Human Impacts on Weather and Climate

WILLIAM R. COTTON
ROGER A. PIELKE
Department of Atmospheric Science, Colorado State University

Published by the Press Syndicate of the University of Cambridge
The Pitt Building, Trumpington Street, Cambridge CB2 1RP
40 West 20th Street, New York, NY 10011–4211, USA
10 Stamford Road, Oakleigh, Melbourne 3166, Australia

Originally published by *ASTeR Press, Fort Collins, 1992

First Published by Cambridge University Press 1995

Printed in the United States of America

Library of Congress Cataloging-in-Publication Data

Cotton, William R., 1940–

Human impacts on weather and climate / William R. Cotton, Roger A. Pielke.

p. cm.

Originally published: Fort Collins : *ASTeR Press, c1992.

Includes bibliographical references and index.

ISBN 0-521-49592-X. – ISBN 0-521-49929-1 (pbk.)

1. Climatology – Social aspects. 2. Weather – Social aspects.
3. Man – Influence on nature. 4. Climatic changes – Environmental aspects. I. Pielke, Roger A. II. Title.
QC981.C72 1995 94-45988
551.6 – dc20 CIP

A catalog record for this book is available from the British Library.

ISBN 0 521 49592 X Hardback
ISBN 0 521 49929 1 Paperback

Contents

Preface

There is increasing evidence that human activities are having a significant impact on weather and climate. Currently the alarm about the impact of so-called 'greenhouse' gases on global warming has attracted the attention of the news media. The aim of this book is to review the basic physical concepts on which human impacts on weather and climate are based and to examine the major uncertainties in determining whether or not those hypotheses are real and if and when those impacts are measurable. As the population of mankind increases on earth, the potential for human impacts on weather and climate becomes increasingly likely. Therefore it is important to examine all potential impacts of human activity on weather and climate and not just one process.

This book is divided into three parts. In Part I we begin by reviewing attempts at purposeful weather modification by cloud seeding. The concepts are of value when considering the total human impact on weather and climate. Moreover, there are many lessons to be learned from the brief history of cloud seeding, both technically and politically. In our enthusiasm for investigating human impact on global climate we must be careful not to repeat the mistakes made during the cloud seeding era. The intent of this book is not only to review the scientific concepts regarding human impacts on weather and climate, but also to consider the entire scientific process including the administration of science and the political/social interactions.

In Part II we review human influences on regional weather and climate. Since people began altering their environment by domesticating animals, cutting down forests, constructing sprawling urban complexes, and planting crops, certain changes in the weather and climate on the regional scale have taken place. We review the evidence that people have inadvertently modified the environment on regional scales and are doing so at increasingly greater rates.

Finally, in Part III we examine the concepts and theories for human effects on global climate including the potential effects of aerosols, large-scale nuclear warfare, greenhouse gases, and global changes in landuse and impacts on the biosphere.

Acknowledgments

This book would not have been possible without the cumulative scientific knowledge that each of the authors gained from sustained research support over the last 20 years from the National Science Foundation, National Oceanic and Atmospheric Administration, Army Research Office, Air Force Office of Scientific Research, and the National Park Service.

William Cotton would like to acknowledge the sabbatical leave support provided by the College of Engineering and Department of Atmospheric Science at the Colorado State University, which allowed him the opportunity to complete extensive literature surveys over the diverse topics in this book.

Roger Pielke would like to acknowledge recently sponsored research by the United States Geological Service in climate change research, as well as numerous discussions on this topic with scientists at the National Resource Ecology Laboratory (NREL) at Colorado State University. Also the CRU/CIRA (Cooperative Research in the Atmosphere) sponsored workshop in 1987, entitled "Monitoring Climate for the Effects of Increasing Greenhouse Gas Concentrations," helped focus issues in climate change research.

David Randall and Harry Orville are thanked for their constructive comments in reviewing a draft of this book.

Brenda Thompson and Dallas McDonald are thanked for their dedicated and professional effort in processing, editing, and typesetting this manuscript. Lucy McCall and Judy Sorbie-Dunn are acknowledged for their assistance with this manuscript and for the drafting of figures.

Part I

The Rise and Fall of the Science of Weather Modification by Cloud Seeding

In Part I we examine human attempts at purposely modifying weather and climate. We also trace the history of the science of weather modification by cloud seeding describing its scientific basis and the rise and fall of funding of weather modification scientific programs, particularly in the United States.

Chapter 1

The Rise of the Science of Weather Modification by Cloud Seeding

Throughout history and probably pre-history man has sought to modify weather by a variety of means. Many primitive tribes have employed witch doctors or medicine men to bring clouds and rainfall during periods of drought and to drive away rain clouds during flooding episodes. Numerous examples exist where modern man has shot cannons, fired rockets, rung bells, etc. in attempts to modify the weather (Changnon and Ivens, 1981).

It was Schaefer's (1948a) discovery in 1946 that the introduction of dry ice into a freezer containing cloud droplets cooled well below zero degrees Celsius (what we call supercooled droplets) resulted in the formation of ice crystals, that launched us into the modern age of the science of weather modification.[1] Working for the General Electric Research Laboratory under the direction of Irving Langmuir on a project investigating ways to combat aircraft icing, Schaefer learned to form a supercooled cloud by blowing moist air into a home freezer unit lined with black velvet. He noted that at temperatures as cold as –23° C, ice crystals failed to form in the cloud. Introducing a variety of substances in the cloud failed to convert the cloud to ice crystals. It was only after a piece of dry ice was lowered into the cloud that thousands of twinkling ice crystals could be seen in the light beam passing through the chamber. He subsequently showed that only small grains of dry ice or even a needle cooled in liquid air could trigger the nucleation of millions of ice crystals.

Motivated by Schaefer's discovery, Vonnegut (1947), also a researcher at the General Electric Research Laboratory, began a systematic search through chemical tables for materials that have a crystalographic structure similar to ice. He hypothesized that such a material would serve

[1] A summary of this early work is given in Havens et al. (1978).

as an artificial ice nucleus. It was well known at that time that under ordinary conditions, the formation (or nucleation) of ice crystals required the presence of a foreign substance called a nucleus or mote which would promote their formation. For some time European researchers such as A. Wegener, T. Bergeron, and W. Findeisen had hypothesized that the presence of supercooled droplets in clouds indicated a scarcity of ice-forming nuclei in the atmosphere. It was believed that the dry ice in Schaefer's experiment cooled the air to such a low temperature that nucleation took place without an available nuclei; the process is referred to as *homogeneous nucleation.* Vonnegut's search through the chemical tables revealed three substances which had the desired crystalographic similarity to ice: lead iodide, silver iodide, and antimony. Dispersal of a powder of these substances in a cold box had little effect. Vonnegut then decided to produce a smoke of these substances by vaporizing the material and as it condensed a smoke of very small crystals of the material was created. Vonnegut found that a smoke of silver iodide particles produced numerous ice crystals in the cold box at temperatures warmer than –20° C similar to dry ice in Schaefer's experiment.

The stage was now set to attempt to introduce dry ice or silver iodide smoke into real supercooled clouds and observe the impact on those clouds. Again, the background of previous research by the Europeans (Wegener, Bergeron, and Findeisen) was important for this stage. They showed that ice crystals once formed in a supercooled cloud could grow very rapidly by deposition of vapor onto them at the expense of supercooled cloud droplets. This is due to the fact that the saturation vapor pressure with respect to ice is lower than the saturation vapor pressure with respect to water at temperatures colder than zero degrees centigrade. As shown in Figure 1.1, the supersaturation with respect to ice increases linearly with decreasing temperature below 0° C for a water-saturated cloud. Thus an ice crystal nucleated in a cloud that is water saturated finds itself in an environment which is supersaturated with respect to ice and can thereby grow rapidly by deposition of vapor. As vapor is deposited on the growing ice crystals the vapor in the cloud is depleted, and the cloud vapor pressure lowers to below water saturation. Thus cloud droplets evaporate providing a reservoir of water vapor for growing ice crystals. The ice crystals, therefore, grow at the expense of the cloud droplets.

It was thus hypothesized that the insertion of dry ice or silver iodide in a supercooled cloud would initiate the formation of ice crystals, which in turn, would grow by vapor deposition into ice crystals. Precipitation could be artificially initiated in such clouds.

Langmuir (1953) calculated theoretically the number of ice crystals that would form from dry ice pellets of a given size. He also predicted that the latent heat released as the ice crystals grew by vapor deposition would warm the seeded part of the cloud, causing upward motion and turbulence

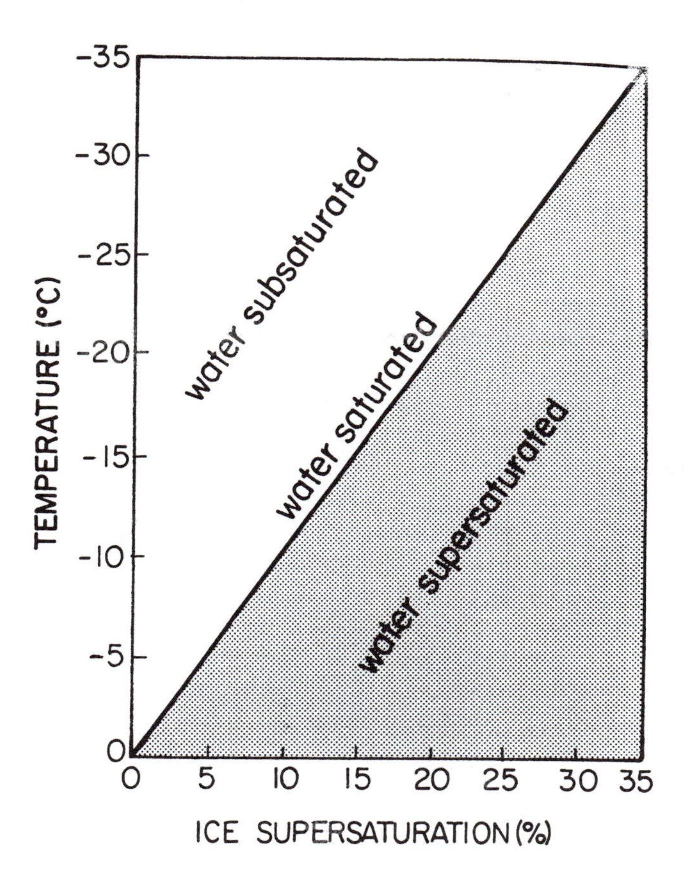

Figure 1.1: *Supersaturation with respect to ice as a function of temperature for a water-saturated cloud. The shaded area represents a water-supersaturated cloud.* [From Cotton, W.R. and R.A. Anthes, 1989: *Storm and Cloud Dynamics.* Academic Press, Inc., San Diego. International Geophysics Series, Vol. 44., 883 pp.]

which would disperse the mist of ice crystals created by seeding over a large volume of the unseeded part of the cloud.

On November 13, 1946, Schaefer (1948b) dropped about 1.4 kg of dry ice pellets from an aircraft flying over a supercooled stratus cloud near Schenectady, New York. Similar to the laboratory cold box experiments, the cloud rapidly converted to ice crystals which fell out as snow beneath the stratus deck. This, as well as a number of other exploratory seeding experiments, led to the formation of Project Cirrus.

1.1 Project Cirrus

Under Project Cirrus, Langmuir and Schaefer performed a number of exploratory cloud seeding experiments including seeding of cirrus clouds, supercooled stratus clouds, cumulus clouds, and even hurricanes. Supercooled stratus clouds yielded the clearest response to seeding. A variety of aircraft patterns were flown over the stratus clouds while dropping dry ice. Patterns included L-shaped, race track, and Greek gammas. The response was the formation of holes in the clouds whose shape mirrored the aircraft flight pattern (see Figure 1.2).

Seeding of supercooled cumulus clouds produced more controversial results. Dry ice and silver iodide seeding experiments were carried out at a variety of locations with the most comprehensive experiments being over New Mexico. Based on four seeding operations near Albuquerque, New Mexico, Langmuir claimed that seeding produced rainfall over a quarter of the area of the state of New Mexico. He concluded that "The odds in favor of this conclusion as compared to the rain was due to natural causes are millions to one." Langmuir was even more enthusiastic about the consequences of silver iodide seeding over New Mexico. The explosive growth of a cumulonimbus cloud and the heavy rainfall near Albuquerque and Santa Fe were attributed to the direct results of ground-based silver iodide seeding. In fact Langmuir concluded that nearly all the rainfall that occurred over New Mexico on the dry ice seeding day and the silver iodide seeding day were the result of seeding.

One of the most controversial experiments performed during Project Cirrus was the periodic seeding experiment. In this experiment a ground-based silver iodide generator was operated on a seven-day periodic schedule with the generator being operated 8 hours a day on Tuesday, Wednesday, and Thursday and turned off the rest of the week. A total of 1000 grams of silver iodide were used per week and the experiment was carried out from December 1949 to the middle of 1951. The analysis of precipitation and other weather records over the Ohio river basin and other regions to the east of New Mexico revealed a highly significant seven-day periodicity. Langmuir and his colleagues were convinced that this periodicity in the rainfall records were a direct result of their seeding in New

Figure 1.2: *Racetrack pattern approximately 20 miles long produced by dropping crushed dry ice from an airplane. The safety pin-like loop at the near end of the pattern resulted when the dry ice dispenser was inadvertently left running as the airplane began climbing to attain altitude from which to photograph results.* [From Havens, B. S., J. E. Jiusto, and B. Vonnegut, 1978: Early history of cloud seeding. Project Cirrus Fund, SUNY-ES/328, 1400 Washington Ave., Albany, NY 12203, 75 pp. Photo courtesy of Dr. Vincent Schaefer.]

Mexico. Other scientists were not so convinced (Lewis, 1951; Wahl, 1951; Wexler, 1951; Brier, 1955; Byers, 1974). They showed that large amplitude seven-day periodicities in rainfall and other meteorological variables, though not common, had occurred during the period 1899-1951. Thus they felt the rainfall periodicity was due to *natural variability* rather than to a direct consequence of cloud seeding.

Convinced that cloud seeding was a miraculous cure to all of nature's evils, Langmuir and his colleagues carried out a trial seeding experiment of a hurricane with the hope of altering the course of the storm or reducing its intensity. On October 10, 1947, a hurricane was seeded off the east coast of the United States. About 102 kg of dry ice were dropped in clouds in the storm. Due to logistical reasons, the eye wall region and the dominate spiral band were not seeded. Observers interpreted visual observations of snow showers as evidence that seeding had some effect on cloud structure. Following seeding the hurricane changed direction from a northeasterly to a westerly course, crossing the coast into Georgia. The change in course may have been a result of the storm's interaction with the larger-scale flow field. Nonetheless, General Electric Corporation became the target of lawsuits for damage claims associated with the hurricane.

While the main focus of research during Project Cirrus was the dry ice and silver iodide seeding of supercooled clouds, some theoretical and experimental effort was directed toward stimulated rain formation in non-freezing clouds or what we will refer to as warm clouds. In 1948, Langmuir (1948) published his theoretical study of rain formation by chain reaction. According to his theory once a few raindrops grew by colliding and coalescing with smaller drops to such a size that they would breakup, the fragments they produced would serve as embryos for further growth by collection. The smaller sized embryos would then ascend in the cloud updrafts while growing by collection and also breakup creating more raindrop embryos. Langmuir hypothesized that insertion of only a few raindrops in a cloud could infect the cloud with raindrops through the chain reaction process. Some attempts were made to initiate rain in warm clouds by water drop seeding in Puerto Rico, though no suitable clouds were found. Subsequently Braham et al. (1957) and others at the University of Chicago demonstrated that one could initiate rainfall by water drop seeding. This experiment will be discussed more fully in a later section.

In summary, Project Cirrus launched the United States and much of the world into the age of cloud seeding. The impact of this project on the science of cloud seeding, cloud physics research, and the entire field of atmospheric science was similar to the effects of the launching of Sputnik on the United States aerospace industry.

Chapter 2

The Glory Years of Weather Modification

2.1 Introduction

The exploratory cloud seeding experiments performed by Langmuir and Schaefer and Project Cirrus personnel fueled a new era in weather modification research as well as basic research in the microphysics of precipitation processes, cloud dynamics, and small-scale weather systems, in general. At the same time commercial cloud seeding companies sprung up worldwide practicing the art of cloud seeding to enhance and suppress rainfall, dissipate fog, and decrease hail damage. Armed with only rudimentary knowledge of the physics of clouds and the meteorology of small-scale weather systems, these weather modification practitioners sought to alleviate all the symptoms of undesirable weather by prescribing cloud seeding medication. The prevailing view was 'cloud seeding is good'!

At the same time scientists were faced with the major challenge of proving that cloud seeding did indeed result in enhancement of precipitation or produce some other desired response, as well as unravel the intricate web of physical processes responsible for both natural and artificially stimulated rainfall. We, therefore, entered the era where scientists had to get down in the trenches and sift through every little piece of physical evidence to unravel the mysteries of cloud microphysics and precipitation processes.

As the science of weather modification developed, two schools of cloud seeding methodology emerged. One school embraced what is called the *static mode* of seeding while the other is called the *dynamic mode* of seeding. In the next few sections we will review these two approaches as well as review the application of cloud seeding to hail suppression, hurricane modification, and precipitation enhancement in warm clouds.

2.2 The "Static Mode" of Cloud Seeding

We have seen that the pioneering experiments of Schaefer and Langmuir suggested that the introduction of dry ice or silver iodide into supercooled clouds could initiate a precipitation process. The underlying concept behind the *static mode* of cloud seeding is that natural clouds are deficient in ice nuclei.[1]

As a result many clouds contain an abundance of supercooled liquid water which represents an underutilized water resource. Supercooled clouds are thus viewed to be inefficient in precipitation formation, where precipitation efficiency is defined as the ratio of the rainfall rate or flux of rainfall on the ground to the flux of water substance entering the base of a cloud. The major focus of the static mode of cloud seeding is to increase the precipitation efficiency of a cloud or cloud system.

In its simplest form the static mode of cloud seeding was based on the Bergeron-Findeisen concept in which ice crystals nucleated either naturally or through seeding in a water-saturated supercooled cloud will grow by vapor deposition at the expense of cloud droplets. Figure 2.1 illustrates schematically the Bergeron-Findeisen process. Seeding therefore can convert a naturally inefficient cloud containing supercooled cloud droplets into a precipitating cloud in which the precipitation is in the form of vapor-grown ice crystals or raindrops formed from melted ice crystals. The 'seedability' of a cloud is thus primarily a function of the availability of supercooled water. Because laboratory cloud chambers predicted that natural ice nuclei concentrations increased exponentially with the degree of supercooling (i.e., degrees colder than 0° C) and because the amount of water vapor available for condensation increases with temperature, it was generally believed that the availability of supercooled water was greatest at warm temperatures, or between 0° C and –20° C.

Cloud seeding experiments and research on the basic physics of clouds during the 1950's through the early 1980's revealed that this simple concept of static seeding is only applicable to a limited range of clouds. It was found that in many supercooled clouds, the primary natural precipitation process was not growth of ice crystals by vapor deposition but growth of precipitation by collision and coalescence, or collection (see Figure 2.2). It was found that clouds containing relatively low concentrations of cloud condensation nuclei (CCN) were more likely to produce rain by collision and coalescence among cloud droplets than clouds containing high concentrations of CCN. If a cloud condenses a given amount of supercooled liquid water, then a cloud containing low CCN concentrations will produce fewer cloud droplets than a cloud containing high CCN

[1] For an excellent, more technical review of static seeding, see Bernard Silverman's review titled *Static Mode Seeding of Summer Cumuli–A Review* in Braham, R.R., Jr., 1986: Precipitation Enhancement–A Scientific Challenge. *Meteor. Monogr.* **21,** 171 pp.

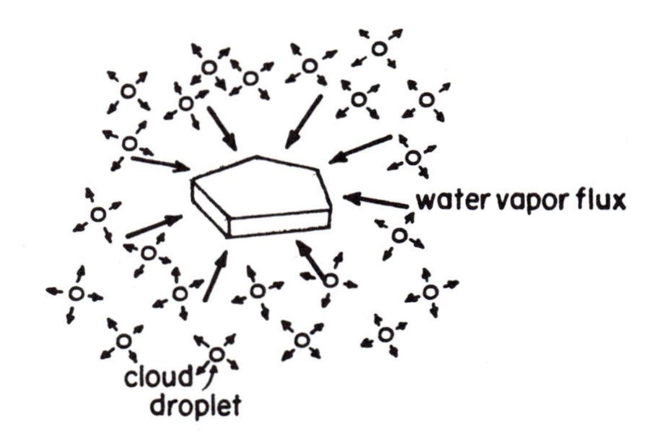

Figure 2.1: *Schematic illustration of the Bergeron-Findeisen process.*

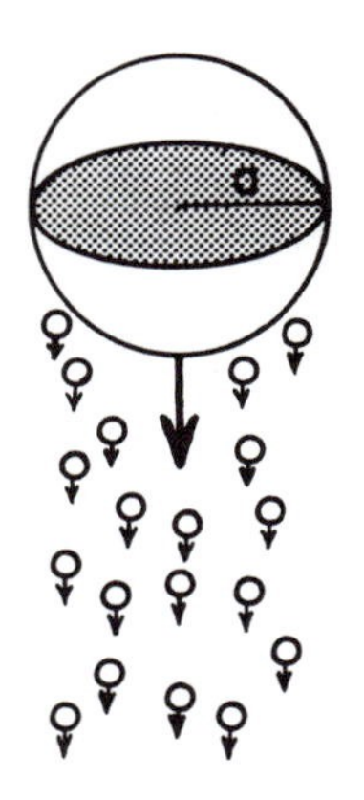

Figure 2.2: *Illustration of growth of a drop by colliding and coalescing with smaller, slower-settling cloud droplets.* [From Cotton, W.R., 1990: *Storms.* Geophysical Science Series, Vol. 1. ASTeR Press, Fort Collins, CO, 158 pp.]

concentrations. As a result, in a cloud containing fewer cloud droplets, the droplets will be bigger on the average and fall faster than a cloud containing numerous, slowly-settling cloud droplets. Because some of the bigger cloud droplets will settle through a population of smaller droplets more readily in a cloud containing low CCN concentrations, a cloud containing low CCN concentrations is more likely to initiate a precipitation process by collision and coalescence among cloud droplets than a cloud with a high CCN concentration. Generally clouds forming in a maritime airmass have lower concentrations of CCN than clouds forming in continental regions, often differing by an order of magnitude or more, and in polluted air masses the CCN concentrations can be 40 times that found in a clean maritime airmass.

It was also found that clouds having relatively warm cloud base temperatures were richer in liquid water content than clouds having cold cloud base temperatures. This is because the saturation vapor pressure increases exponentially with temperature. As a result clouds with warm cloud base temperatures have much more water vapor entering cloud base available to be condensed in the upper levels of the cloud than a cloud with cold base temperatures. What this means is that clouds forming in a maritime airmass with low CCN concentrations and having warm cloud base temperatures have a high potential of being very efficient natural rain producers by collision and coalescence of cloud droplets.

The collision and coalescence process is not limited to just liquid drops colliding with liquid drops. Once ice crystals become large enough and begin to settle through a cloud of small supercooled droplets, the ice crystals can grow by collecting those droplets as they rapidly fall through a population of cloud droplets to form what we call *rimed* ice crystals or graupel particles (see Figure 2.3). Frozen raindrops can also readily collide with supercooled cloud droplets to form hailstones or large graupel particles. The larger the liquid water content in clouds, the more likely that precipitation will form by one of the above collection mechanisms. Therefore, natural clouds can be far more efficient precipitation producers than would be expected from the simple concept of precipitation formation primarily by vapor growth of ice crystals.

Research during the same period revealed that laboratory ice nucleus counters were not always good predictors of ice crystal concentrations. Observations of ice crystal concentrations showed that in many clouds the observed ice crystal concentrations exceeded estimates of ice crystal concentrations by 4 to 5 orders of magnitude! The greatest discrepancies between observed ice crystal concentrations and concentrations diagnosed from ice nucleus counters occurred in clouds with relatively warm cloud top temperatures (i.e., warmer than –10° C) and those having significant concentrations of heavily rimed ice particles such as graupel and frozen raindrops. These are the clouds that contain relatively high liquid water contents and/or an active collection process. In other words clouds that

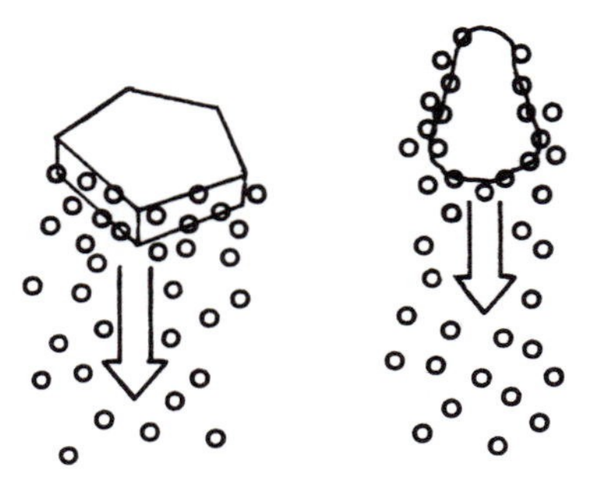

Figure 2.3: *Riming of ice crystals or graupel particles.*

are warm-based and maritime are most likely to contain much higher ice crystal concentrations than ice nuclei concentrations. On the other hand, clouds in which ice crystal growth by vapor deposition prevails and in which riming is modest, generally exhibit ice crystal concentrations comparable to ice nuclei concentrations.

The reasons for the discrepancy between ice crystal concentrations and ice nuclei concentrations are not fully understood today. Some researchers concluded from observational studies that temperature has little influence on the ice crystal concentrations (Hobbs and Ragno, 1985). Instead, it is argued that the droplet size-distribution in clouds has the major controlling influence on ice crystal concentrations.

In recent years several laboratory experiments have revealed that under certain cloud conditions, ice crystal concentrations can be greatly enhanced by an ice multiplication process (Hallett and Mossop, 1974; Mossop and Hallett, 1974). The laboratory studies suggest that over the temperature range –3° to –8° C, copious quantities of secondary ice crystals are produced when an ice crystal or graupel particle collects or rimes supercooled cloud droplets. The secondary production of ice crystals is greatest when the supercooled cloud droplet population contains a significant number of large cloud droplets ($r > 12$ microns). Figure 2.4 illustrates the rime-splinter, secondary ice crystal production process. The presence of large cloud droplets would be greatest in clouds which are warm-based and maritime. Moreover, warm-based maritime clouds are more likely to contain supercooled raindrops which, when frozen, can serve as active sites for riming growth and secondary particle production. Thus, the Hallett-Mossop rime-splinter process is consistent with many field observations which suggest that clouds that are maritime and warm-based are more likely to contain ice crystal concentrations greatly in excess of ice nuclei concentrations.

The rime-splinter secondary ice crystal production process may not explain all the observations of high ice crystal concentrations relative

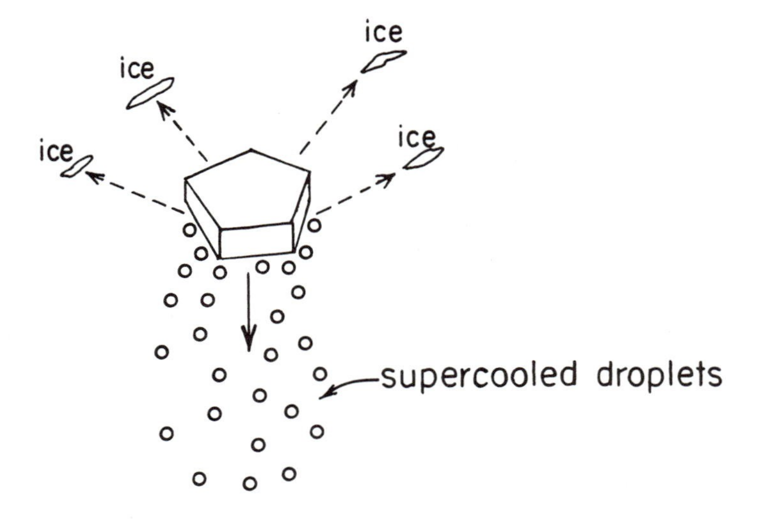

Figure 2.4: *Illustration of secondary ice particle production by ice particle collection of supercooled cloud droplets at temperatures between -4 to -8 ° C.* [From Cotton, W.R., 1990: *Storms.* Geophysical Science Series, Vol. 1. ASTeR Press, Fort Collins, CO, 158 pp.]

to ice nuclei estimates but it is consistent with many of them. Still other processes not understood at this time may be operating in some cases of observed high ice crystal concentrations relative to ice nuclei concentrations.

The implication of these physical studies is that the 'window of opportunity' for precipitation enhancement by cloud seeding is much smaller than it was originally thought. Clouds which are warm-based and maritime have a high natural potential for producing precipitation. On the other hand, clouds which are cold-based and continental have reduced natural potential for precipitation formation and, hence, the opportunity for precipitation enhancement by cloud seeding is much greater, although the total water available would be less than in a warm-based cloud.

This is consistent with the results of field experiments testing the static seeding hypothesis. The Israeli I and II Experiments were quite successful in producing positive yields of precipitation in seeded clouds (Gagin and Neumann, 1981). The clouds that were seeded over Israel had relatively cold bases (5 to 8° C) and were generally continental such that there was little evidence of a vigorous warm rain process or the presence of large quantities of heavily rimed graupel particles. Other cloud seeding experiments were not so fortunate and either no effects of seeding or even decreases in precipitation were inferred (Tukey et al., 1978; Kerr, 1982). Presumably the opportunities for vigorous warm rain processes and secondary ice particle formation were greater in those clouds.

In those clouds, seeding could not compete effectively with natural precipitation formation processes or natural precipitation processes masked the seeding effects so that they could not be separated from the ***natural variability*** of precipitation.

A number of observational and theoretical studies have also suggested that there is a cold temperature 'window' of opportunity as well. Studies of both orographic and convective clouds have suggested that clouds colder than –25° C have sufficiently large concentrations of natural ice crystals such that seeding can either have no effect or even reduce precipitation (Grant and Elliot, 1974; Grant, 1986; Gagin and Neumann, 1981; Gagin et al., 1985). It is possible that seeding such cold clouds could reduce precipitation by creating so many ice crystals that they compete for the limited supply of water vapor and result in numerous, slowly settling ice crystals which evaporate before reaching the ground. Such clouds are said to be '*overseeded*'.

There are also indications that there is a warm temperature limit to seeding effectiveness (Gagin and Neumann, 1981; Grant and Elliot, 1974; Cooper and Lawson, 1984). This is believed to be due to the low efficiency of ice crystal production by silver iodide at temperatures approaching – 4° C and to the slow rates of ice crystal vapor deposition growth at warm temperatures. Thus there appears to be a 'temperature window' of about –10° C to –25° C where clouds respond favorably to seeding (i.e., exhibit seedability).

There also seems to be a 'time window' of opportunity for seeding in many clouds; especially convective clouds. It is well known that the life cycle of convective clouds is significantly affected by the entrainment of dry environmental air. As dry environmental air is entrained into a cumuli, cloud droplets formed in moist updraft air are evaporated, causing cooling that then forms downdrafts which terminate the life cycle of the cumuli. It was found during the HIPLEX-1 Experiment (Cooper and Lawson, 1984) that the time scale for which sufficient supercooled cloud water was available for seeding in ordinary cumuli was less than 14 minutes. In towering cumuli and small cumulonimbi the time scale is not so limited by entrainment as in smaller cumuli, but those clouds are more likely to produce precipitation naturally thus competing for the available cloud water. The 'time window' of opportunity in larger cumuli is thus much more variable and uncertain since it depends not only on dynamic time scales which control entrainment, but also on the time scales of natural precipitation formation.

Thus, physical studies and inferences drawn from statistical seeding experiments suggest that there exists a more limited window of opportunity for precipitation enhancement by the static-mode of cloud seeding than was originally thought. The window of opportunity for cloud seeding appears to be limited to:

1) clouds which are relatively cold-based and continental;

2) clouds having top temperatures in the range –10 to –25° C;

3) a time scale limited by the availability of significant supercooled water before depletion by entrainment and natural precipitation processes.

This limited scope of the opportunities for rainfall enhancement by the static-mode of cloud seeding that has emerged in recent years may explain why some cloud seeding experiments have been successful while other seeding experiments have yielded inferred reductions in rainfall from seeded clouds or no effect. A successful experiment in one region does not guarantee that seeding in another region will be successful unless all environmental conditions are replicated as well as the methodology of seeding. This, of course, is highly unlikely.

We must also recognize that implementing a seeding experiment or operational program that operates only in the above listed windows of opportunity is extremely difficult and costly. It means that in a field setting we must forecast the top temperatures of clouds to assure that they fall within the –10 to –25° C temperature window. We must determine the extent to which clouds are maritime and warm-based, versus continental and cold-based, or the likelihood that clouds will naturally contain broad droplet spectra and an active warm-rain process. Because there are no routine measurements of cloud condensation nuclei spectra nor forecast models with a demonstrated skill in predicting the particular modes of precipitation formation, the potential of successfully implementing a seeding strategy in the field in which consideration is given to the natural widths of cloud droplet spectra and to the natural modes of precipitation formation is not very good. Furthermore, consideration of a 'time window' complicates implementation of an operational seeding strategy even more. Seeding material would have to be targeted at the right time in a cumulus cloud before either entrainment depletes the available supercooled water or natural precipitation processes deplete the available liquid water. This would require airborne delivery of seeding material, which is expensive, and a prediction of the time scales of liquid water availability in clouds of differing types.

The success of a cloud seeding experiment or operation, therefore, requires a cloud forecast skill that is far greater than is currently in use. As a result, such experiments or operations are at the mercy of the *natural variability* of clouds. The impact of *natural variability* may be reduced in some regions where the local climatology favors clouds which are in the appropriate temperature windows and are more continental. The 'time window' will still exist, however, and this will yield uncertainty to the results unless the field personnel are particularly skillful in selecting suitable clouds.

Orographic clouds are less susceptible to a 'time window' as they are steady clouds so they offer a greater opportunity for successful precipitation enhancement than cumulus clouds. A time window of a different type does exist for orographic clouds which is related to the time it takes a parcel of air to condense to form supercooled liquid water and ascend to the mountain crest. If winds are weak, then there may be sufficient time for natural precipitation processes to occur efficiently. Stronger winds may not allow efficient natural precipitation processes but seeding may speed up precipitation formation. Even stronger winds may not provide enough time for even seeded ice crystals to grow to precipitation before being blown over the mountain crest and evaporating in the sinking subsaturated air to the lee of the mountain. A time window related to the ambient winds, however, is much easier to assess in a field setting than the time window in cumulus clouds.

In summary, the 'static' mode of cloud seeding has been shown to cause the expected alterations in cloud microstructure including increased concentrations of ice crystals, reductions of supercooled liquid water content, and more rapid production of precipitation elements in both cumuli (Cooper and Lawson, 1984) and orographic clouds (Reynolds, 1988; Super and Boe, 1988; Super et al., 1988; Super and Heimbach, 1988; Reynolds and Dennis, 1986). The documentation of increases in precipitation on the ground due to static seeding of cumuli, however, has been far more elusive with the Israeli experiment (Gagin and Neumann, 1981) providing the strongest evidence that static seeding of cold-based, continental cumuli can cause significant increases of precipitation on the ground. The evidence that orographic clouds can cause significant increases in snowpack is far more compelling, particularly in the more continental and cold-based orographic clouds (Mielke et al., 1981; Super and Heimbach, 1988). It is clear, however, that we still do not have the ability to produce statistically significant increases in surface precipitation from all supercooled cumuli or orographic clouds. At the very least we conclude that we do not yet have the ability to discriminate seeding-induced increases in surface precipitation from the background 'noise' created by the high *natural variability* of surface precipitation for many cloud systems. The stronger evidence for positive seeding effects on orographic clouds versus cumuli is due in large measure to the lower *natural variability* of wintertime orographic precipitation than in summertime cumuli.

2.3 The Dynamic Mode of Cloud Seeding

2.3.1 Introduction

We have seen that the fundamental concept of the '*static mode*' of cloud seeding is that precipitation can be increased in clouds by enhancing their precipitation efficiency. While alterations in the dynamics or air motion in clouds due to latent heat release of growing ice particles, redistribution of condensed water, and evaporation of precipitation is inevitable with *static mode* seeding, it is not the primary aim of the strategy. By contrast, the focus of the '*dynamic mode*' of cloud seeding is to enhance the vertical air currents in clouds and thereby vertically process more water through the clouds resulting in increased precipitation.[2] In this section we examine the concepts behind the dynamic mode strategy and discuss the physical/statistical evidence supporting the concept.

2.3.2 Fundamental concepts

We noted earlier that Langmuir postulated that the latent heat released as ice crystals grow by vapor deposition would warm the seeded part of a cloud and cause upward motion and turbulence. The concept is a simple one. As ice crystals grow by vapor deposition a phase transition takes place in which water vapor molecules deposit on an ice crystal lattice. During the phase transformation the latent heat of sublimation, $2.83 \times 10^6\ J\ kg^{-1}$, is released warming the immediate environment of the ice crystals. If the cloud contains cloud droplets, however, the growth of ice crystals causes the lowering of the cloud saturation pressure below water saturation, resulting in the evaporation of cloud droplets to restore the cloud to water saturation. The evaporation of cloud droplets absorbs the latent heat of vaporization or $2.50 \times 10^6\ J\ kg^{-1}$ resulting in a net warming of the cloud of $0.33 \times 10^6\ J\ kg^{-1}$ for the vapor deposited on the ice crystals. Only if all condensed liquid water is evaporated and deposited on ice crystals, will the cloud experience the full warming effects of the latent heat of sublimation.

Moreover, if supercooled cloud droplets or raindrops freeze by contacting an ice crystal or ice nuclei, the phase transformation from liquid to ice will release the latent heat of fusion or $0.33 \times 10^6\ J\ kg^{-1}$ of water frozen. In some instances, so much supercooled water may freeze that the cloud can become subsaturated with respect to ice causing the sublimation of ice crystals and partially negating the positive heat released by freezing.

---

[2]For an excellent, more technical review of dynamic seeding see Harold Orville's review titled *A Review of Dynamic-Mode Seeding on Summer Cumuli* in Braham, R.R., Jr., 1986: Precipitation Enhancement–A Scientific Challenge. *Meteor. Monogr.* **21,** 171 pp.

Why is this heating important to clouds? Many clouds such as cumulus clouds are buoyancy-driven. When a small volume of air, which we shall call an air parcel, becomes warmer than its environment it expands and displaces a volume of environmental air equal to the weight of the warm air. According to Archimedes' principle, the warmed air will be 'buoyed up' with a force that is equal to the weight of the displaced environmental air. This upward-directed buoyancy force will then accelerate a cloud parcel upwards similar to the upward acceleration one can experience in a hot air balloon when the air inside the balloon is heated with a propane burner. The simple addition of heat to atmospheric air parcels, however, does not guarantee that the air will become buoyant.

The buoyancy of a cloud is determined not only by how warm a cloud is with respect to its environment, but also by how much water is condensed in a cloud. Condensed water produces negative buoyancy, such that a cloud that is warmer than its environment can actually become negatively buoyant due to the load of condensed water it must carry. One consequence of a precipitation process is that it unloads the upper portions of a cloud from its burden of condensed water (see Figure 2.5a). Unleashed from its burden of condensed water, the top of the cloud can penetrate deeper into the atmosphere. Of course, the water that settles from the upper part of the cloud transfers the burden of condensed water to lower levels, causing a weakening of updrafts or formation of downdrafts at lower levels.

Once the raindrops settle into the subsaturated, subcloud layer, they begin to evaporate. Evaporation of the raindrops absorbs latent heat from the surrounding air, thereby cooling the air. The denser, evaporatively-chilled air sinks towards the surface, spreading horizontally as it approaches the ground (see Figure 2.5). The dense, horizontally-spreading air undercuts the warm, moist air, often elevating it to the lifting condensation level (LCL) and perhaps the level of free convection (LFC). Thus, the settling of raindrops below cloud base can promote the development of new cumulus clouds or help sustain existing ones by causing lifting of warm, moist air into the cloud base level.

Because towering cumulus clouds are taller than fair-weather cumulus clouds, they often extend to heights that are colder than 0° C, or the freezing level. Before significant precipitation occurs, these clouds are called cumulus congestus. Ice particles can therefore form by either the freezing of *'supercooled'* drops or by nucleation on ice nuclei (IN). As far as the overall behavior of a cumulus cloud is concerned, the important consequence of droplet freezing and vapor deposition growth of ice crystals is that additional latent heat is added to the cloudy air. The latent heat liberated during the freezing and vapor deposition growth of ice particles therefore contributes to the buoyancy of the cloud, giving the cloud a boost in its vertical ascent. As a result, towering cumulus clouds often exhibit *explosive* vertical development once ice-phase precipitation

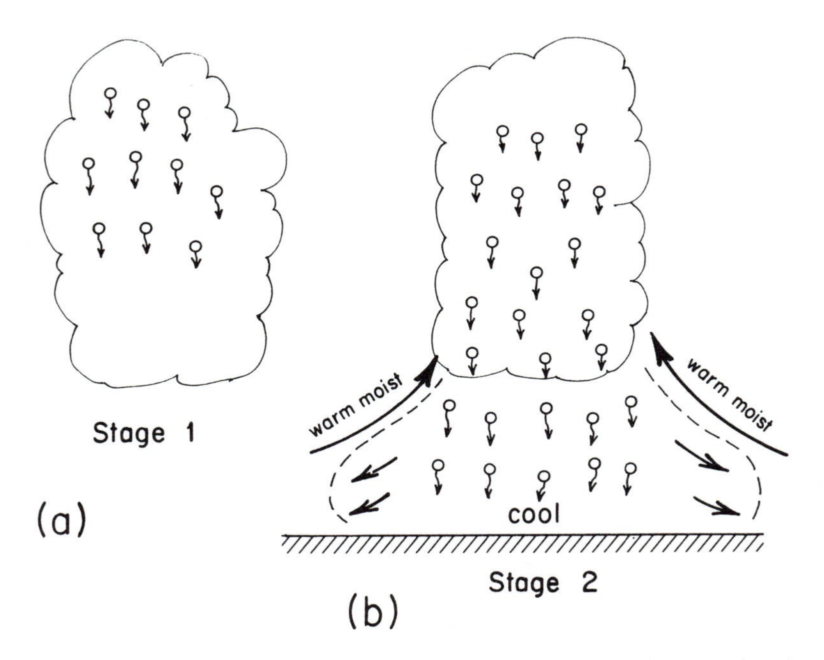

Figure 2.5: *(a) Illustration of droplets settling from the upper levels of a cloud, thus reducing the amount of liquid water content or water-loading burden on the cloud. (b) Illustration of the formation of an evaporatively-chilled layer near the surface which can lift surrounding moist air sometimes to the lifting condensation level (LCL) and level of free convection (LFC).* [From Cotton, W.R., 1990: *Storms.* Geophysical Science Series, Vol. 1. ASTeR Press, Fort Collins, CO, 158 pp.]

processes take place. The taller cumulus clouds typically produce more rainfall and perturb the stably-stratified environment more, thus producing gravity waves which may impact the development of other cumulus clouds (see Cotton and Anthes, 1989).

An important step in the transition of cumulus congestus clouds to thunderstorms or cumulonimbus clouds is the merger of a number of neighboring towering cumulus clouds. Figure 2.6 illustrates the merger of two cumulus clouds due to the interaction of the low-level, cool outflows from the neighboring clouds. As the merger process proceeds, a 'bridge' of smaller cumuli forms between the two clouds. The bridge of clouds eventually deepens and fills the gap between the clouds, resulting in wider and often taller clouds. The clouds resulting from merger generally produce larger rainfall rates, last longer, and are bigger, so that the volume of rainfall from merged clouds is sometimes a factor of ten or more greater than the sum of the rain volumes from similar, non-merged clouds (Simpson et al., 1980).

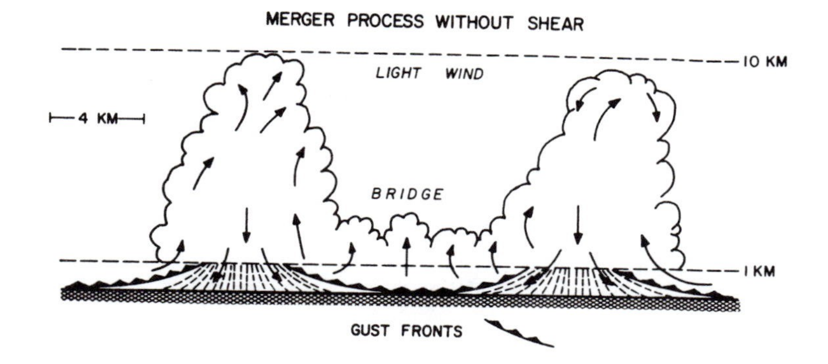

Figure 2.6: *Schematic illustration relating downdraft interaction to bridging and merger in case of light wind and weak shear.* [From Simpson, J., N.E. Westcott, R.J. Clerman, and R.A. Pielke, 1980: On cumulus mergers. *Arch. Meteor. Geophy. Bioklim., Series A,* **29**, 1-40.]

There are many factors which influence the merger of cumulus clouds. We know that merger takes place more frequently in regions where there exists convergence of warm, moist air at low levels on a scale greater than the individual cumulus clouds. The convergence of warm, moist air provides the fuel necessary to sustain convection on the scale of the merged system.

Merger is often accompanied by the explosive growth of at least one of the neighboring clouds. Explosive growth of a cumulus cloud, perhaps due to the release of additional latent heat from the growth of ice particles, generally results in greater precipitation which, in turn, causes stronger subcloud cooling and outflow. Also, the more vigorously, growing clouds create a region of low pressure beneath their bases, which draws warm, moist air into the cloud base, and perhaps along with it, draws in neighboring cumulus clouds. Moreover, explosively rising cumulus clouds perturb the stably-stratified, surrounding environment, triggering gravity waves which can enhance the growth of some clouds and weaken others.

One may ask, if the latent heat released by freezing supercooled drops is only about one-eighth the latent heat released during the condensation of an equivalent mass of vapor onto droplets, why are we interested in its impact on cloud growth? The reason is that at cold temperatures where the ice phase becomes prevalent, the saturation vapor pressure with respect to water is relatively small and varies more slowly with temperature. As a result, as a cloud volume rises and becomes colder, the amount of water available to be condensed in a cloud and correspondingly the latent heat released, becomes less and less. Moreover, unless the cloud is raining heavily, the water vapor that has condensed in the cloud to form water drops at warmer temperatures is available in large amounts for freezing. If this stored water is then frozen by seeding or spontaneously through natural ice nucleation processes, the cloud will experience a boost in buoyancy at precisely those levels where the latent heat liberated during condensation is lessened. In addition, since at colder temperatures the saturation vapor pressure becomes small in magnitude, the differences between environmental vapor pressures and the saturation vapor pressure in the cloudy region become smaller. As a result, entrainment of dry environmental air into the cloud causes less evaporative cooling and the consequences of entrainment are less of a brake on cloud vertical development. Thus the artificial stimulation of the ice phase in a cloud by seeding can cause a boost in the buoyancy of a cloud that is less likely to be destroyed by entrainment of dry environmental air.

All these factors must be considered when estimating whether or not the latent heat released by freezing or vapor deposition growth of ice crystals created by seeding will boost a cloud upwards in the atmosphere. We must examine the local environment or each individual sounding in the neighborhood of a cloud to see if it will support deep convection and if natural cloud vertical growth will be limited by a stable layer or inversion or by the effects of entrainment. Will the cloud experience a sufficient boost in buoyancy when seeded to overcome the effects of entrainment or a stable inversion layer so that its vertical growth will be enhanced? To answer these and other questions about cloud behavior, we must simulate the behavior of natural and seeded clouds on a computer.

The computer simulation of clouds involves the use of a mathematical or numerical model. Such a model simulates a cloud by solving or integrating a prescribed set of equations numerically on a computer. The earliest cloud models (and those used most extensively for simulating dynamic seeding) are based on the hypothesis that clouds behave similar to buoyant laboratory thermals or jets (see Figure 2.7). The laboratory studies suggest that thermals or jets are primarily buoyancy-driven, and that the rate of rise can be mathematically described in terms of the cloud buoyancy, vertical momentum, and the rate at which buoyancy is eroded as dry, cooler environmental air is entrained into the bubble or jet. Different entrainment laws were hypothesized for jets and thermals based on laboratory tank calibrations.

Application of these models to atmospheric clouds involves the use of a thermodynamic energy equation along with the vertical rise rate equation. The models are typically initialized with a local atmospheric sounding of temperature, relative humidity, and winds, as well as prescribing some initial ascent at an estimated or prescribed cloud base height. As illustrated by the square in Figure 2.7a and b, a small parcel of air is then integrated upward while calculating the changes in cloud buoyancy and rise rate due to condensation of vapor, freezing of raindrops, and

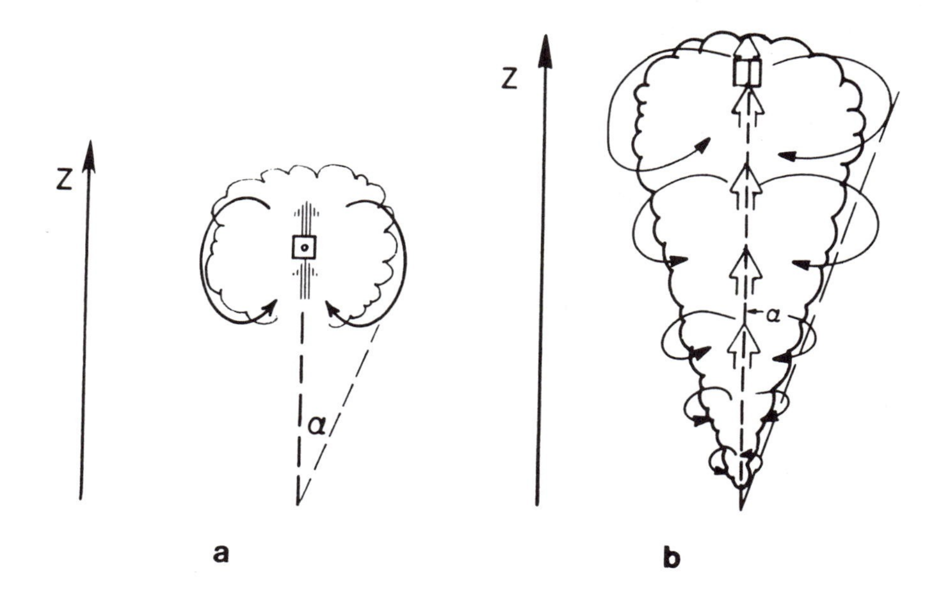

Figure 2.7: *(a) Schematic view of the "bubble" or "thermal" model of lateral entrainment in cumuli. (b) Schematic view of the "steady-state jet" model of lateral entrainment in cumuli.* [From Cotton, W.R. and R.A. Anthes, 1989: *Storm and Cloud Dynamics.* Academic Press, Inc., San Diego. International Geophysics Series, Vol. 44., 883 pp.]

vapor deposition on ice crystals as well as removal of condensate products by precipitation. The calculations are terminated when the modeled cloud loses all positive buoyancy. To simulate the effects of dynamic seeding, the calculations are first done for a *natural* cloud in which natural ice nucleation processes are simulated, and then they are repeated for a seeded cloud in which enhanced ice particle nucleation is simulated for an assumed amount of seeding material. The difference in height between *natural* and *seeded* clouds is defined as the dynamic seeding potential or *seedability* of clouds that develop in such an environment.

Application of such models to the semi-tropical and tropical atmosphere often resulted in seedability predictions of 2 to 3 km, while in midlatitudes the predicted height changes due to seeding are generally less though some days exhibit large values of predicted explosive growth. Simple models such as these were used to support field experiments by predicting the potential for obtaining significant increases in cloud growth on a given day. They have been also used for identifying how effective seeding actually was. Figure 2.8 illustrates an example of values of seedability predicted versus the observed heights of both seeded and unseeded clouds. These results strongly suggest there is a significant difference between the heights of seeded versus unseeded clouds, at least in the semi-tropics.

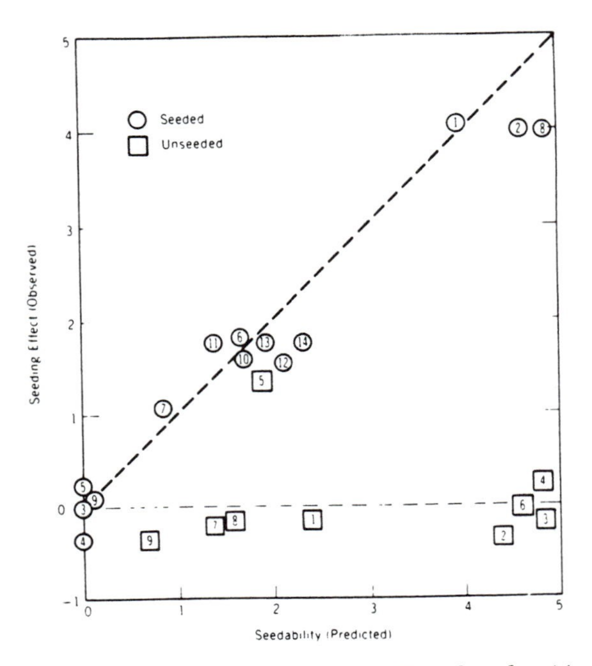

Figure 2.8: *Seedability versus seeding effect for the 14 seeded (circles) and 9 control (squares) clouds studied in 1965. Note that seeded clouds lie mainly along a straight line with slope 1 (seeding effect is close to seedability), while control clouds lie mainly along a straight horizontal line (showing little or no seeding effect regardless of magnitude of seedability). Units of each axis are in kilometers.* [From Simpson, J., G.W. Brier, and R.H. Simpson, 1967: STORMFURY cumulus seeding experiment 1965: Statistical analysis and main results. *J. Atmos. Sci.*, **24**, 508-521. ©American Meteorological Society.]

The higher cloud top heights does not necessarily mean that the desired goal of greater rainfall on the ground has been achieved. It is generally well known that in a population of natural clouds, taller clouds produce more rain on the average (Gagin et al., 1985). Because seeded clouds are altered microphysically, it does not necessarily follow that taller seeded clouds rain more. Some limited exploratory field experiments have been conducted that suggest that seeding clouds for dynamic effects can increase rainfall (Woodley, 1970). Woodley speculated that the seeded clouds were larger, longer lasting, and processed more moisture than their unseeded counterparts resulting in an increase in precipitation. Extensive area-wide, randomized statistical experiments have not been able to confirm the earlier exploratory studies (Barnston et al., 1983;

Woodley et al., 1982a, 1983; Meitín et al., 1984; Dennis et al., 1975). The reasons for the failure of the confirmatory seeding experiments are not fully known but they may be due to: (1) the simple model relating increased cloud growth to enhanced surface rainfall may not work for all clouds and in some environments (i.e., certain wind shear profiles, some mesoscale weather regimes); (2) large *natural variability* of rainfall over fixed targets of large areal extent and inadequate models (physical or statistical) to account for that variability; (3) the size of the sample of clouds seeded and not-seeded was not large enough to accommodate the *natural variability* in rainfall (i.e., a single, heavy rainfall day swamped the natural rainfall statistics).

Lacking in the dynamic seeding research is an identification of the hypothesized chain of physical processes that lead to enhanced rainfall on the ground over a target region. Observations in clouds seeded for dynamic effects showed that seeding did indeed glaciate the clouds (convert the cloud from liquid to primarily ice) [Sax, 1976; Sax et al., 1979; Sax and Keller, 1980; Hallett 1981]. The one-dimensional models clearly predict that artificial glaciation of a cloud should result in increased vertical development of the cloud. Those one-dimensional models, however, cannot simulate the consequences of increased vertical growth. A chain of physical responses to dynamic seeding has been hypothesized (Woodley et al., 1982b) that includes: *(1) pressure falls beneath the seeded cloud towers and convergence of unstable air in the cloud will as a result develop, (2) downdrafts are enhanced, (3) new towers will therefore form, (4) the cloud will widen, (5) the likelihood that the new cloud will merge with neighboring clouds will therefore increase, and (6) increased moist air is processed by the cloud to form rain.*

Few of these hypothesized responses to dynamic seeding have been observationally documented in any systematic way. A few exploratory attempts to identify some of the hypothesized links in the chain of responses were attempted using multiple Doppler radars, but they were largely unsuccessful since they occurred at a time that multiple Doppler technology was still in its infancy. Likewise, two- and three-dimensional numerical prediction models were applied to simulate dynamic effects. These models have the potential for simulating pressure perturbations caused by seeding throughout the cloud, as well as the formation of downdrafts, new towers, cloud merger, and increased rainfall. Only a few attempts were made to simulate dynamic seeding with multidimensional cloud models (Orville and Chen, 1982; Levy and Cotton, 1984) but these simulations did not produce the hypothesized sequence of responses including enhanced rainfall on the ground. This could have been due to the inadequacies of the models at that time, or to the fact the the soundings selected were not ideal for dynamic responses, or the chain of hypothesized events did not occur. More research is needed to determine which is indeed the case.

In summary, the concept of dynamic seeding is a physically plausible hypothesis that offers the opportunity to increase rainfall by much larger amounts than simply enhancing the precipitation efficiency of a cloud. It is a much more complex hypothesis, however, requiring greater quantitative understanding of the behavior of cumulus clouds and their interaction with each other and with larger-scale weather systems. Fundamental research in dynamic seeding essentially terminated at the time that new cloud observing tools and multidimensional cloud models became available to the community. Systematic application of these tools to the study of dynamic seeding would help in evaluating if the hypothesized chain of events is possible and perhaps the conditions under which such responses are likely to occur.

2.4 Modification of Warm Clouds

2.4.1 Introduction

Attempts to augment precipitation by cloud seeding has not been limited to supercooled clouds. In tropical and semi-tropical regions, in particular, clouds too shallow to extend into freezing levels are prevalent. During drought periods, if any clouds form at all, they are generally warm, non-precipitating clouds. The motivation is therefore strong to develop techniques for extracting rainfall from non-ice-phase clouds.

In this section we review the basic physical concepts governing the formation of precipitation in warm clouds. We then discuss the various concepts for enhancing precipitation in warm clouds and describe the physical/statistical experiments and modeling studies attempting to refine and develop warm cloud seeding strategies.

2.4.2 Basic physical concepts of precipitation formation in warm clouds

The process of precipitation formation in warm clouds begins with the nucleation of cloud droplets on hygroscopic aerosol particles. These are airborne dust particles that are normally quite small (i.e., less than one micrometer in diameter) and have a natural affinity for water vapor. Examples are salts such as ammonium sulfate and sodium chloride particles. A rising volume of cloud-free air will cool dry-adiabatically resulting in an increase in relative humidity. At relative humidities greater than about 78%, hygroscopic particles begin taking on water vapor and swell in size. Eventually the relative humidity will exceed 100% and we say the cloud is *supersaturated.* In a supersaturated environment the hygroscopic aerosol particles (CCN) will allow deposition of water vapor on the particle surface eventually forming a cloud droplet. As long as a supersaturation in a cloud is maintained (generally by continued ascent and adiabatic cooling) vapor will deposit on the surface of the droplet, allowing the cloud droplet to grow bigger (see Figure 2.9). Because vapor deposition

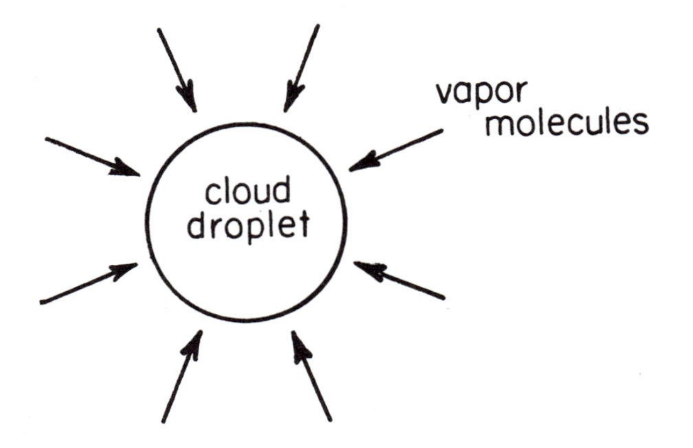

Figure 2.9: *Schematic illustration of a droplet growing by deposition of vapor molecules.*

occurs on the surface of cloud droplets, the rate of growth (in terms of droplet radius) diminishes as the droplet gets bigger. This is because the surface to volume ratio of a nearly spherical droplet gets less the bigger the droplet. That is, the surface area of the droplet diminishes relative to the amount of vapor mass that must be added to the droplet to make it expand. Cloud droplets therefore grow quite rapidly to a size of 4 to 12 μm in radius, but then grow very slowly to radii exceeding 20 μm. In fact, for a droplet to grow by vapor deposition to raindrop size (1000 μm or 1 mm) in a smooth ascending updraft takes more than a day. Vapor deposition is, therefore, not the major process causing the formation of warm rain.

Instead, warm rain formation is dominated by collision and coalescence among droplets. Collision and coalescence refers to the process in which a large drop settling through the air at a high terminal velocity overtakes a small drop with a smaller settling velocity. Figure 2.2 illustrates a large drop that is settling through a population of smaller droplets. The large drop sweeps out a cross-sectional area ($\pi\ a^2$) indicated by the shaded region. Thus the larger the drop relative to a smaller drop, the greater will be its fall velocity through the air relative to a smaller drop, and the greater the cross-sectional area the large drop will sweep out. Likewise the more numerous the smaller drops and the more mass they collectively have (i.e., the higher the liquid water content or mass of water per unit mass of air), the faster will the larger drop grow by colliding and coalescing with smaller droplets.

Unfortunately the problem is a bit more complicated since not all the drops in front of the larger drop shown in Figure 2.2 actually collide and coalesce with the larger droplet. As illustrated in Figure 2.10, streamlines of the air flowing about a drop are quite curved near the drop surface. Some smaller drops near the center line of the big drop depart slightly

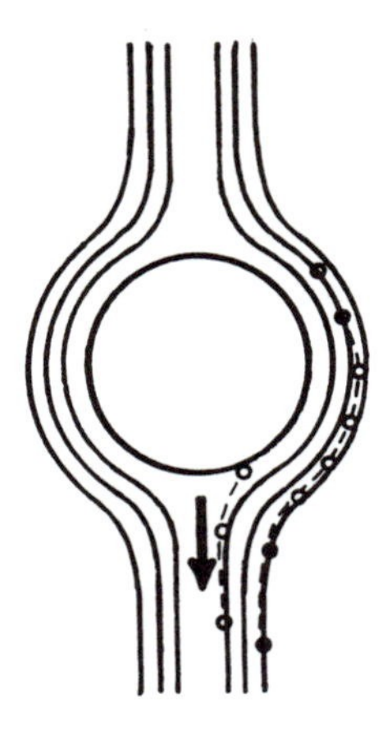

Figure 2.10: *Schematic of streamlines of airflow around a falling droplet.*

from the airflow streamline and make an impact with the larger drop. On the other hand, drops further from the center line of the bigger drop also follow the airflow streamline, departing from it only slightly, and are not able to collide with the larger drop. The likelihood that a small drop will cross streamlines and impact the larger drop is greater the larger the smaller drop. This likelihood of a collision is also greater if the larger drop falls faster through the air.

Moreover, not all the drops that make impact with a smaller drop actually coalesce with the bigger drop. If two drops are of similar size and thus settle through the air relatively slowly, their impact speeds may not be large enough to break the surface tension that keeps a drop intact or squeeze out any air that may get trapped between the drops. While we don't know the numbers precisely, the indications are that a large fraction of drops smaller than 100 μm that make impact actually coalesce. Larger raindrops of similar size, on the other hand, have a higher probability of colliding but not coalescing.

Summarizing all these complicated effects, we find that if all the drops in a cloud are very small, say less than 15 μm in radius, the rate of collision and coalescence among those drops is almost negligible. This is because the relative velocities among the drops is small, the cross-sectional areas swept out by the slightly bigger drops relative to the small ones is also small, and the efficiency of collision and coalescence is nearly zero. On the other hand once there exists a few drops greater than 50 or 100 μm in radius in a cloud containing numerous 5 to 15 μm radius drops (i.e., a cloud containing high liquid water content) those few big drops fall rapidly through the population of small drops, sweep out large cross-sectional areas, and their efficiency of colliding and coalescing with the smaller drops is large. These few larger drops grow rapidly to raindrop size.

The problem is how do those favored drops, which are large enough to grow by collision and coalescence, get to a size of 25 μm and greater? Remember that the rate of change of droplet radius by vapor deposition growth in smooth unmixed updrafts diminishes as droplets become larger. It takes a long time for droplets to grow from 15 μm to 25 μm in radius. We also know that some clouds produce warm rain very rapidly, particularly clouds which reside in a tropical oceanic environment. As noted previously the reasons why warm cloud precipitation is favored in a tropical marine environment are: droplet concentrations are few and liquid water contents are high.

Thus, clouds most favored to create rain by collision and coalescence of warm cloud are clouds that are warm-based and maritime, while those least likely are cold-based and continental.

There are many clouds between these two extremes, however, some produce rain by collision and coalescence and others do not. The reasons why some do and others do not is still being debated by scientists. Some theorize that the particular properties of the turbulence in clouds favors a more rapid growth of few droplets by vapor deposition to a size that collision and coalescence can operate efficiently. Others theorize that many clouds sweep up very large dust particles called ultra-giant particles that can function as coalescence embryos. It is generally difficult to determine which if either of these processes are operational in any given cloud.

2.4.3 Strategies for enhancing rainfall from warm clouds

The goal of most attempts at enhancing rainfall from warm clouds by cloud seeding has been to introduce particles in the cloud that are large enough to function as embryos for collision and coalescence growth. The approach has been to seed clouds with actual water drops or to seed with hygroscopic particles such as sodium chloride which take on water vapor at rapid rates in a supersaturated cloud to become droplets larger in size than the natural population of droplets. Braham et al. (1957) carried out a water drop seeding experiment in maritime tropical trade wind cumulus clouds. They released about 450 gallons of water per mile from an aircraft flying at middle levels in cumulus clouds. The initiation of precipitation was determined from radar observations of the clouds. They found that seeded clouds initiated precipitation 6 minutes earlier than unseeded clouds on the average. Furthermore, twice as many of the seeded clouds observed produced radar-detectable rainfall than the unseeded clouds. Their experiments provided convincing evidence that water spray seeding can enhance the rate of formation of precipitation and in some cases produce rainfall where naturally occurring rain would not occur. Direct measurements of rainfall at the surface were not made, therefore one cannot be certain how effective this technique is in increasing the amount of rain reaching the surface. The primary limitation of this approach is that large amounts of water must be carried aloft by an aircraft. In fact the costs of hauling the water aloft is so great that the technique has not been viewed as being economically feasible.

A more popular approach to seeding warm clouds has been to introduce hygroscopic aerosol particles (salt) ranging in size from 5 to 100 μm or greater. Once the salt particles experience a supersaturated cloudy environment, they grow rapidly in size by vapor deposition becoming droplets 25 to 30 μm in radius. Mason (1971) calculated that 100 grams of salt would be equivalent to seeding with one gallon of water ($\sim$ 3600 grams) in 25 μm radius drops. Thus hygroscopic seeding has a much greater potential for economic payoff in enhancing rainfall. Furthermore, a number of investigators have performed ground-based hygroscopic seeding experiments thus eliminating the use of costly aircraft.

Unfortunately most of the hygroscopic seeding experiments have been 'black box' type experiments in which clouds are randomly seeded from aircraft or the ground and rainfall is measured. These experiments yield little insight into the actual physical responses to seeding and can provide only the answer to the question – does seeding increase rainfall by a measurable amount? Many of those experiments have been inconclusive while a few suggested strong increases in rainfall (Roy et al., 1961; Biswas et al., 1967; Murty and Biswas, 1968) and others decreases in rainfall (Fournier d'Albe and Aleman, 1976). Why some experiments yielded increases while others decreases in rainfall is not known. It is quite possible that the large *natural variability* of rainfall could be a factor. That is the experiments with decreases in rainfall could have been strongly influenced by a few natural, heavy rainfall events in the unseeded population of clouds. Conversely, those experiments with apparent large increases in rainfall could have experienced well below normal rainfall in the unseeded population of clouds.

The salt seeding experiments carried out in northwestern India are another excellent example of a 'black box' experiment (Roy et al., 1961; Biswas et al., 1967; Murty and Biswas, 1968). These ground-based seeding experiments suggested that a 41.9% increase in rainfall occurred on seed days in the downwind direction. Many scientists remain skeptical about the results (Mason, 1971; Simpson and Dennis, 1972; Warner, 1973) largely due to the lack of supporting physical evidence that the seeded clouds exhibited a significantly different microstructure from unseeded clouds. This illustrates that even a well-designed statistical experiment will not be accepted by the scientific community as being credible unless that experiment is supported by physical evidence that: (1) the seeding material actually entered the clouds, (2) the seeded clouds exhibit broader droplet spectra than unseeded clouds, (3) raindrops are initiated earlier, lower in the cloud, with higher drop concentrations, and (4) larger amounts of rainfall actually reach the ground.

Some scientists have hypothesized that hygroscopic seeding will initiate a premature Langmuir chain reaction process (Biswas and Dennis, 1971; Dennis and Koscielski, 1972). If hygroscopic seeding initiates a Langmuir chain reaction, then measurably larger concentrations of raindrops should be observed in seeded clouds. Unfortunately no documentation of such higher raindrop concentrations have been reported.

While the main thrust of most hygroscopic seeding experiments has been to increase rainfall by enhancing the collision and coalescence of liquid drops, there also exists the possibility that hygroscopic seeding may enhance ice phase precipitation in taller, colder clouds. It has been known for some time that the speed of conversion of a cloud from an all liquid supercooled cloud to an all ice cloud (what we call a glaciated cloud) is greater if the cloud contains a broad droplet spectrum (Cotton, 1972b; Koenig and Murray, 1976; Scott and Hobbs, 1977). Supercooled raindrops speed up the glaciation process by rapidly collecting ice crystals which causes the raindrops to freeze. The frozen raindrops, in turn, collect supercooled cloud droplets which freeze on impact with the frozen raindrop.

A broad droplet spectrum also aids faster glaciation of a cloud by favoring secondary ice crystal production by the rime-splinter mechanism (Chisnell and Latham, 1976a,b; Koenig, 1977; Lamb et al., 1981). As noted previously, Hallett and Mossop (1974) and Mossop and Hallett (1974) showed in laboratory experiments that when a frozen raindrop or graupel particle collects supercooled raindrops, secondary ice crystals are produced (see Figure 2.4). The process occurs most vigorously between – 4 and – 8° C and when a broad droplet spectrum is present including drops smaller than about 7 μm and drops greater than 12.5 μm radius.

Hygroscopic seeding, therefore, has the potential for creating a broad droplet spectrum and therefore initiating, or enhancing, rime-splinter production of small ice crystals, and of enhancing the formation of supercooled raindrops which can accelerate the glaciation of a cloud. Hygroscopic seeding could also create dynamic responses such as postulated with silver iodide seeding described in Section 2.3. No observational confirmation of such responses to hygroscopic seeding has been made, however.

In summary, there appears to be a real opportunity to enhance rainfall through hygroscopic seeding in some clouds. It has not been determined how open the 'window of opportunity' actually is. In warm-based, maritime clouds the rate of natural production of rainfall may be so great that there is little opportunity to beat nature at its own game. On the other hand, some cold-based continental clouds may have so many small droplets that seeding-produced big drops cannot collect them owing to very small collection efficiencies. Thus there probably exists a spectrum of clouds between these two extreme types that have enough liquid water to support a warm cloud precipitation process that can be accelerated by hygroscopic seeding. The problem is to identify those clouds, and deliver the right amount of seeding material to them at the right time.

2.5 Hail Suppression

2.5.1 Introduction

For hundreds if not thousands of years man has sought techniques for suppressing hail. This has ranged from ringing bells, to firing cannons, to the modern era of cloud seeding (Changnon and Ivens, 1981). The motivation for suppressing hail is strong since hail can devastate a wheat or grape crop, and when a major hail storm occurs over a modern urban area millions of dollars of property damage can result and humans and livestock can be severely injured or even killed (i.e., the Fort Collins, Colorado hailstorm of 1983). Scientific programs aimed at developing a hail suppression effort using cloud seeding strategies began in the 1950's in the Soviet Union and in Alberta, Canada. In the early 1960's reports of major success in hail suppression in the Soviet Union were made by visiting United States scientists. Subsequently several delegations of U.S. scientists visited the various hail suppression programs in the Soviet Union (e.g., Battan, 1969; Marwitz, 1972).

Not to be outdone by the Soviet Union during the cold war era, in May 1965 the Interdepartmental Committee on Atmospheric Sciences (ICAS), which serves as a coordinating group for U.S. federal meteorological activities, asked the National Science Foundation to prepare plans for a national program of hail suppression. This led to the formation of a pilot program called Hailswath which was carried out in South Dakota and Colorado in the summer of 1966. This was followed by the formation of the National Hail Research Experiment (NHRE) that operated both a randomized cloud seeding program patterned to some degree on the Soviet hail suppression program and basic physical studies of hailstorms and hail growth mechanisms. NHRE operated field programs during the period 1972 to 1976.

In this section we summarize basic concepts of hailstorms and hail formation processes, review the concepts for suppressing hail by cloud seeding, and examine the status of our ability to suppress hail.

2.5.2 Basic concepts of hailstorms and hail formation

In order to understand how hail forms, one must first understand the varying behavior of thunderstorms which are the factories where hailstones are produced. A thunderstorm produces the liquid water content upon which hailstones grow. The updrafts then suspend the embryo hailstones, which ultimately determines how long a growing hailstone will remain in a water-rich environment.

The primary energy driving a thunderstorm is the buoyancy experienced by updrafts as latent heat is released as vapor condenses to form cloud droplets, supercooled drops freeze, and vapor deposits on ice crystals. The buoyancy, which is determined by the temperature excess of an updraft relative to its environment multiplied by the acceleration due to gravity, is a local measure of the acceleration of the updraft. The actual magnitude of the updraft strength at any height in the atmosphere is largely determined by the integrated buoyancy that an updraft experiences as it ascends from cloud base to a given height in the atmosphere. The integrated buoyancy through the atmosphere is called convective available potential energy (CAPE). In general, the greater the CAPE, the stronger the strength of updrafts in a thunderstorm.

The likelihood that hail will form, the size of hailstones, and the extent of the hailswath, is not only determined by how much CAPE there is in the atmosphere at any given time. Other environmental factors also are important in initiating thunderstorms and determining the particular flow structure of the storm system. For example, as an updraft in a thunderstorm rises through the atmosphere, it carries with it the horizontal momentum that is characterized by the winds at the updraft source level. As the updraft ascends it encounters air having differing horizontal momentum (i.e., different wind speeds and direction). The vertical variation in horizontal wind speed and direction is called *wind shear*. The interaction of the updraft with ambient air having different horizontal momentum causes the updraft to tilt from the vertical and creates pressure anomalies in the air that can also accelerate the air. Thus the complicated interactions of updraft and downdraft air with an environment having vertical shear of the horizontal wind can alter the storm structure markedly. For example, *ordinary thunderstorms* develop in an atmosphere containing moderate amounts of CAPE and weak to moderate vertical wind shear.

The ordinary cumulonimbus cloud or thunderstorm is distinguished by a well-defined life cycle as shown in Figure 2.11, that lasts 45 minutes to one hour. Figure 2.11a illustrates the growth stage of the system. The growth stage is characterized by the development of towering cumulus clouds, generally in a region of low-level convergence of warm moist air. Often during the growth stage, towering cumulus clouds merge to form a larger cloud system or a vigorous cumulus cloud expands to a larger cloud. During the growth stage, updrafts dominate the cloud system and precipitation may be just beginning in the upper levels of the convective towers.

The mature stage of a cumulonimbus cloud commences with rain settling in the subcloud layer. Upon encountering the surface, the downdraft air spreads horizontally, where it can lift the warm moist air into the cloud system. At the interface between the cool, dense downdraft air and the warm, moist air, a gust front forms. Surface winds at the gust front are squally; rapidly changing direction and speed. The warm, moist air lifted by the gust front provides the fuel for maintaining vigorous updrafts. Upon encountering the extreme stability at the top of the troposphere, called the tropopause, the updrafts spread laterally, spewing ice crystals and other cloud debris horizontally to form an anvil cloud. In many cases, the updrafts are strong enough that they penetrate into the lower stratosphere creating a cloud dome. Often the stronger updrafts form a thin layered cloud that caps the cloud and is separated from the main body of the cloud which is called *pileus*. The presence of pileus is a visual clue that strong updrafts exist in the storm.

Water loading and the entrainment of dry environmental air into the storm generate downdrafts in the cloud interior, which rapidly transport precipitation particles to the subcloud layer. The precipitation particles transported to the subcloud layer partially evaporate, further chilling the subcloud air and strengthening the low-level outflow and gust front. Thus, continued uplift of warm, moist air into the cloud system is sustained during the mature stage. Lowering of pressure at middle-levels in the storm as a result of warming by latent heat release and diverging air flow results in an upward-directed, pressure-gradient force which helps draw the warm, moist air lifted at the gust front up to the height of the level of free convection. Thus, the thunderstorm becomes an efficient "machine" during its mature stage, in which warming aloft and cooling at low levels sustains the vigorous, convective cycle.

The intensity of precipitation from the storm reaches a maximum during its mature stage. Therefore, the mature cumulonimbus cloud is characterized by heavy rainfall and gusty winds, particularly at the gust front (see Figure 2.12). The propagation speed of the gust front increases as the depth of the outflow air increases and the temperature

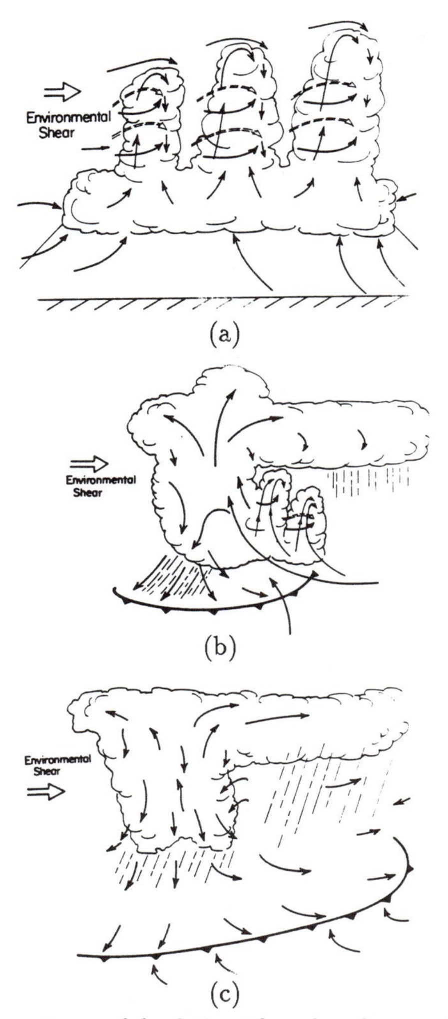

Figure 2.11: *Schematic model of the lifecycle of an ordinary thunderstorm. (a) The cumulus stage is characterized by one or more towers fed by low-level convergence of moist air. Air motions are primarily upward with some lateral and cloud top entrainment depicted. (b) the mature stage is characterized by both updrafts and downdrafts and rainfall. Evaporative cooling at low levels forms a cold pool and gust front which advances, lifting warm moist, unstable air. An anvil at upper levels begins to form. (c) The dissipating stage is characterized by downdrafts and diminishing convective rainfall. Stratiform rainfall from the anvil cloud is also common. The gust front advances ahead of the storm preventing air from being lifted at the gust front into the convective storm.* [From Cotton, W.R. and R.A. Anthes, 1989: *Storm and Cloud Dynamics.* Academic Press, Inc., San Diego. International Geophysics Series, Vol. 44., 883 pp.]

of the outflow air decreases. The optimum storm system is one in which the speed of movement of the gust front is closely matched to the speed of movement of the storm as a whole.

Once the gust front advances too far ahead of the storm system, air lifted at the gust front does not enter the updraft air of the storm, but may only form fair weather cumulus clouds. This marks the beginning of the dissipating stage of the thunderstorm shown in Figure 2.11c. During the dissipating stage, updrafts weaken and downdrafts become predominant, rainfall intensity subsides often turning into a period of light steady rainfall.

Many hail-producing thunderstorms are composed of a number of cells, each undergoing a life cycle of 45 to 60 minutes. The thunderstorm system, called a *multicell storm*, may have a lifetime of several hours. Figure 2.13 illustrates a conceptual model of a severe multicell storm as depicted by radar. At 3 minutes, the vertical cross section through a vigorous cell, C1, shows a weak echo region (WER) at low levels, where there is an absence of precipitation particles large enough to be seen on radar. Updrafts in this region are so strong that there is not enough time for precipitation to form. The region of intense precipitation at middle levels is due to a cell that existed previous to C1. By 9 minutes, Cell C1 exhibits a well defined precipitation maximum at middle levels, while the weak echo region has disappeared. By 15 minutes, precipitation from C1 has reached the surface and a new cell, C2, is evident on the right forward flank of the storm (relative to storm motion). By 21 minutes, the maximum of precipitation associated with C1 has lowered and C2 has grown in horizontal extent. Cell C2 will soon become the dominant cell of the storm, only to be replaced by another cell shortly.

Multicell storms, where updrafts reach 25 to 35 m s^{-1}, can produce extensive moderate-sized hailstones (i.e., golfball-sized). Severe multicell storms typically occur in regions where the atmosphere is quite unstable and where the vertical wind shear is moderate in strength.

The grandaddy of all thunderstorms is the *supercell* thunderstorm. It is noted for its persistence, lasting for two to six hours, in a single cell structure. Figure 2.14 is a schematic illustration of the dominant airflow branches in a supercell storm. The storm is often characterized by a broad, intense updraft entering its southeast flank, rising vertically and then, in the northern hemisphere, turning clockwise in the anvil outflow region. Updrafts in supercell storms may exceed 40 m s^{-1}, capable of suspending hailstones as large as grapefruit. Horizontal and vertical cross sections of a supercell storm as viewed by radar are shown in Figure 2.15a and 2.15b. A distinct feature of the supercell storm is the region that is free of radar echo that can be seen on the southeast flank of the storm at the 4 and 7 km levels and in the vertical cross section. This persistent feature, called a bounded weak echo region (BWER), or echo free vault, is a result of the strong updrafts in that region which do not provide

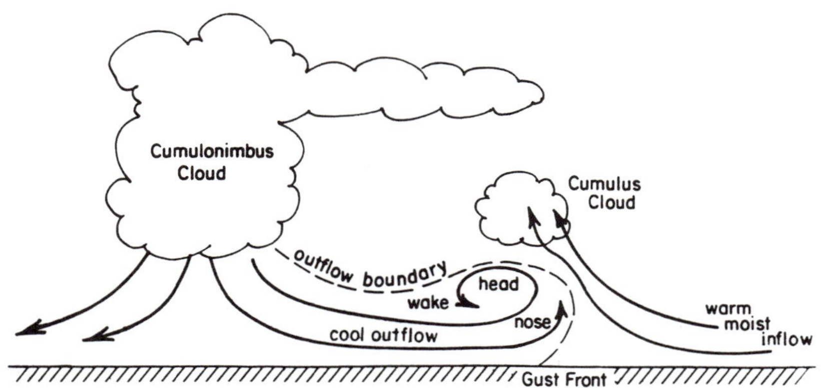

Figure 2.12: *Illustration of a gust front formed at the leading edge of the downdraft outflow from a thunderstorm.* [From Cotton, W.R., 1990: *Storms.* Geophysical Science Series, Vol. 1. ASTeR Press, Fort Collins, CO, 158 pp.]

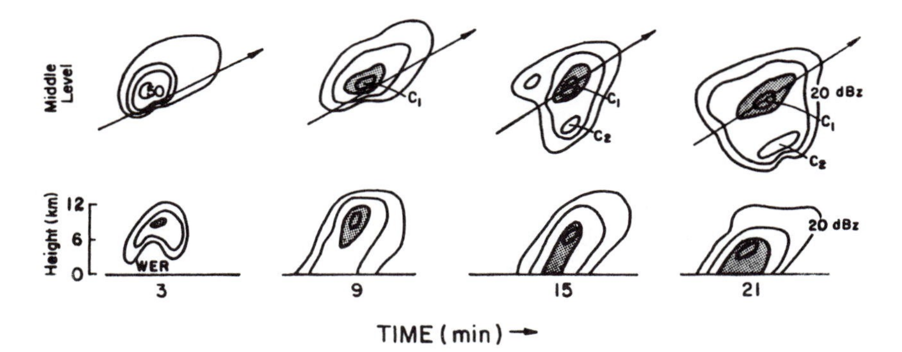

Figure 2.13: *Conceptual model of horizontal and vertical radar sections for a multicell storm at various stages during its evolution, showing reflectivity contours at 10 dBZ intervals. Horizontal section is at middle levels (~6 km) and the vertical section is along the arrow depicting cell motion. The lifecycle of Cell C_1 is depicted on the right flank of the storm beginning at 15 minutes.* [Adapted from Chisholm, A.J., and J.H. Renick, 1972: The kinematics of multicell and supercell Alberta hailstorms, Alberta Hail Studies, 1972. Research Council of Alberta Hail Studies Rep. No. 72-2, 24-31.]

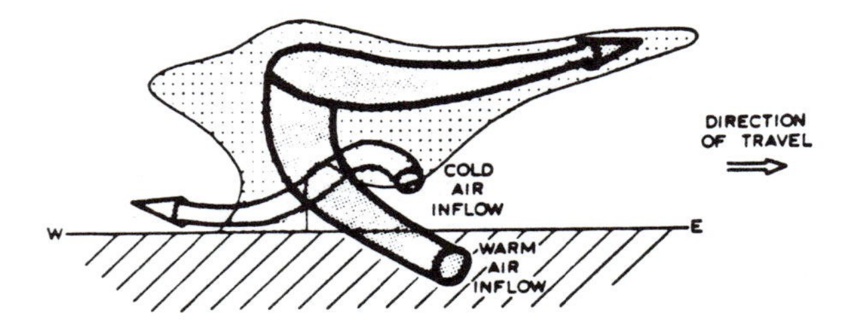

Figure 2.14: *Model showing the airflow within a 3-dimensional severe right-turning (SR) storm traveling to the right of the tropospheric winds. The extent of precipitation is lightly stippled and the up- and downdraft circulations are shown more heavily stippled. Air is shown entering and leaving the updraft with a component into the plane of the diagram. However, the principal difference of this organization is that cold air inflow, entering from outside the plane of the vertical section, produces a downdraft ahead of the updraft rather than behind it.* [From Browning, K.A., 1968: The organization of severe local storms. *Weather,* **23,** 429-434. ©Royal Meteorological Society.]

enough time for precipitation to form in the rapidly-rising air. As noted previously, severe multicell thunderstorms exhibit weak echo regions, but they are neither as persistent as in supercell storms, nor do they maintain the characteristic bounded structure of supercells. Cyclonic rotation (or counter-clockwise in the northern hemisphere) of the updrafts may also contribute to the BWER by causing any precipitation elements that do form to be centrifuged laterally out of the rotating updraft region. Supercell storms are also known as severe right-moving storms because, in the northern hemisphere, most supercells move to the right of the mean flow due to the interaction of the updraft with environmental shear. The rotational characteristic of supercells, as well as their strong updrafts, results in a storm system which produces the largest hailstones.

Between the severe multicell thunderstorm type and the supercell-type storm exist a continuum of thunderstorm types. Some of these are quite steady, exhibiting a dominant cell for a time and a rotating updraft, but also contain transient cell groups for part of the storm lifetime or at the flanks of the dominant cell and thus do not meet all the criteria of a supercell storm. The more a thunderstorm resembles a supercell thunderstorm, the more likely it is to produce very large hailstones and long, continuous swaths of hail.

Hailstorms generally occur in an environment with large values of CAPE. In such an environment, thunderstorms develop significant positive buoyancy and associated strong updrafts capable of suspending hail-

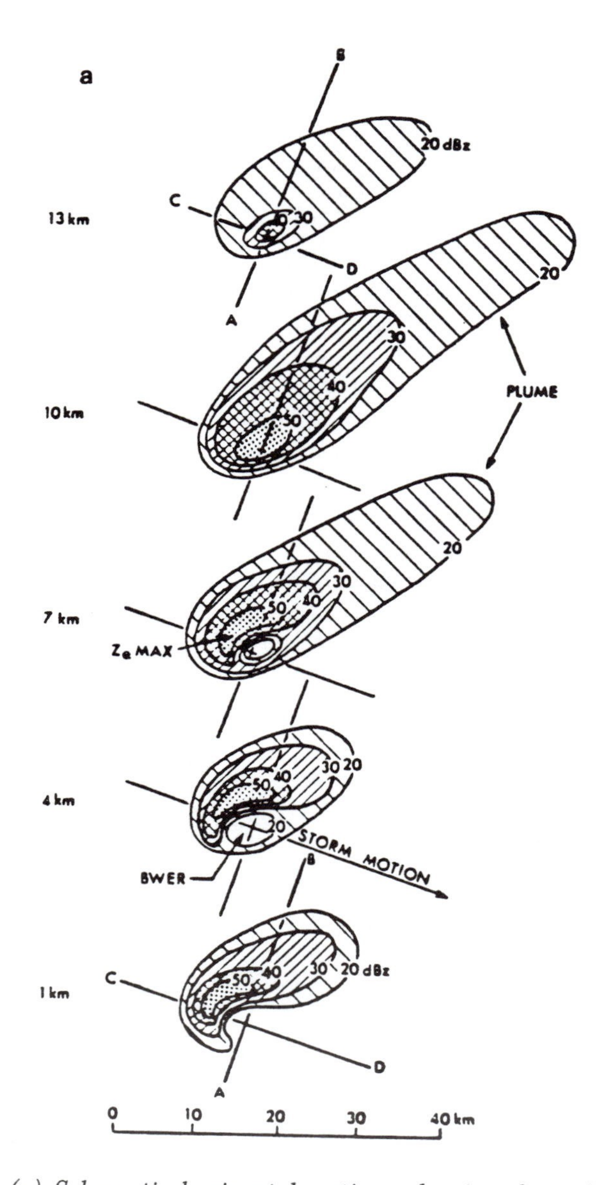

Figure 2.15: *(a) Schematic horizontal sections showing the radar structure of a unicellular supercell storm at altitudes of 1, 4, 7, 10, and 13 km AGL. Reflectivity contours are labeled in dBZ. Note the indentation on the right front quadrant of the storm at 1 km, which appears as a weak-echo vault (or BWER, as it is labeled here) at 4 and 7 km. On the left rear side of the vault is a reflectivity maximum extending from the top of the vault to the ground (see Fig 2.15b.)* [From Chisholm, A.J., and J.H. Renick, 1972: The kinematics of multicell and supercell Alberta hailstorms, Alberta Hail Studies, 1972. Research Council of Alberta Hail Studies Rep. No. 72-2, 24-31.]

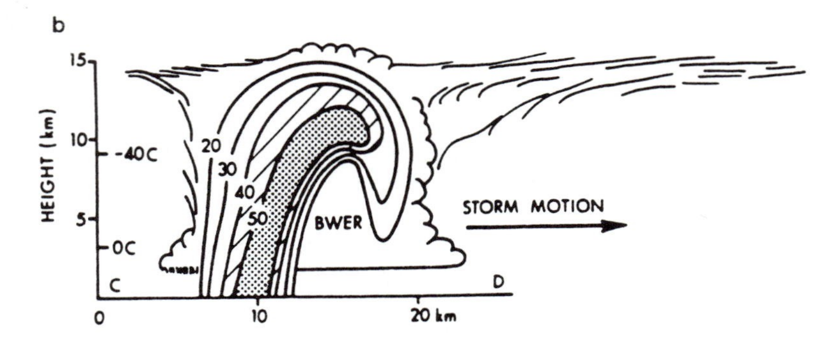

Figure 2.15: *(b) Schematic vertical section through a unicellular supercell storm in the plane of storm motion (along CD in Figure 2.15a). Note the reflectivity maximum, referred to as the hail cascade, which is situated on the (left) rear flank of the vault (or BWER, as it is labeled here). The overhanging region of echo bounding the other side of the vault is referred to as the* embryo curtain, *where it is shown to be due to millimetric-sized particles some of which are recycled across the main updraft to grow into large hailstones.* [From Chisholm, A.J., and J.H. Renick, 1972: The kinematics of multicell and supercell Alberta hailstorms, Alberta Hail Studies, 1972. Research Council of Alberta Hail Studies Rep. No. 72-2, 24-31.]

stones falling through the air at speeds of 15 to 25 m s^{-1}. Often, severe thunderstorms producing large hail also produce tornadoes and even flash floods. Storms producing the largest hailstones normally develop in an environment with strong wind shear which favors the formation of supercell thunderstorms. The height of the melting level is also important in determining the size of hailstones that will reach the surface. It has been estimated that as many as 42% of the hailstones falling through the 0° C level melt before reaching the ground over Alberta, Canada, while it may be as high as 74% over Colorado and 90% in southern Arizona. This is consistent with observations indicating that the frequency of hail is greater at higher latitudes. Any thunderstorm with a radar echo top over 8 km near Alberta, Canada has a significant probability of producing damaging hail!

Hailstone growth is a complicated consequence of the interaction of the airflow in thunderstorms and the growth of precipitation particles. Hailstones grow primarily by collection of supercooled cloud droplets and raindrops. At temperatures colder than 0° C, many cloud droplets and raindrops do not freeze and can remain unfrozen to temperatures as cold as –40° C. A few ice particles do freeze, however, perhaps by collecting an aerosol particle that can serve as a freezing nucleus. If the frozen droplet is small, it will grow first by vapor deposition forming snowflakes such as dendrites, hexagonal plates, needles, or columns. After some time, perhaps five to ten minutes, the ice crystals become large enough to settle relative to small cloud droplets, which immediately freeze when they impact the surface of the ice particle. If enough cloud droplets are present or the supercooled liquid water content of the cloud is high, the ice particle can collect enough cloud droplets so that the original shape of the vapor grown crystal becomes obscured and the ice particle becomes a graupel particle of several millimeters in diameter. At first, the density of the graupel particle is low as the collected frozen droplets are loosely compacted on the surface of the graupel particle. As the ice particle becomes larger, it falls faster, sweeps out a larger cross-sectional area, and its growth by collection of supercooled droplets increases proportionally (see Figure 2.16). As the growth rate increases, the collected droplets may

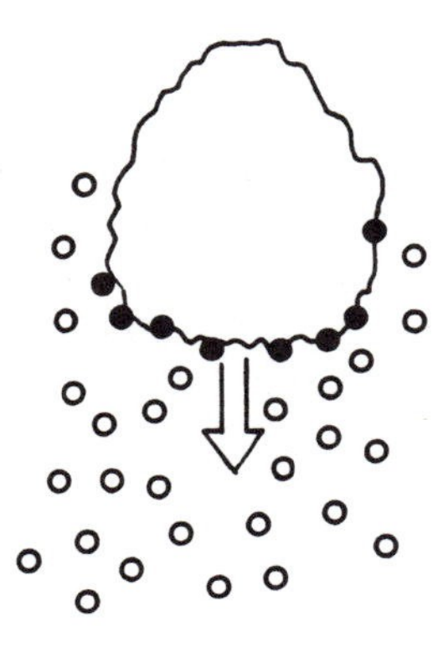

Figure 2.16: *Illustration of a hailstone growing by collecting supercooled droplets.* [From Cotton, W.R., 1990: *Storms.* Geophysical Science Series, Vol. 1. ASTeR Press, Fort Collins, CO, 158 pp.]

not freeze instantaneously upon impact, and thus flow over the surface of the hailstone filling in the gaps between collected droplets. The density of the ice particle, therefore, increases close to that of pure ice as the dense hailstone falls still faster, growing by collecting supercooled droplets as long as the cloud liquid water content is large. The ultimate size of the hailstone is determined by the amount of supercooled liquid water in the cloud and the time that the growing hailstone can remain in the high rainwater region. The time that a hailstone can remain in the high liquid water content region, in turn, is dependent on the updraft speed and the fall speed of the ice particle. If the updraft is strong, say 35-40 m s^{-1}, and the particle fall-speed through the air is only of the order of 1-2 m s^{-1}, then the ice particle will be rapidly transported into the anvil of the cloud before it can take full advantage of the high liquid water content region. The ideal circumstance for hailstone growth is that the ice particle reaches a large enough size as it enters the high liquid water content region of the storm so that the ice particle fall speed nearly matches the updraft speed. In such a case, the hailstone will only slowly ascend or descend while it collects cloud droplets at a very high rate. Eventually, the hailstone fall speed will exceed the updraft speed or it will move into a region of weak updraft or downdraft. The size of the hailstone reaching the surface will be greatest if the large airborne hailstone settles into a vigorous downdraft, as the time spent below the 0° C level will be lessened and the hailstone will not melt very much. Thus, a particular combination of airflow and particle growth history is needed to produce large hailstones. Let us now examine several conceptual models of hailstone growth in the different thunderstorm models we have described previously.

The Soviet Hail Model

The Soviet hail model builds on the ordinary thunderstorm model illustrated in Figure 2.17. If the storm develops particularly vigorous updrafts and high liquid water contents during the growth stage, raindrops may form by collision and coalescence with smaller cloud droplets. As the growing raindrops are swept aloft, they continue to grow and eventually ascend into supercooled regions. If the updraft exhibits a vertical profile as shown in the insert of Figure 2.17, with a maximum updraft speed (w) in the layer between –10 to –20° C, many raindrops may become suspended just above the updraft maximum. This is because the fall speed of raindrops approaches a maximum value slightly greater than 9 m s^{-1} for raindrops larger than 2.3 mm in radius (see lower insert Figure 2.17). The region just above the updraft maximum serves as a trap for large raindrops, and rainwater accumulates in the region as long as an updraft speed greater than 9 m s^{-1} persists. Supercooled liquid water contents greater than 17 g m^{-3} have been reported in such regions. Normal values of supercooled liquid content rarely exceed 2-4 g m^{-3}.

If now a few supercooled raindrops freeze in the zone of accumulated liquid water content, they will experience a liquid-water-rich environment and hailstone growth can proceed quite rapidly. The frozen supercooled raindrops serve as very effective hailstone embryo, greatly accelerating the rate of formation of millimeter-sized ice particles. Recent observations of thunderstorms near Huntsville, Alabama, U.S.A., with multiparameter Doppler radar revealed regions of very high radar reflectivity where ice particles were not detected. Light hailfall was observed at the surface. These observations are consistent with the Soviet hail model.

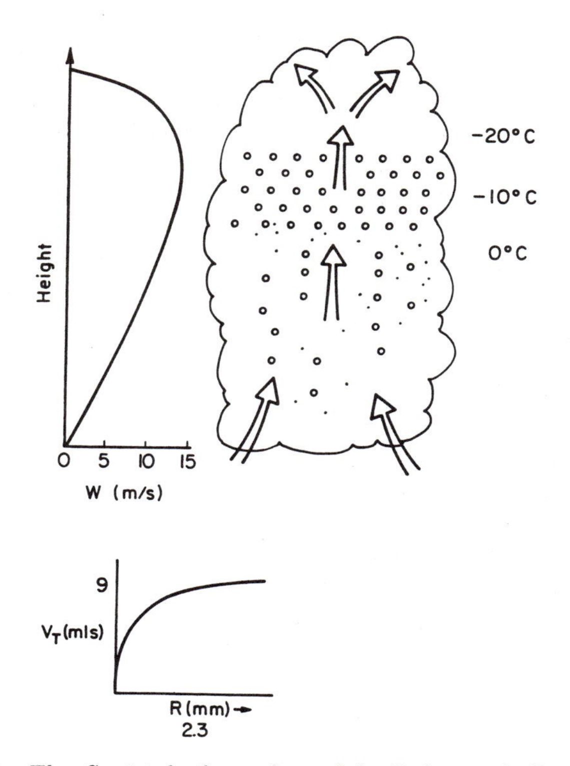

Figure 2.17: *The Soviet hailgrowth model. Left panel illustrates a favorable updraft profile. Right panel illustrates the formation of an accumulation zone. Bottom panel illustrates variation in terminal velocity of raindrops with size.* [From Cotton, W.R., 1990: *Storms.* Geophysical Science Series, Vol. 1. ASTeR Press, Fort Collins, CO, 158 pp.]

Conceptual Model of Hail Formation in Ordinary Multicell Thunderstorms

Previously we described a multicell thunderstorm as a storm containing several cells each undergoing a life cycle of 45-60 minutes. Over the High Plains of the United States and Canada, multicell thunderstorms are found to be prolific producers of hailstones; if not the very largest in size, at least the most frequent. Large hailstones grow during the mature stage of the cells illustrated in Figure 2.13 when updrafts may exceed 30 m s^{-1}. In such strong updrafts, the time available for the growth of hailstones from small ice crystals to lightly rimed ice crystals, to graupel particles or aggregates of snowflakes, to hailstone embryos, is only 5 or 6 minutes! This time is far too short, as it takes some 10-15 minutes for an ice particle to grow large enough to begin collecting supercooled cloud droplets or aggregating with other ice crystals to form an embryonic hailstone. The mature stage of each thunderstorm cell provides the proper updraft speeds and liquid water contents for mature hailstones to grow, but they must be sizeable precipitation particles at the time they enter the strong updrafts in order to take advantage of such a favored environment. Here is where the growth stage of each cell is very important to hailstone growth, as the weaker, transient, updrafts provide sufficient time for the growth of graupel particles and aggregates of snow crystals, which can then serve as hailstone embryos as the cell enters its mature stage. The growth stage of the multicellular thunderstorm thereby preconditions the ice particles and allows them to take full advantage of the high water contents of the mature stage of the storm. Upon entering the mature stage, the millimeter-sized ice particles settle through the air at 8-10 m s^{-1} and therefore rise slowly as the updrafts increase in speed from 10-15 m s^{-1} at low levels to 25-35 m s^{-1} at higher levels.

As mentioned previously, the optimum growth of a hailstone occurs when the updraft speed exceeds the particle fall-speed by only a few meters per second. In that case, the hailstone rises slowly only a few kilometers as it collects supercooled droplets. Fortunately, not all multicellular thunderstorms develop hailstone embryos of the appropriate sizes during the growth stage nor do the embryos enter the updrafts of the mature cell at the right location for optimum hailstone growth.

Conceptual Model of Hailstone Growth in Supercell Thunderstorms

We have seen that the supercell thunderstorm is a steady thunderstorm system consisting of a single updraft cell that may exist for 2-6 hours. Updraft speeds are so strong that they are characterized by having a bounded weak echo region (see Figure 2.15a and 2.15b) in which precipitation particles of a radar-detectable size do not form. Nonetheless, the supercell thunderstorm produces the largest hailstones, sometimes over very long swaths. Consider for example, the Fleming hailstorm that occurred on 21 June 1972. Figure 2.18 illustrates that this storm first reached supercell proportions in northeast Colorado and produced a nearly continuous swath of damaging hail 300 km long over eastern Colorado and western Kansas. How can a storm system consisting of a single, steady updraft with speeds in excess of 30 m s^{-1} develop hailstones before the ice particles are thrust into the anvil of the storm?

Browning and Foote (1976) visualized hail growth in a supercell thunderstorm as a three-stage process illustrated in Figure 2.19. During Stage 1, hail embryos form in a relatively narrow region on the edge of the updraft, where speeds are on the order of 10 m s^{-1}, allowing time for the growth of millimeter-sized hail embryos. Those particles forming on the western edge of the main updraft have a good chance of sweeping around the main updraft and entering the region called the *embryo curtain* on the right-forward flank of the storm. These particles will follow the tra-

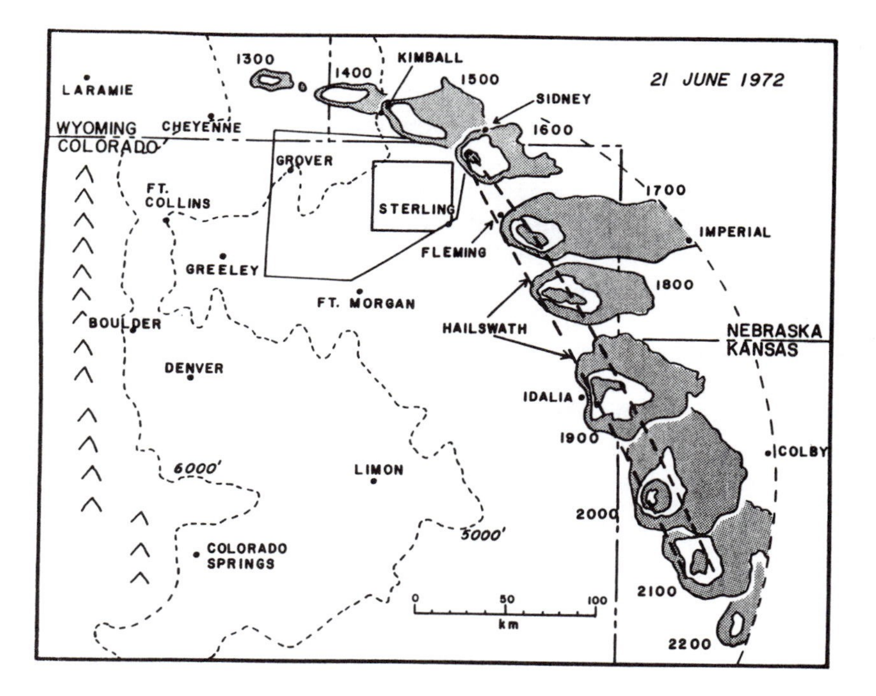

Figure 2.18: *Hourly positions of the Fleming hailstorm as determined by the NWS Limon radar (CHILL radar data used 1300-1500 MDT). The approximate limits of the hailswath are indicated by the bold, dashed line. Continuity of the swath is not well established but total extent is. Special rawinsonde sites were located near the towns of Grover, Ft. Morgan, Sterling and Kimball. Contour intervals are roughly 12 dBZ above 20 dBz.* [From Browning, K.A., and G.B. Foote, 1976: Airflow and hail growth in supercell storms and some implications for hail suppression. *Quart. J. Roy. Meteor. Soc.,* **102,** 499-533. ©Royal Meteorological Society.]

jectory labeled 1 in Figure 2.19. By contrast, particles that enter the main updraft directly follow the trajectory labeled 0 and do not have sufficient time to grow to hailstone size. They are thrust out into the storm anvil.

During Stage 2, the embryos formed on the western edge of the main updraft are carried along the southern flank of the storm by the diverging flow field. Some of the larger embryos settle into the region of weak updrafts that characterizes the embryo curtain. The particles following the trajectory labeled 2 experience further growth as they descend in the embryo-curtain region. Some of the particles settle out of the lower tip of the embryo curtain and re-enter the base of the main updraft, commencing Stage 3.

Stage 3 represents the mature and final stage of hail growth, in which the hailstones experience very high liquid water concentrations and grow by collecting numerous cloud droplets during their ascent in the main updraft. The growth of hailstones from embryos is viewed as a single up-and-down cycle. Those embryos which enter the main updraft at lowest levels, where the updraft is weakest, are likely to have their fall speed nearly balanced by the updraft speed. As a result of their slow rise rate, they will have plenty of time to collect the abundant supercooled liquid water. Eventually their fall velocities will become large enough to overcome the large updraft speeds and/or they will move into the downdraft region and descend to the surface on the northern flank of the storm.

Some researchers argue that hail embryos are not formed on the flanks of a single, main updraft, where updraft speeds may be weaker, but, instead the embryos form in towering cumulus clouds that are flanking the main updraft of the storm. They argue that the towering cumulus clouds are obscured by the precipitation debris falling out of the main updraft. Embryos can form in the transient, weaker towering convective elements and 'feed' the main updraft with millimeter-sized embryos. In some thunderstorms, such transient, flanking cumulus towers are visually separated from the dominant parent cell. Such thunderstorms are a hybrid between the single supercell storm and the ordinary multicell storm and are called *organized multicell thunderstorms* or *weakly evolving thunderstorms.* Such thunderstorms are characterized by a single, dominant cell which maintains a steady flow structure similar to a supercell thunderstorm, including a BWER and a rotating updraft. They also contain distinct flanking towering cumulus clouds that can serve as the manufacturing plant for hailstone embryos that can settle into the main updraft of the steady cell. The so-called *'feeder'* cells have to be relatively close to the parent cell in order to be effective suppliers of hailstone embryos. Whether or not a supercell actually contains such embedded feeder clouds or is just a single cell is not known at the present time.

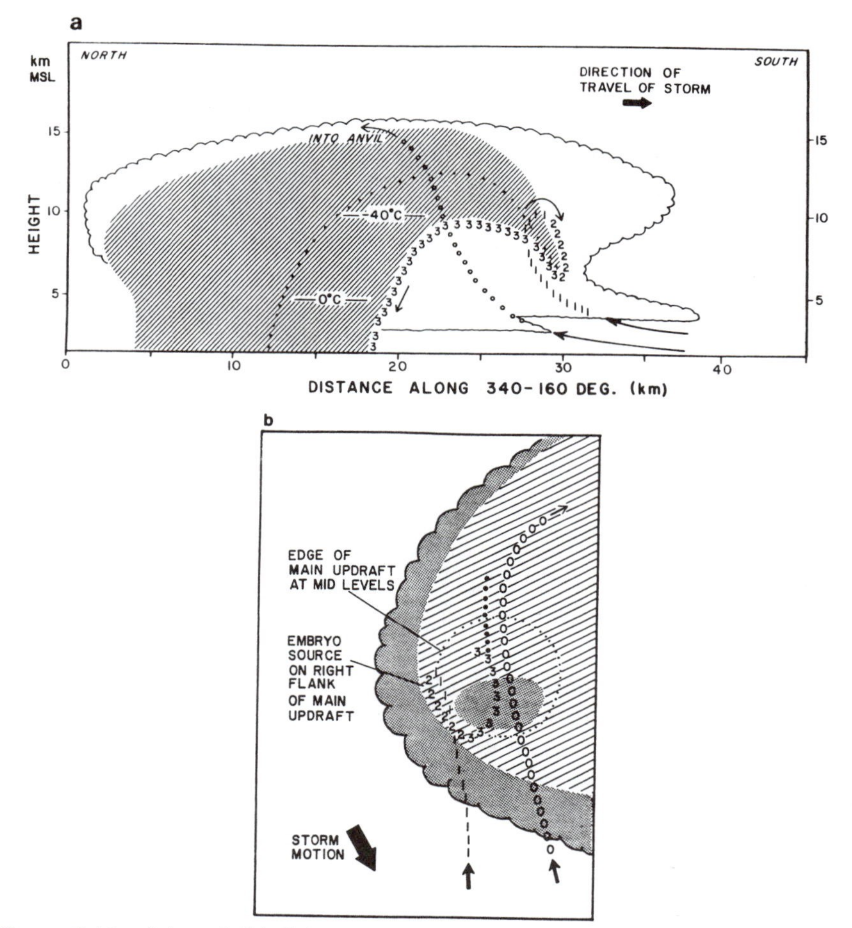

Figure 2.19: *(a) and (b) Schematic model of hailstone trajectories within a supercell storm based upon the airflow model inferred by Browning and Foote (1976). (a) Shows hail trajectories in a vertical section along the direction of travel in the storm. (b) Shows these same trajectories in plan view. Trajectories 1, 2, and 3 represent the three stages in the growth of large hailstones discussed in the text. The transition from stage 2 to 3 corresponds to the re-entry of a hailstone embryo into the main updraft prior to a final up-and-down trajectory during which the hailstone may grow large, especially if it grows close to the boundary of the vault. Other, slightly less favored, hailstones will grow a little farther away from the edge of the vault and will follow trajectories resembling the dotted trajectory. Cloud particles growing "from scratch" within the updraft core are carried rapidly up and out into the anvil along trajectory 0 before they can attain precipitation size.* [From Browning, K.A., and G.B. Foote, 1976: Airflow and hail growth in supercell storms and some implications for hail suppression. *Quart. J. Roy. Meteor. Soc.,* **102**, 499-533. ©Royal Meteorological Society.]

2.5.3 Hail suppression concepts

The approaches to suppressing hail vary considerably depending on the conceptual model one considers the most appropriate for the storms in a region of interest. Differences in conceptual storm models result from differing environmental conditions and from different perceptions of storm structure. Perceptions of storm structure, in turn, vary from one scientist to another and are dependent on the storm observing systems and numerical models available to a researcher. For example, one's perception of a storm structure changes as one moves from observing a storm with a radar providing only reflectivity values, to observing a storm with multiple Doppler radar capable of defining storm air motions, to observing a storm with multiple Doppler radar and a multiparameter radar capable of also identifying ice versus liquid precipitation elements.

The Soviet hail suppression scheme

Hail suppression techniques in the Soviet Union are largely based on the Soviet conceptual model of a hailstorm described above. In that conceptual model hailstones grow from embryos of frozen supercooled raindrops. In a region of high liquid water contents resulting from a accumulation of supercooled raindrops, a few drops freeze naturally and grow rapidly to hailstones in the water-rich environment. The Soviets, therefore, developed several techniques for direct injection of silver iodide seeding material in the region of high radar reflectivity and presumably large amounts of supercooled water. The techniques involve the use of several types of rockets and cannons which carried seeding material to heights as much as 8 km where the material is explosively dispersed over the target region (Bibilashvili et al., 1974). The dispersed silver iodide is then hypothesized to be swept up by the supercooled raindrops (or promote freezing of small ice crystals which are swept up by the supercooled raindrops) and promote their freezing. The numerous frozen raindrops then compete "beneficially" for the available supercooled liquid water, thus inhibiting the formation of hailstones. As described by Battan (1969) the projects required the coordinated use of radars to detect regions of high rain water content and ground-based rocket launchers or cannons.

The results reported from a number of seeding operations in the Soviet Union suggested 50 to 100% reduction in hail damage (Battan, 1969; Burtsev, 1974; Sulakvelidze et al., 1974). Attempts to replicate the Soviet hail suppression scheme in Switzerland and the United States have been unsuccessful (Federer et al., 1986; Atlas, 1977). The failure in the United States however, may have been due to the inability of the scientists to successfully deploy rockets. Most of the seeding was done by burning flares on aircraft flying in updrafts. It is also possibly a result of the fact that the storms in the NHRE experimental area were quite different from the hail-producing storms over the Soviet Union. First, there is clear evidence that the embryos for hail stones in northeastern Colorado are primarily graupel particles rather than frozen drops (Knight and Squires, 1982). Therefore, it is unlikely that the Soviet concept that a accumulation zone of supercooled raindrops serves as the main source region of hailstone growth in the High Plains of the United States. Secondly, the most severe hail-producing storms over northeast Colorado are supercell storms whereas supercells appear to be rather rare in the Soviet hail regions.

The reasons for the failure of the Swiss hail experiment is less obvious since the characteristics of hailstorms in that region more closely resemble the Soviet storms and the Swiss more faithfully modeled their experiment on the Soviet hail suppression scheme.

The Glaciation Concept

The aim of hail suppression by glaciation is to introduce so many ice crystals via seeding that the ice crystals consume all the available supercooled liquid water as they grow by vapor deposition and riming of cloud droplets. To be effective this technique requires the insertion of very large amounts of seeding materials in the storm updrafts. Modeling studies (Weickmann, 1964; Dennis and Musil, 1973; English, 1973; Young, 1977) have suggested that unless very large amounts of seeding material are used, the strongest updrafts remain all liquid and hail growth is not substantially affected. Therefore, the glaciation concept is generally thought not to be a feasible approach to hail suppression. The glaciation concept is also not popular because many scientists think that it may result in a reduction in rainfall along with hail. Since most hail-prone areas are semi-arid, the loss of rainfall can have a greater adverse impact on agriculture than economic gains from hail suppression.

The Embryo Competition Concept

The competing embryo concept, first introduced by Iribarne and dePena (1962), involves the introduction of modest concentrations of hailstone embryos (on the order of 10 per cubic meter) in the regions of major hailstone growth. The idea is that millimeter-sized ice particles will then compete 'beneficially' for the available supercooled water and result in numerous small hailstones or graupel particles rather than a few large, damaging hailstones. Because it is not economically feasible to introduce hailstone embryos directly in the cloud, one must use a seeding strategy which utilizes the storm's natural hailstone embryo manufacturing process. For example, the Soviet concept of hail suppression can be considered a embryo competition strategy. In their case the hypothesized hailstone embryos are frozen supercooled raindrops. By dispersing seeding material into a region containing supercooled raindrops, the raindrops readily freeze and immediately become millimeter-sized hailstone embryos. The numerous hailstone embryos then beneficially compete for the available supercooled water resulting in the formation of numerous small hailstones, many of which would melt before reaching the ground.

Now consider a cloud in which supercooled raindrops are not present. In such clouds millimeter-sized ice particles first must form by vapor deposition until ice crystals of the size of a few hundred micrometers (0.1 mm) form. This takes a significant amount of time, on the order of 5 to 10 minutes. The larger vapor-grown ice crystals can then settle through a population of cloud droplets and grow rapidly by riming those droplets to form graupel particles. The larger ice crystals can also collide with each other to form clusters of ice crystals called ***aggregates.*** Both the graupel particles and aggregates can serve as hailstone embryos since they have significant fall velocities and cross-sectional areas to enable them to grow rapidly by accreting supercooled cloud droplets to form hailstones.

As a result of the significant amount of time for hailstone embryos to form, seeding intense updrafts, such as exist in supercell storms and the mature cell of severe multicell storms with weak echo regions, is unlikely to have any significant effect on hail growth. The ice crystals formed from seeding would probably be swept aloft into the anvil before becoming large enough to serve as embryos of hailstones. In the case of multicell storms, the recommended approach is to seed in the flanking towering cumulus clouds where updrafts are weaker and transient. If the cell is a daughter cell or a cell that eventually becomes a mature cell, it may be laden with numerous artificially produced hailstone embryos. Likewise, if the flanking cell is in the right location to serve as a feeder cell, then the natural and artificially produced hailstone embryos will be entrained into the mature cell to beneficially compete for the supercooled liquid water and reduce the size of hailstones.

The problem is how can one implement an embryo competition strategy in supercell storms? Remember that supercells have steady updraft speeds of 15 to 40 m s^{-1} and that there do not appear to be any flanking towering cumuli associated with them. Modeling studies (e.g., Young, 1974) have suggested that significant embryo growth is only possible in regions with cloud base updraft speeds less than 3 m s^{-1}. A conclusion drawn from the NHRE is that it is probably not feasible to suppress hail growth in supercell storms (Atlas, 1977).

Liquid Water Depletion by Salt Seeding

Another proposed approach to hail reduction is to seed the base of clouds with salt particles or some other hygroscopic material and thereby initiate a warm rain process in the lower levels of the cloud. The concept, also called the *trajectory-lowering technique* is that the precipitation settling out of the lower part of the cloud will deplete the liquid water in the cloud and therefore limiting hailstone growth. Some cloud modeling studies (Young, 1977) suggest that this may be a feasible approach in regions such as the High Plains of the United States or Canada where cloud base temperatures are cold and cloud droplet concentrations are large.

It has also been proposed to use a combination of salt seeding and ice phase seeding to both deplete supercooled liquid water and to promote beneficial embryo competition (Dennis and Musil, 1973). Only limited exploratory field studies have been performed to examine the feasibility of this approach.

2.5.4 Field confirmation of hail suppression techniques

As noted previously the reports of major success in hail suppression in the Soviet Union spawned a number of operational and scientific research programs on hail suppression in the United States, Canada, Switzerland, South Africa and elsewhere. The Swiss and U.S. programs were attempts to replicate the Soviet strategy. Unfortunately, the U.S. NHRE did not replicate the Soviet strategy for a variety of reasons. As mentioned previously, the meteorology over the High Plains of the U.S. is quite different than in the Soviet hail regions with the storms being cold-based and continental so that a warm rain process is not very active. An accumulation zone of supercooled raindrops is, therefore, not likely in that region. Furthermore, there appears to be a greater preponderance of supercell storms in the NHRE area than in the Soviet hail regions (Battan, 1969; Marwitz, 1972). The storms in NHRE were thus less amenable to hail suppression. Finally, scientists in NHRE were never able to successfully implement a rocket based seeding strategy. Instead, most of the seeding was conducted as broadcast seeding from aircraft flying beneath the base of the strongest updrafts; a procedure that is least likely to produce competing embryos. The NHRE seeding experiment was therefore curtailed prematurely without definitive results (Atlas, 1977).

In contrast, the Swiss experiment (Federer et al., 1986) had everything going for it. The meteorology in Switzerland is relatively similar to the Soviet hail regions. They successfully implemented a rocket seeding strategy and carried out a well designed and well implemented field program. Again, this experiment did not yield positive results of hail suppression. Similarly hail suppression experiments in Canada and South Africa have been inconclusive. Only long-term statistical analyses of non-randomized, operational programs have suggested that seeding can significantly reduce hail frequency (e.g., Mesinger and Mesinger, 1992). One can not rule out the possibility that natural climatic variations in hail frequency lead to the 'apparent' reductions in hail frequency in such studies. The scientific basis of hail suppression, therefore, remains unresolved.

2.6 Modification of Tropical Cyclones

As mentioned in Chapter 1, the first attempt at modifying a tropical cyclone or hurricane occurred during Project Cirrus in 1947. Because of the extensive damage produced by hurricanes, interest in developing an economical technique to modify hurricanes developed to a high level. For example Gentry (1974) stated: "*For scientists concerned with weather modification, hurricanes are the largest and wildest game in the atmospheric preserve. Moreover, there are urgent reasons for 'hunting' and taming them.*" Therefore in 1962 a joint project, between the U.S. Weather Bureau and the U.S. Navy, called Project STORMFURY was created. Before we examine the results of that study, let us review the basic concepts of tropical cyclones.

2.6.1 Basic conceptual model of hurricanes

The tropical cyclone or hurricane is a large cyclonically circulating (counter-clockwise rotation in the northern hemisphere) weather system that is composed of bands of deep convective clouds (see Figure 2.20). Inside a radius of about 400 km, the low-level flow is convergent and asso-

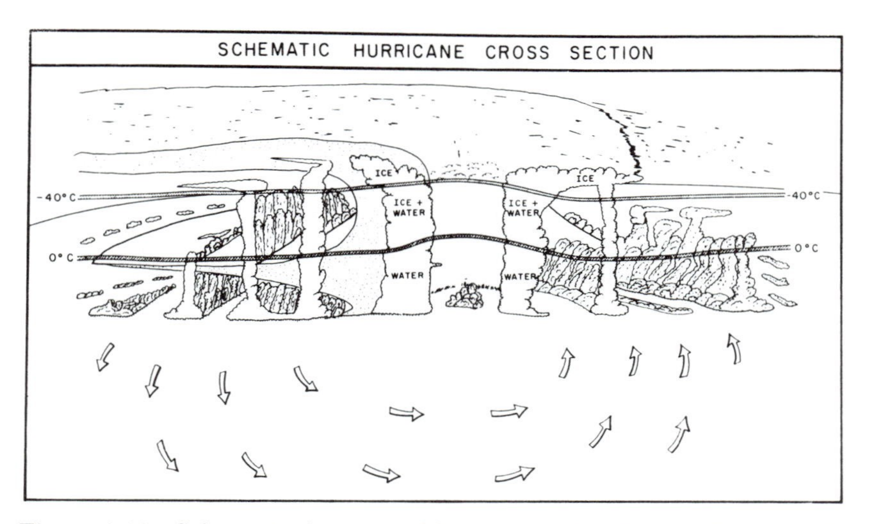

Figure 2.20: *Schematic diagram of hurricane, showing low-level circulation and cloud types. The highest clouds, composed of cirrus and cirrostratus, occur at the tropopause, which is about 16 km.* [From STORMFURY, 1970: Project STORMFURY Annual Report 1969, National Hurricane Research Laboratory, NOAA, AOML/Hurricane Research Division, Miami, FL, 20 pp.]

ciated lifting of warm, moist air produces extensive cumulonimbus clouds and precipitation. The storm is composed of an eye wall (a generally circular ring of intense cumulonimbus convection surrounding the often cloud free eye), a region of stratiform cloud and precipitation outside the eyewall, and, beyond the eye wall, spiral bands of convective clouds and rainfall. Like a giant flywheel, the region within 100 km of the storm center is inertially stable and is not affected strongly by outside weather systems. Only through enhancements or reductions in the divergence pattern associated with jet streaks is the strength of the storm affected by larger-scale weather systems. Likewise it does not respond rapidly to small changes in heat release but only to sustained, large amplitude changes in heating.

The primary energy driving the tropical cyclone comes from the sea. Air flowing over a warm ocean surface receives energy primarily in the form of sensible and latent heat. Small-scale turbulent eddies near the ocean surface transfer the heat and moisture upwards to levels where it becomes saturated and cumulus clouds form. By condensing water to form cloud droplets and consequently releasing latent heat, the moisture and latent heat transferred from the ocean surface warms the cloudy air at a rate which is roughly proportional to the precipitation rate in the clouds. Thus cumulus clouds and subcloud eddies transfer the sensible and latent heat from the ocean surface to the middle and upper troposphere. As the air moves laterally outward from the central region of the storm at upper levels in stratiform anvil clouds, much of the energy gained at the ocean surface is radiated to space by infrared radiation.

These features of a tropical cyclone have motivated Emanuel (1988) to suggest that a tropical cyclone is like an idealized heat engine called a Carnot engine. In a Carnot engine heat is input at a single high temperature and all the heat output is ejected at a single low temperature. The amount of work produced by the Carnot engine is proportional to the difference between the input and output temperatures, and is the maximum amount of energy that can be extracted from a heat source. In the case of a tropical storm, the amount of work or the maximum strength of the winds in a storm is proportional to the difference in temperature between the heat input level (i.e., at the ocean surface), and the heat output level, or the tops of the stratiform-anvil clouds. However, this is not the total story. If the sea surface temperature and temperature at the tropopause were the only determinant factors on hurricane formation and strength, there would be more than ten times as many hurricanes as normally occur. Other factors which determine the extent to which larger scales of motion and convective scales interact in an optimum way to utilize the energy flowing from the ocean surface must also be important as discussed in Pielke (1990).

2.6.2 The STORMFURY modification hypothesis

The original STORMFURY hypothesis was first advanced by Simpson et al. (1963) and Simpson and Malkus (1964). They proposed that the additional latent heat released by seeding the supercooled water present in the eyewall cloud would produce a hydrostatic pressure drop that would modestly reduce the surface pressure gradient and as a consequence the maximum wind speed.

The original hypothesis was subsequently modified following a series of numerical hurricane simulations (Gentry, 1974). Those numerical experiments suggested that application of the individual cloud dynamic seeding hypothesis to towering cumuli immediately outward of the eyewall would cause enhanced vertical development of the towering cumuli and removal of low-level moisture from the boundary layer immediately beneath them. The loss of moisture outward from the eyewall would starve the clouds of moisture in the eyewall region, causing a shift in the eyewall convection outward to greater radii (as illustrated in Figure 2.21). Like ice skaters extending their arms, the storm should rotate slower and the winds diminish appreciably.

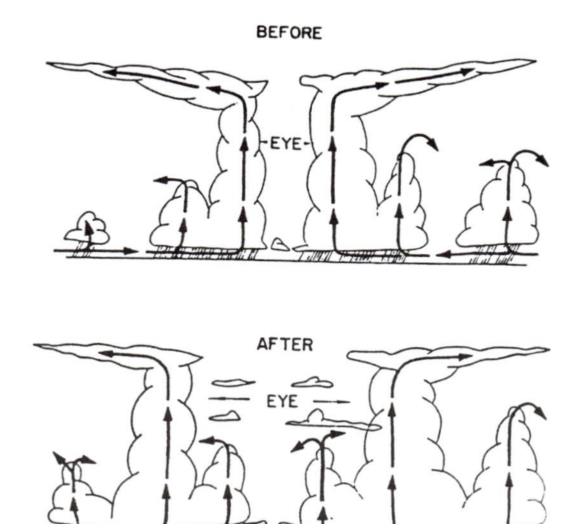

Figure 2.21: *Hypothesized vertical cross sections through a hurricane eye wall and rain bands before and after seeding. Dynamic growth of seeded clouds in the inner rain bands provides new conduits for conducting mass to the outflow layer and causes decay of the old eye wall.* [From Simpson, R.H., et al., 1978: TYMOD: Typhoon Modification Final Report. Prepared for the Government of the Philippines. Virginia Technology, Arlington, Virginia.]

2.6.3 STORMFURY field experiments

For nearly a decade STORMFURY performed seeding experiments in an attempt to reduce the intensity of hurricanes. Only a few storms were actually seeded, however, due to the fact the hurricanes are a relatively infrequent phenomena and the experiment was constrained to operate in a limited region of the north Atlantic well away from land. Some encouraging results were obtained from seeding Hurricane Debbie in 1969 with 30% and 15% reductions in wind speeds following seeding on two days separated by a no-seed day. This led to greatly expanded field programs for a few years but the program was eventually curtailed in the late 1970's with no definitive results. STORMFURY succumbed to the very large *natural variability* of hurricanes including a period of very low hurricane frequency from the middle 1960's to the middle 1980's. The conclusions of the project were summarized in Sheets (1981) in which 10–15% decreases in the maximum wind speed with associated damage reductions of 20–60% should occur if the STORMFURY hypothesis (Gentry, 1974) were implemented. No significant changes in storm motion or storm averaged rainfall at any specific location would be expected.

Chapter 3

The Fall of the Science of Weather Modification by Cloud Seeding

For nearly two decades vigorous research in weather modification was carried out in the United States and elsewhere. As shown in Figure 3.1 federal funding in the United States for weather modification research peaked in the middle 1970's at nearly $19 million per year. Even at its peak, funding for weather modification research was only 6% of the total federal spending in atmospheric research (Changnon and Lambright, 1987) and this amount included considerable support for basic research on the physics of clouds and of tropical cyclones. Nonetheless, research funding in cloud physics, cloud dynamics, and mesoscale meteorology was largely justified based on its application to development of the technology of weather modification. Research on the basic microphysics of clouds particularly benefited from the political and social support for weather modification.

By 1980 the funding levels in weather modification research began to fall appreciably and by 1985 they had fallen to the level of $12 million. After 1985 funding in weather modification research became so small and fragmented that no federal agency kept track of it. Currently the Bureau of Reclamation has only about $0.25 million per year that can be identified as weather modification. They are, however, operating a program in Thailand that is supported by the Agency for International Development. Basic research in the National Science Foundation that can be linked to weather modification is on the order of $1 million. Likewise the Department of Commerce has no budgeted weather modification program, but has supported a cooperative state/federal program at about the $3.5 million level. This "pork barrel" program is supported by annual congressional write-ins rather than a line item in the NOAA budget. As a result, states having strong political lobbying support for weather modification are earmarked for support in this program. Overall the to-

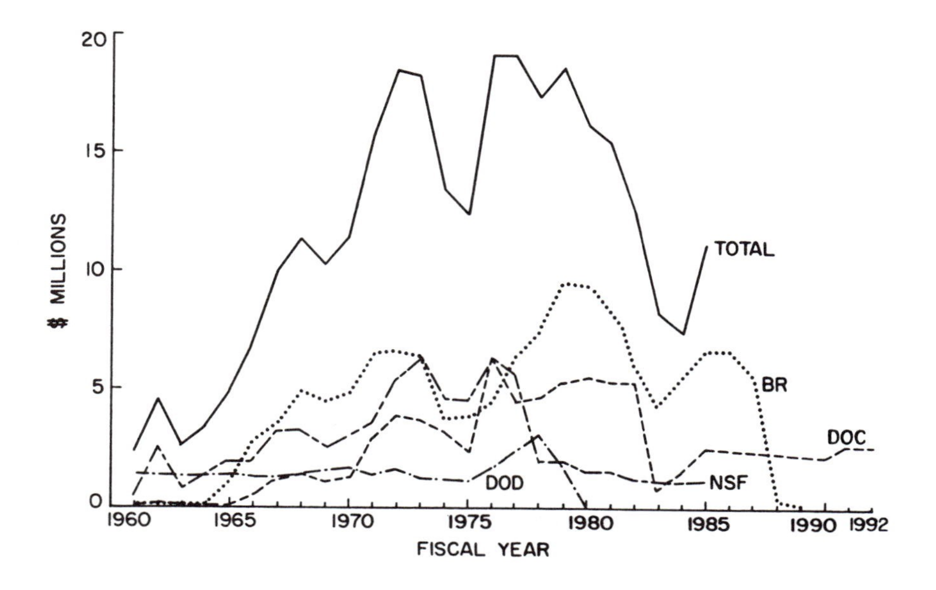

Figure 3.1: *Estimates of federal spending levels in the United States for weather modification research. The agencies are BR (Bureau of Reclamation), DOC (Department of Commerce), DOD (Department of Defense), and NSF (National Science Foundation). Data provided courtesy of the National Science Foundation.*

tal federal program for weather modification in the United States is on the order of 10% of its peak level in the middle 1970's. What caused this virtual crash in weather modification research?

Changnon and Lambright (1987) listed the following reasons for this reduction in funding:

- poor experimental designs;
- widespread use of uncertain modification techniques;
- inadequate management of projects;
- unsubstantiated claims of success;
- inadequate project funding; and
- wasteful expenditures.

Changnon and Lambright concluded that the primary cause of the rapid decline in weather modification funding was the lack of a coordinated federal research program in weather modification. However, there are other factors that also must be considered:

- Weather modification was oversold to the public and legislatures.
- Demands for water resource enhancement declined due to an abnormal wet period.
- Impact of Reaganomics.
- Change in public attitude toward the environment.
- The limited period of government and public interest in specific environmental problems and the resultant diversion of public attention to other weather and climate issues.

Changnon and Lambright argued that Congress' decision to terminate NSF's lead agency role in weather modification research in 1968 was a major blunder. Instead, support for weather modification research in the United States became fragmented between the U.S. Bureau of Reclamation (BR) of the Department of the Interior, the Department of Defense (DOD), the National Oceanic and Atmospheric Administration (NOAA), the National Science Foundation (NSF), and the Department of Agriculture (DOA). With the exception of NSF, these agencies were mission oriented and as such the primary support was not for basic research, but development of a technology to be applied to water resource enhancement, severe weather abatement, agricultural uses, and military applications. This led to a great deal of interagency rivalry and to research programs having *short-term* goals of establishing a weather modification technology. As a result most of the research programs were of a 'black box' nature in which clouds were seeded in a randomized statistical design where the only measurements were the amount of rain on the ground or the amount of hail damage, etc. There were few attempts to design studies in which the entire sequence of hypothesized responses to seeding were measured. Did seeding produce more numerous ice crystals? Is the precipitation particle size-spectra different in seeded clouds? What are the sequence of dynamic responses of a cloud to seeding? It was only in very recent years, about the time of the crash in weather modification funding, that major attempts to answer these fundamental questions were performed in weather modification programs. Clearly the lack of a coordinated federal research program had a major adverse impact on weather modification research.

Another factor affecting the decline in weather modification research was that from the early days of Project Cirrus onward, weather modification was oversold to the funding agencies and to the public. The cry was weather modification is good! It can enhance rainfall, suppress hail, weaken winds in hurricanes, inhibit lightning, and put out forest fires. Many scientists and program managers argued that only a few years of research were needed to put cloud seeding on a sound scientific basis and be ready for routine applications. Such overselling was often undoubtedly performed by a few unscrupulous scientists, program managers, and commercial seeders. However, more importantly, this behavior reflected the rather naïve perceptions of many in the scientific community of the difficult problems faced by the weather modification community. Not only are the physical problems faced by weather modification scientists extremely complex, but scientists generally underestimated the impact of the *natural variability* of precipitation and weather on discerning a seeding signal from the natural background. Thus this overselling has led to a lack of scientific credibility since after more than 30 years, scientists are still disagreeing about the outcome of cloud seeding projects and the scientific status of the field. This loss of scientific credibility came to a head in a meeting of atmospheric scientists organized by the United States National Academy of Sciences to assess research themes of the 1980's (NAS, 1980). The consensus of the attendees was that weather modification should not be given a high priority for federal research funding in the 1980's. Many were vigorously opposed to support for weather modification research. They argued that climate, atmospheric chemistry, and mesoscale meteorology should be given the highest priority.

Major support for weather modification research has traditionally come from the semi-arid western states of the United States where the demand for water often exceed the supply. However, in the middle 1970's to the 1980's, precipitation was often above normal so that demand for cloud seeding projects and weather modification research dwindled. During the same period, tropical cyclones striking the east coast of the United States also reached a low level (Gray, 1990; 1991). These factors plus the Reagan administrations attempts at reducing federal expenditures, combined to make it easy for Congress to cut federal funding in weather modification research.

Finally, enthusiasm for weather modification developed during a time when the prevailing attitude was to restructure the environment to suit the needs of mankind by building dams, cutting trees, building sea walls, and other means of altering the environment. This philosophy has been replaced, by and large, with an environmental awareness ethic. No longer is it acceptable to build nuclear power plants, dams, or highways without a major assessment of the total environmental impact. Weather modification, with its primary aim to change the weather, no longer fits the environmental ethic that prevails in many developed nations.

As a consequence of the above factors, weather modification research experienced a sharp decline in funding in the middle 1980's in the United States. A similar decline in funding was also experienced in many developed countries during the same period, largely due to the loss of scientific credibility of the field. Some countries such as China and the Union of South Africa (Garstang et al., 1987), however, have maintained vigorous weather modification research programs. More than 100 operational cloud seeding programs currently exist throughout the world including the United States.

There is also evidence of a rebirth of weather modification research. In Australia, for example, several operational and research programs for precipitation enhancement have been instigated. With the drought of 1988 and renewed enhanced hurricane activity along the east coast of the United States will demands for weather modification research be increasing? Certainly several operational programs were begun in the drought of 1988 in the United States.

Clearly there is a great need to establish a more credible, stably-funded scientific program in weather modification research, one that emphasizes the need to establish the physical basis of cloud seeding rather than just a 'blackbox' assessment of whether or not seeding increased precipitation. We need to establish the complete hypothesized physical chain of responses to seeding by observational experiments and numerical simulations. We also need to assess the total physical, biological, and social impacts of cloud seeding. Recently, for example, E.K. Bigg (Bigg and Turton, 1988; Bigg, 1988; 1990b) suggested that silver iodide seeding can trigger biogenic production of secondary ice nuclei. His research suggests that fields sprayed with silver iodide release secondary ice nuclei particles at 10 day intervals and that such releases could account for inferred increases in precipitation 1 to 3 weeks following seeding in several seeding projects (e.g., Bigg and Turton, 1988). If Bigg's hypothesis is verified, an implication of biogenic production of secondary ice nuclei is that many seeding experiments have thus been contaminated such that the statistical results of seeding are degraded. This effect would be worst in randomized crossover designs and in experiments in which one target area is used and seed days and non-seed days are selected over the same area on a randomized basis. Thus, not only is the weather modification community faced with very difficult physical problems and large *natural variability* of the meteorology, but they also are faced with the possibility of responses to seeding through biological processes.

We shall see later that scientists dealing with human impacts on global change are also faced with very difficult physical problems, large *natural variability* of climate, and the possibility of complicated feedbacks through the biosphere. There is also a great deal of overselling of what models can deliver in terms of prediction of human impacts over time scales of decades or centuries.

Part II

Human Impacts On Regional Weather and Climate

In Part I, we discussed man's purposeful attempts at altering weather and climate by cloud seeding. We saw that there is strong evidence indicating that clouds and precipitation processes, can be altered through cloud seeding. In general, however, our knowledge about clouds is still not sufficient to enable cloud seeders to alter precipitation processes and severe weather in anything but the simplest weather systems (e.g., orographic clouds, supercooled fogs and stratus, and some cumuli). In Part II we examine the mechanisms and evidence indicating there have been changes in regional weather and climate through anthropogenic emissions of aerosols and gases, and through alterations in landscape. Regional scale refers to horizontal scales of less than a few thousand kilometers. On these scales we examine the possible changes in rainfall, severe weather, temperatures, and cloud cover caused by anthropogenic activity.

Chapter 4

Anthropogenic Emissions of Aerosols and Gases

A variety of human activities result in the release of substantial quantities of aerosol particles and gases which may influence cloud microstructure, precipitation processes and other weather phenomena. In this section we examine evidence that these particulate and gaseous releases are influencing regional weather and climate. In this chapter, however, we will not focus on urban emissions of particulates and gases, that discussion will be reserved for Chapter 5.

4.1 Cloud Condensation Nuclei and Precipitation

In our discussion of purposeful modification of clouds in Part I, we described attempts to enhance precipitation from warm clouds by seeding them with hygroscopic materials. The hygroscopic particles which are called cloud condensation nuclei (CCN) can alter the microstructure of a cloud by changing the concentrations of cloud droplets and the size spectrum of cloud droplets. There are numerous examples of anthropogenic sources of CCN, including automobile emissions (Squires, 1966), certain urban industrial combustion products, and the burning of vegetative matter, especially sugar cane.

Warner and Twomey (1967) observed substantial increases in CCN concentrations beneath the base of cumulus clouds and increases in cloud droplet concentrations above their bases downwind of areas in which the burning of sugar cane fields was taking place. The practice of burning sugar cane fields to remove leaf and trash before harvesting is quite common in most areas where sugar cane is grown. Warner and Twomey hypothesized that the larger numbers of CCN and cloud droplets would slow down collision and coalescence growth of precipitation by virtue of their smaller size and, consequently, small collection efficiencies and small collection kernels. Slower coalescence growth should therefore lead to less rainfall, at least from smaller clouds which do not contain large amounts of liquid water. Warner (1968) performed an analysis of precipitation records downwind of sugar cane burning areas near Bunderburg, Queensland, Australia and those in upwind 'control' areas. The analysis was performed only during the three month burning periods and over a 60 year record. He found substantial reductions in rainfall amounting to a decrease of approximately 25% downwind of the burning areas. Attempts to confirm these findings in other areas of Australia (Warner, 1971) and over the Hawaiian Islands (Woodcock and Jones, 1970) have been unsuccessful. Nonetheless, Warner's findings strongly suggest that enhanced concentrations of CCN particles can lead to reductions in rainfall, at least in some clouds and in some regions.

There is also evidence suggesting that changes in the size spectrum of CCN by anthropogenic emissions may be responsible for enhancing precipitation (Hobbs et al., 1970; Mather, 1991). Hobbs et al. found that pulp and paper mills and certain other industries are prolific sources of CCN. Observations of some of the clouds downwind of the source factories suggest that they actually produce precipitation more efficiently than other clouds in the region. The analysis of precipitation and stream flow records for a period before construction of the mills to after the mills were in operation suggested that rainfall downwind of the pulp and paper mills was 30% greater in the later period than before the mills were put in operation (Hobbs et al., 1970). Subsequent studies (Hindman et al., 1977a) indicated that large (>0.2 μm) and giant (> 2 μm) CCN were increased in concentration downwind of the plants, while small CCN (<0.2 μm) were not significantly altered in concentration. It was concluded by Hobbs et al. and Hindman (1977a,b) that the large and giant nuclei emitted by the mills increased the efficiency of the collision and coalescence process thereby enhancing precipitation.

Attempts to simulate the response of cumulus clouds and stratocumulus clouds to injections of large and ultra-giant CCN having similar concentrations to those emitted by paper mills (Hindman et al., 1977b) revealed little change in simulated precipitation. They speculated that heat and moisture emitted by the mills, in combination with the CCN, may have been responsible for increased rainfall. This conclusion is consistent with the finding by Hindman et al. (1977a) that liquid water contents in clouds downwind of the paper mills were 1.3 to 1.5 times greater than surrounding clouds, though the results were not statistically significant.

At this time it cannot be determined if the models used in these studies were too crude to realistically simulate the response of precipitation processes to injections of large and ultra-giant nuclei by paper mills, or if those nuclei have little effect by themselves on rainfall. More realistic models of cloud microphysics and cloud dynamics are required to further investigate this problem. In addition, observational studies of other potential or known sources of CCN should be done. A study performed by Mather (1991) suggests that emissions from a paper mill in South Africa are also causing an enhancement of rainfall. The mechanisms responsible for that anomaly can only be speculated. In the South African case the storms are much more complex, being small cumulonimbi.

Another example of a possible link between anthropogenic emissions of CCN and precipitation is evident from ship track trails viewed on satellite imagery (Coakley et al., 1987; Scorer, 1987; Porch et al., 1990). The ship track trails appear as a line of enhanced brightness in satellite imagery particularly at 3.7 micrometer wavelength. Figure 4.1 shows clear evidence of the much brighter ship trails. The implications of those brighter clouds to global climate will be discussed in Part III. Here, we will concentrate mainly on evidence of the relationship of ship CCN emissions on cloud structure.

The prevailing hypothesis is that the ship's trails appear brighter on satellite imagery because the effluent from the ships is rich in CCN particles. The more numerous CCN particles create larger concentrations of cloud droplets which then reflect more solar energy than the surrounding clouds. Aircraft observations in the ship track clouds as well as surrounding clouds (Radke et al., 1989) reveal that the ship track clouds exhibit higher droplet concentrations, smaller droplet sizes, and higher liquid water content than surrounding clouds. The higher droplet concentrations and smaller droplet sizes are consistent with the hypothesis that the higher cloud brightness is due to a higher concentration of CCN in the ship effluent. The greater liquid water content in the ship trails, however, is a surprising result.

Albrecht (1989) hypothesized that the higher droplet concentration in ship track clouds reduced the rate of formation of drizzle drops by collision and coalescence similar to Warner and Twomey's (1967) hypothesis for cloud microphysical structure downwind of sugar cane fields. The reduced rate of drizzle formation then resulted in higher liquid water content and higher droplet concentration in ship track clouds compared to surrounding clouds. Radke et al. (1989) found that the concentration of drizzle drops (droplets of diameter greater than or equal to 200 μm) in the ship track was only 10% of that in surrounding clouds. This supported Albrecht's hypothesis that the liquid water content in the tracks was higher because the higher droplet concentration and smaller droplet sizes limited collision and coalescence.

Another hypothesis is that the ship track clouds exhibit higher liquid water content because the heat and moisture emissions from the ships invigorated the air motions in the clouds, thereby creating deeper and wetter (hence brighter) clouds. Porch et al. (1990) examined this hypothesis and showed that ship tracks are not only characterized by greater brightness but also clear bands on the edges of the cloud tracks (see Fig-

Figure 4.1: *Images constructed from 1 km by NOAA-9 AVHRR data 3.7 μm radiance for a 500 km by 500 km region of ocean off the coast of California. The data were taken at 2246 UTC on 3 April 1985. The ship tracks evident at 3.7 μm may be due to a shift toward smaller droplets for the contaminated clouds. The shift causes an increase in reflected sunlight at 3.7 μm.* [From Coakley, J.A., R.L. Bernstein, and P.A. Durkee, 1987: Effect of ship-stack effluents on cloud reflectivity. *Science,* **237,** 1020-1022. ©American Association for the Advancement of Science.]

ure 4.2). They speculated and provided some modeling evidence that the heat and moisture fluxes from the ship effluent excited a dynamic mode of instability which, in some marine stratocumulus environments, led to enhanced upward and downward motion associated with the cloud circulations.

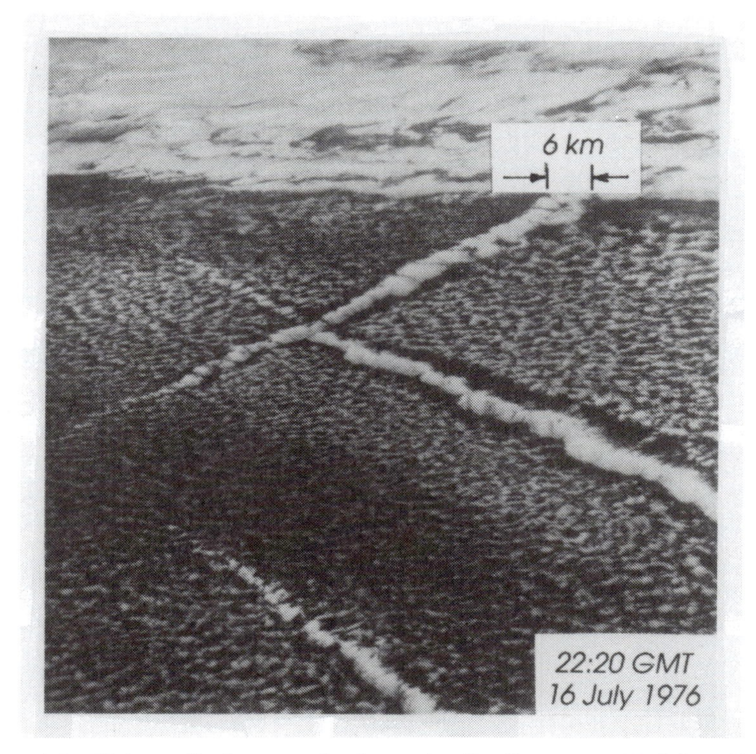

Figure 4.2: *Ship-trail photography from Apollo-Soyuz on 16 July 1975 at 2221 GMT.* [From Porch, W. M., C-Y. J. Kao, and R. G. Kelley, Jr., 1990: Ship trails and ship induced cloud dynamics. *Atmos. Environ.,* **24A,** 1051-1059. ©Pergamon Press PLC.]

In summary, the evidence is compelling that anthropogenic emissions of CCN can result in decreases or increases of precipitation, depending on their size distribution. Research to date, however, is not sufficiently definitive to confirm the hypothesis that anthropogenic emissions do indeed alter precipitation substantially.

4.2 Aircraft Contrails

It has become quite common to observe jet contrails covering extensive parts of the sky in regions of heavy jet traffic. Often several independent contrails merge to form an almost solid overcast of thin clouds, much like natural cirrus. The primary source of contrail formation is the water vapor emitted by the jet aircraft engine during combustion. Murcray (1970) noted that a typical medium-sized commercial jet such as a Boeing 727 burns 3100 kg of fuel per hour while cruising and produces over 1.2 times as much water (more than a kilogram per second) as a result of the combustion in the presence of atmospheric oxygen. At very cold temperatures such as exist in the upper troposphere, the saturation vapor pressure with respect to water and ice are very small in magnitude and since there is little difference between those saturation values and typical water vapor mixing ratios, small additions of water vapor to the air can lead to water saturation and also supersaturation with respect to ice. Thus the water emitted by jets is the primary cause of contrail formation.

When the moisture laden jet exhaust enters the very cold atmosphere, it is rapidly cooled forming a cloud of small droplets. Evidence suggests that the liquid phase is very short-lived (on the order of a second) and the droplets freeze to form a cloud composed of nearly spherical ice crystals a few micrometers in diameter (Murcray, 1970). As a result of their small size, and hence small settling velocities, and because the atmosphere may be slightly supersaturated or only weakly subsaturated with respect to ice, these crystals can survive for periods of several hours or more.

We now examine what impact such clouds of ice crystals in the upper troposphere can have on regional weather and climate. The most obvious impact is that they can alter the radiation budget that affects surface temperatures. That is, the contrails reflect incoming solar radiation back to space and absorb and re-radiate upwelling terrestrial or longwave radiation back toward the ground. (For a more detailed discussion of atmospheric radiation the reader is referred to Part III). Figure 4.3 illustrates that a portion of the incoming solar energy (Rsd) incident on the earth's surface is reflected back to the atmosphere (Rsu), a part is conducted into the ground (G0), some is transferred into the atmosphere by sensible heat convection (S0) and by latent heat transfer (L0), and a portion is emitted by terrestrial radiation (Rlu). Part of the upwelling terrestrial radiation is transmitted through the atmospheric window (see Part III) while the remainder of the terrestrial radiation is absorbed in the atmosphere by water vapor, carbon dioxide and other trace gases, and by clouds. A percentage of this terrestrial radiation absorbed in the atmosphere is re-radiated back towards the ground (Rld) and is absorbed at the earth's surface. The surface temperature at any time is controlled by a balance between these contributions to the energy transfer. If incoming solar radiation is reduced (and all other contributions remain the same), then surface temperatures will be cooler. If Rld is increased and all other contributions remain the same, surface temperatures will rise.

Kuhn (1970) observed the radiation budget of contrails from an aircraft. He found a 500 m thick contrail depleted incoming solar radiation (Rsd) by 15% and increased downward terrestrial radiation (Rld) by 21%.

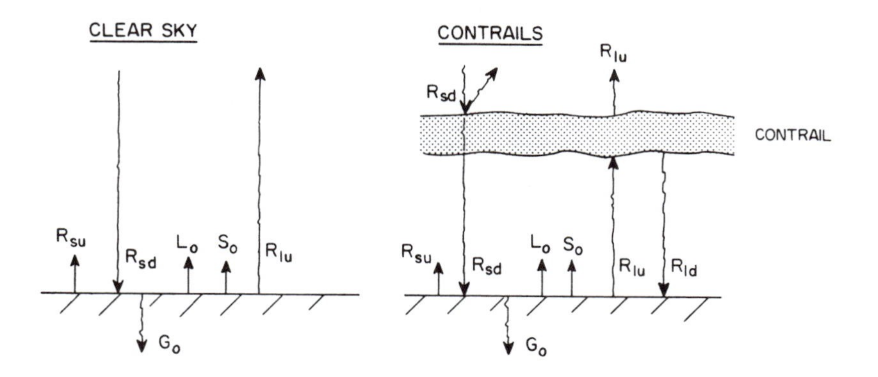

Figure 4.3: *Energy budget at earth's surface, where* R_{sd} *is incoming solar radiation,* R_{su} *is reflected solar radiation* $\left(\text{albedo} = \frac{R_{su}}{R_{sd}}\right)$, G_o *is heat diffusion into the ground,* L_o *is latent heat transfer to the atmosphere,* S_o *is sensible heat transfer to the atmosphere,* R_{lu} *is upwelling terrestrial radiation, and* R_{ld} *is downward terrestrial radiation.*

The enhanced terrestrial radiation, however, could not make up for the loss of solar radiation, so that surface temperatures are reduced by the presence of contrails during the daytime. At nighttime, when there is no incident solar radiation, the enhanced downward flux of terrestrial radiation leads to a net warming. Thus the presence of contrails reduces afternoon maximum temperatures, and raises nighttime minimums, causing a moderation of local climate. Kuhn (1970) calculated that if the contrails were persistent over a 24-hour period then the net effect would be a 5° C to 6° C cooling of surface temperatures. This result, however, depends on the latitude and length of day, and the optical thickness of the cloud. Warming is the net effect for shorter days and higher latitudes and optically thin contrails.

Changnon (1981a) performed a climatological analysis of cloudiness, sunshine, and surface temperatures over the midwestern United States during the period 1901 to 1977. He found that in the period since 1960, there was more cloud cover, a decrease in sunshine, and surface temperature extremes were moderated (less extreme minimum and maximum monthly averages), especially in the area of the midwest where jet traffic is greatest. In a study of sky cover over the United States, Seaver and Lee (1987) found a decrease in cloudless days over large regions of the United States since 1936. The results are consistent with the hypothesis that increased contrail formation led to the changes, but *natural fluctuations* in climate trends cannot be ruled out.

It has also been proposed that contrails can seed lower-level clouds with ice crystals and thereby enhance surface precipitation (Murcray,

1970). Natural cirrus ice crystals are known to survive long fall distances and potentially seed lower-level clouds (Braham and Spyers-Duran, 1967). Whether or not contrail crystals, which are quite likely smaller than natural cirrus crystals, can survive falling through great depths of the troposphere is not known. If, indeed, high concentrations of contrail crystals can survive descent into low-level, water-rich clouds such as orographic clouds and frontal clouds, they could seed those clouds with ice crystals and enhance surface precipitation in a 'seeder-feeder' type process (e.g., Bergeron, 1965; Browning et al., 1975; Hobbs et al., 1980; see Cotton, 1990 for a review of the seeder-feeder process). This hypothesis has not been supported by any further studies, however.

In summary, contrails have the potential of altering regional and possibly even global climate if they become extensive enough. Their biggest potential impact is on moderating surface temperature extremes (i.e., maximum and minimum temperatures). It is also possible that they could impact daily average surface temperatures, but whether they will cause a net increase or decrease in surface temperatures depends on their thickness, persistence, latitude, and time of the year.

4.3 Ice Nuclei and Precipitation

In Part I we examined cloud seeding strategies in which artificial ice nuclei (IN) were purposely inserted in clouds to enhance precipitation. We have seen that it is no simple matter to relate enhanced concentrations of ice nuclei with increases in surface precipitation. Although cloud seeding clearly increases ice crystal concentrations, the impact of those higher ice crystal concentrations on surface precipitation has only been identified in a limited number of cloud types and locations. The same can be said for inadvertent emissions of IN.

Schaefer (1966) showed that automobile exhaust emissions from lead-burning gasoline were prolific sources of IN, especially when those exhaust products react with iodine vapor. Many industries, especially those associated with steel production and refining of metal ores, are well known sources of IN. Schaefer (1969) described measurements of IN downwind of several eastern United States cities in which IN concentrations were substantially enhanced above those found in pollution free regions. In some cases Schaefer measured concentrations in excess of 1000 per liter, which would likely create a stable cloud of very small ice crystals leading to reduced precipitation; a phenomena called *overseeding.* Schaefer also described examples of plumes of ice crystal clouds extending downwind from major industrial effluents while no such ice crystal clouds could be seen in the surrounding countryside. One author (Cotton) has also observed such localized plumes of ice crystal clouds downwind of Buffalo, N.Y. during research flights in that region in the 1960's. Whether those enhanced ice crystal concentrations have any significant impact on surface precipitation, however, is unknown. Schaefer (1969) also suggests that dust from plowed fields can serve as ice nuclei, again causing enhanced ice crystal emissions by anthropogenic means. Recently Bigg (1990a) has suggested that there has been a systematic decrease in IN concentrations at several sites in the southern hemisphere as well as Hawaii over the last 25 years. It is not known at this time if such observations are a direct result of the increased use of lead-free gasolines or contamination of natural ice nuclei by pollutants, or just a measurement anomaly. Nonetheless, we agree with Bigg (1990b) that *persistent, baseline measurements of IN along with CCN should be routinely made at a number of locations throughout the world.*

4.4 Other Pollution Effects

Besides being sources of CCN and IN, air pollution can be high in concentrations of total aerosol particles. These particles are sufficiently numerous in urban areas that they can deplete direct solar radiation in cities by about 15%, sometimes more in winter and less in summer (Landsberg, 1970). Welch and Zdunkowski (1976) used a model to show that solar radiative heating of a polluted boundary layer can cause warming of 4° C hr^{-1} in the pollution layer with a zenith angle of 45° . Hänel et al. (1982) reported observed heating rates from absorption of solar radiation by aerosols as large as about 0.5° C hr^{-1} during the middle of the day under clear sky conditions over Frankfurt, Germany. Welch et al. (1978) applied a two-dimensional model to a polluted urban area for stagnant synoptic conditions and found temperatures at the ground to be reduced by 2° C because of low-level pollution sources and up to 7° C when the pollution was situated higher above the surface. During the Yellowstone National Park fires of 1988, as the plume spread over Golden, Colorado, the daily total global horizontal irradiance was 91% of the solar flux measured on a clear day (Hulstrom and Stoffel, 1990). While this reduction in solar radiation would lead to cooling of the ground, it is usually dominated by the urban heat island effect which we will discuss in Chapter 5. The main impact of these pollutants is a degradation in visibility and this is greatest at higher relative humidities where the hygroscopic particles swell in size and thereby attenuate visible wavelengths more. The swelled particles, called haze particles, many of which serve as CCN at cloud supersaturations, are hygroscopic and begin absorbing small amounts of water at relative humidities greater than 70%. *Smog,* which is largely composed of haze particles, is not a cloud, but haze particles. Haze has optical properties which are distinct from pure water clouds. There is some evidence that the higher concentrations of haze particles contributes to a higher incidence of fogs (Landsberg,

1970), but whether that is due to a confusion between smog and fog, or urban sources of moisture, or actually due to pollutants has not been determined. Another factor is that urban areas warm more slowly in the morning hours because the depth of the nocturnal stable layer is greater. Since fog is preferred in the morning hours, the cooler cities would favor more persistent sustained fog. At relative humidities greater than 90%, the haze particles could contain enough water to cause substantial radiative cooling at the top of the haze layer and thereby prematurely initiate the onset of a mature fog through radiative-convective interactions (see Cotton and Anthes, 1989). Further modeling and observational studies are needed to determine if that indeed takes place.

Many industries are also major sources of moisture. Power plant steam plumes and cooling ponds (Murray and Koenig, 1979; Orville et al., 1980; 1981) release sufficient amounts of moisture into the atmosphere to cause cloud and fog formation. Especially in the cold winter months when saturation mixing ratios are small in magnitude, these moisture sources can lead to the persistence of cloud or fog. Highway departments often place fog caution signs along roads near power plants to warn motorists of the increased likelihood of fog in the area. Occasionally, these persistent plumes have been observed to produce snowfall and streaks of snow-covered ground downwind of the moisture sources when air temperatures are less than 0° C (Kramer et al., 1976). These supercooled cloud plume ice crystals are probably nucleated on natural IN, although some coal-burning power plant steam plumes could be strong sources of IN as well as moisture.

Moisture emissions due to anthropogenic activity in winter months at high latitudes can also lead to persistent ice fog. Cities such as Fairbanks, Alaska (Ohtake and Huffman, 1969) and many industrialized cities in Siberia are prolific sources of ice fog, which affect aircraft operations and automobile travel. The physics of formation of ice fog is similar to contrail formation. At very cold temperatures, small additions of moisture to the air by burning fossil fuels, automobile exhaust emissions, and even human respiration can lead to persistent clouds of small ice crystals.

Overall, the evidence is certainly suggestive that anthropogenic emissions of aerosols and gases are having an impact on cloud microstructure, precipitation processes, and cloud/fog occurrence. Further research is needed, however, to identify physical mechanisms responsible for observed or inferred effects of those emissions.

Chapter 5

Urban-Induced Changes in Precipitation and Weather

5.1 Introduction

In Chapter 4, we examined the possible effects of particulate and gaseous emissions on precipitation and weather on the regional scale in a general sense rather than specific urban-induced changes. In this Chapter we examine the evidence suggesting that pollutants as well as other urban effects are causing changes in the weather and climate, in and immediately surrounding, urban areas.

There is considerable evidence which suggests that major urban areas are causing changes in surface rainfall, increased occurrences of severe weather, especially hailfalls, and alterations to surface temperatures (Ashworth, 1929; Kratzer, 1956; Landsberg, 1956; 1970; Changnon, 1968; 1981a; Changnon and Huff, 1977; 1986). Some of the hypothesized causes of those changes include:

- urban increases in CCN concentrations and spectra, and IN concentrations;
- changes in surface roughness and low-level convergence;
- changes in the atmospheric boundary layer and low-level convergence caused by urban heating and landuse changes; and
- addition of moisture from industrial sources.

A major cooperative experiment was carried out in the St. Louis, Missouri area in the 1970's to identify urban-induced changes in weather and climate and to identify the primary causes of those changes. A comprehensive review and summary of the experiment and its results are described in the monograph *METROMEX: A Review and Summary* (Changnon, 1981b). In this section we draw heavily on those findings to discuss the potential mechanisms causing urban-induced changes in weather and climate.

First of all METROMEX and related studies showed that St. Louis exhibits a major summertime precipitation anomaly relative to the surrounding rural area. The area-average urban related increase is about 25%. Much of the enhanced rainfall occurs during the afternoon (1500 to 2100 local daylight time (LDT)), over the city and the close-in area east and northeast. The clouds producing those changes are deep convective clouds and thunderstorms. In fact the frequency of thunderstorms is enhanced in that region by 45% and hailstorms increased by 31%. Not only is the hailstorm frequency higher, but hailstones are larger and of greater number. The rainfall observations also indicated a maximum around midnight extending from approximately 2100 LDT to 0330 LDT located northeast of the city. Changnon and Huff (1986) estimated that the area experienced a 58% increase in nocturnal rainfall relative to the surrounding countryside. The storms responsible for the nocturnal maxima were well organized storms such as squall line thunderstorms that swept across the urban area and moved across the affected region.

How does an urban area cause those changes? Let us examine each of the hypothesized mechanisms and see how well each fits the METROMEX observations.

5.2 Urban Increases in CCN and IN Concentrations and Spectra

Not surprising, anthropogenic activity in the St. Louis urban area caused major increases in CCN concentrations; as much as 94%. Droplet size distributions as a result were found to be narrower with larger concentrations of droplets in the clouds downwind of the city compared to upwind. Large numbers of large, wettable particles, having radii greater than 10 μm with many as large as 30 μm were found over the city. These 'ultra-giant' particles can serve as embryos for initiation of collision and coalescence. This is consistent with the finding that clouds over the city did have a greater number of larger droplets. The METROMEX scientists cautioned, however, that they had less confidence in those observations compared to the observed higher concentrations of small cloud droplets.

Similar to the study of paper pulp mills, the METROMEX modeling studies revealed that the time required to initiate precipitation in upwind and downwind clouds was only different by a few minutes. It was therefore concluded that the anthropogenic CCN do not play a major role in the creation of the urban rainfall anomaly.

It was also found that the concentrations of IN were not greatly altered over and downwind of the urban area. If anything, it was found in the winter months that the IN concentrations were actually less over the urban region. This suggested that the coagulation of the few IN with the more numerous anthropogenic aerosol actually deactivated the IN.

In summary it does not appear that the anthropogenic emissions of aerosols can by themselves cause the observed increases in rainfall. It is possible that changes in the cloud and raindrop spectra can have an impact on the rate of glaciation of a cloud and thereby the subsequent cloud behavior. We will examine this hypothesis next as the '*glaciation*' mechanism.

5.3 The Glaciation Mechanism

As noted in Part I, it is generally accepted that cumuli containing supercooled raindrops glaciate more readily than more continental, cold-based cumuli that do not contain supercooled raindrops. There are several reasons for this. First of all, larger drops freeze more readily than smaller drops by immersion freezing. More importantly, as noted in Part I, the coexistence of large, supercooled drops and small ice crystals, nucleated by some mechanism of primary nucleation, favors the rapid conversion of a cloud from a liquid cloud to an ice cloud (i.e., glaciation) [Cotton, 1972a,b; Koenig and Murray, 1976; Scott and Hobbs, 1977]. Thus the ultra-giant particles observed over St. Louis could produce more supercooled raindrops which would accelerate the glaciation process. This process does not require any change in IN concentrations.

A second factor potentially affecting the rapid glaciation of urban clouds is that the altered drop-size spectra could initiate secondary production of ice crystals. Laboratory studies have indicated that copious quantities of ice splinters are produced when an ice particle collects supercooled cloud droplets when cloud temperatures are within the range of –3° C to –8° C, and when the cloud is composed of a mixture of large drops (greater than 12.5 μm radius) and small droplets (less than 7 μm). All these criteria were met in the clouds observed over St. Louis during METROMEX.

As noted by Keller and Sax (1981), however, in broad, sustained fast rising updrafts, even when all the criteria for secondary ice production are met, the secondary ice particles and graupel particles will be swept upwards out of the limited temperature range favorable for secondary ice crystal production. Until the updrafts weaken and graupel particles settle into the secondary production zone, the positive feedback mechanism of secondary production is broken. Therefore the opportunities for rapid and complete glaciation of a cloud are greatest if the cloud has a relatively weak, steady updraft or the updraft is a pulsating convective tower. We will show that the clouds over the St. Louis urban area had less buoyant energy or CAPE (as evidenced by lower values of θ_e)[1] than rural clouds. Thus the clouds over the urban area would be expected to have weaker updraft strengths as they enter freezing levels than rural clouds, further enhancing the potential for rapid and complete glaciation.

The hypothesis then builds on the dynamic seeding hypothesis described in Part I. That is, the rapidly glaciated, urban clouds would explosively deepen after they penetrate into subfreezing temperatures, process more moisture through their greater depths, live longer, and rain more. Evidence supporting the glaciation hypothesis is as follows. First of all, it was observed during METROMEX that cumuli over the adjacent rural areas exhibited a distribution in cloud top heights that was bimodal, with many clouds terminating at a height of about 6 km and many others rising to 12 km, but with few clouds penetrating just to heights of around 9 km. In contrast, urban cloud top heights had a more continuous distribution with cloud top heights at all levels between 5 and 13 km. One interpretation of these measurements is that enhanced glaciation of the urban clouds allowed more clouds to penetrate upward through an arresting level, such as an inversion in temperature or a dry layer, and thereby rise to greater heights.

The fact that there is a downwind maximum in thunderstorm activity and hail is consistent with more vigorous glaciation of the clouds as well. Finally the finding that merger of clouds was more frequent over the urban area is consistent with the dynamic seeding hypothesis (Simpson, 1980).

Unfortunately enhanced glaciation of urban clouds was never directly observed during METROMEX. This is because the airborne sensors used were not capable of discriminating between glaciated and unglaciated clouds. Thus this mechanism remains an unproven hypothesis.

5.4 Impact of Urban Landuse on Precipitation and Weather

Except for the Ohio River valley in the immediate area of the city, St. Louis is located in a vast farmland on a relatively flat plain. The presence of a major urban area changes the surface properties markedly. First, the presence of buildings, particularly tall downtown structures, alters surface roughness from the relatively smooth cropland and occasional forest to a very rough surface. This rough surface creates surface drag which slows the winds near the ground. As shown in Figure 5.1 air approaching the city would slow down and tend to divert around the city something like flow around an isolated rock in a stream. On the

[1] θ_e, called equivalent potential temperature, is a conservative variable for wet-adiabatic processes. See Cotton and Anthes (1989) and Pielke (1984) for a mathematical definition of θ_e.

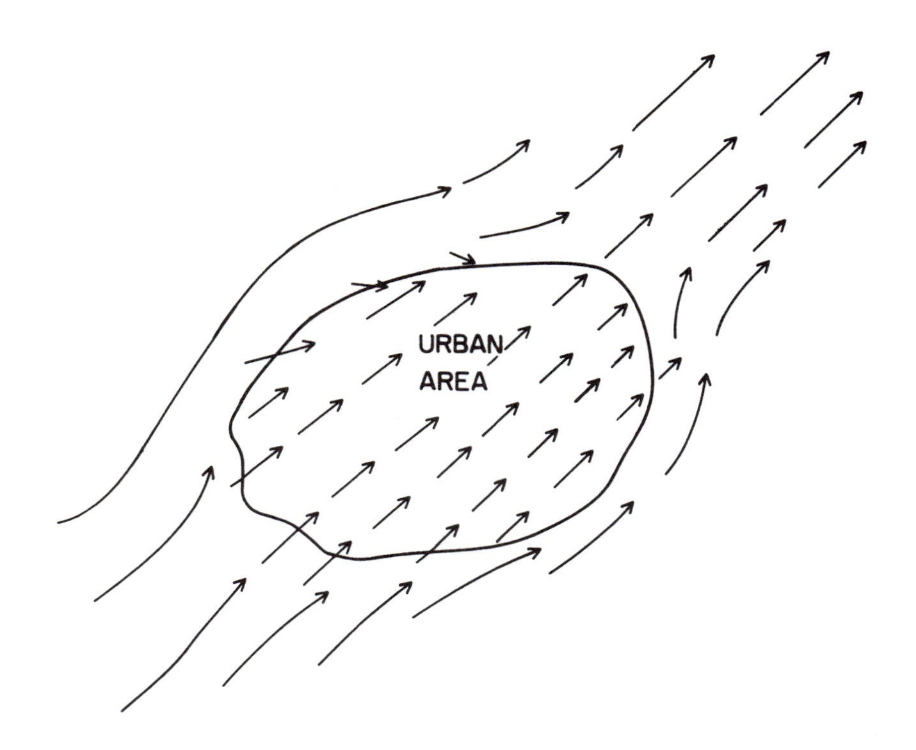

Figure 5.1: *Schematic illustration of low-level air flow over and around a major urban area due to changes in surface roughness.*

downwind side of the city air streaming around the city would tend to converge, causing upward motion in that region. There are documented cases where changes in surface roughness have led to a slowing down of cold fronts upwind of New York City, and acceleration of the front downwind (Loose and Bornstein, 1977).

Even more important are the changes in the heat and water budget at the surface caused by the presence of a city. In the countryside, the earth's surface consists of fallow and plowed fields, grasslands, and small forested areas. The soils are, relative to much of the urban area, rather moist and contain vegetation which can transfer significant amounts of moisture to the atmosphere. By contrast, the surface in the city is a rather impermeable layer, consisting of a mixture of concrete, asphalt, and buildings with a relatively small area of undisturbed soils and vegetation. A greater fraction of rainfall therefore runs off in urban areas than in the countryside.

These changes in surface properties alter the surface energy budget in two ways. First of all, in an urban area such as found in the central United

States, a greater fraction of the incoming solar radiation is reflected over the cities as the concrete and buildings are more reflective than plowed fields and cropland. This greater reflectance, or what we call *albedo*, has a cooling effect over the urban areas. Secondly, the more moist land surfaces over the countryside causes a greater fraction of the solar energy absorbed at the surface to be converted into latent heat release rather than sensible heat transfer. In other words, much of the absorbed energy goes into evaporating water from the soil and evapotranspiration from the vegetative canopy. This causes a cooling effect in rural areas relative to the drier, less vegetated urban areas. Because the impact of creating a drier, less vegetation covered soil in the urban area is much greater than the cooling effect of increased albedo over urban areas, the urban areas in a humid climate such as St. Louis warm more quickly than the surrounding countryside. This causes what is called an *urban heat island.*

During METROMEX, St. Louis was shown to have a well-defined heat island centered over the downwind commercial district, northeast of the core of the urban area. Its maximum size and intensity occurred between midnight and 0600 LDT. It was also found that the air immediately above the urban area was usually drier than over nearby rural areas. Let us consider the hypothetical diurnal variation of the boundary layer of the urban area.

At sunrise, air temperatures begin to rise over both the rural and urban areas. Owing to a shallower nocturnal inversion over the country than the city, air temperatures rise more quickly over the countryside at first. As the ground is heated in both the urban and rural areas, a mixed layer forms which deepens more rapidly over the city than the rural areas. This is because the low-level nocturnal inversion strength is weaker over the city. By midday, heating proceeds more rapidly over the city because more of the absorbed energy goes into sensible heat rather than latent heat. The boundary layer thus becomes increasingly deeper and drier over the city. On typical afternoons, the urban boundary layer was found to be 100 to 400 m deeper over St. Louis than the rural areas.

Figure 5.2a illustrates a late afternoon vertical cross section over the city showing a deeper urban boundary layer, that is warmer and drier than the countryside. Associated with this warmer and drier pool of air over the city is rising motion which produces a sea-breeze-like circulation between the city and the countryside. As seen in Figure 5.2b this rising motion over the city draws low-level air into the city causing low-level convergence. Such low-level convergence has been found to be favorable for producing deep, precipitating cumulus clouds and also increases the likelihood that those clouds will merge in this low-level convergence zone to produce bigger, heavier raining clouds (Pielke, 1974a; Ulanski and Garstang, 1978a,b; Chen and Orville, 1980; Simpson et al., 1980; Tripoli and Cotton, 1980). The maximum convergence would occur somewhat

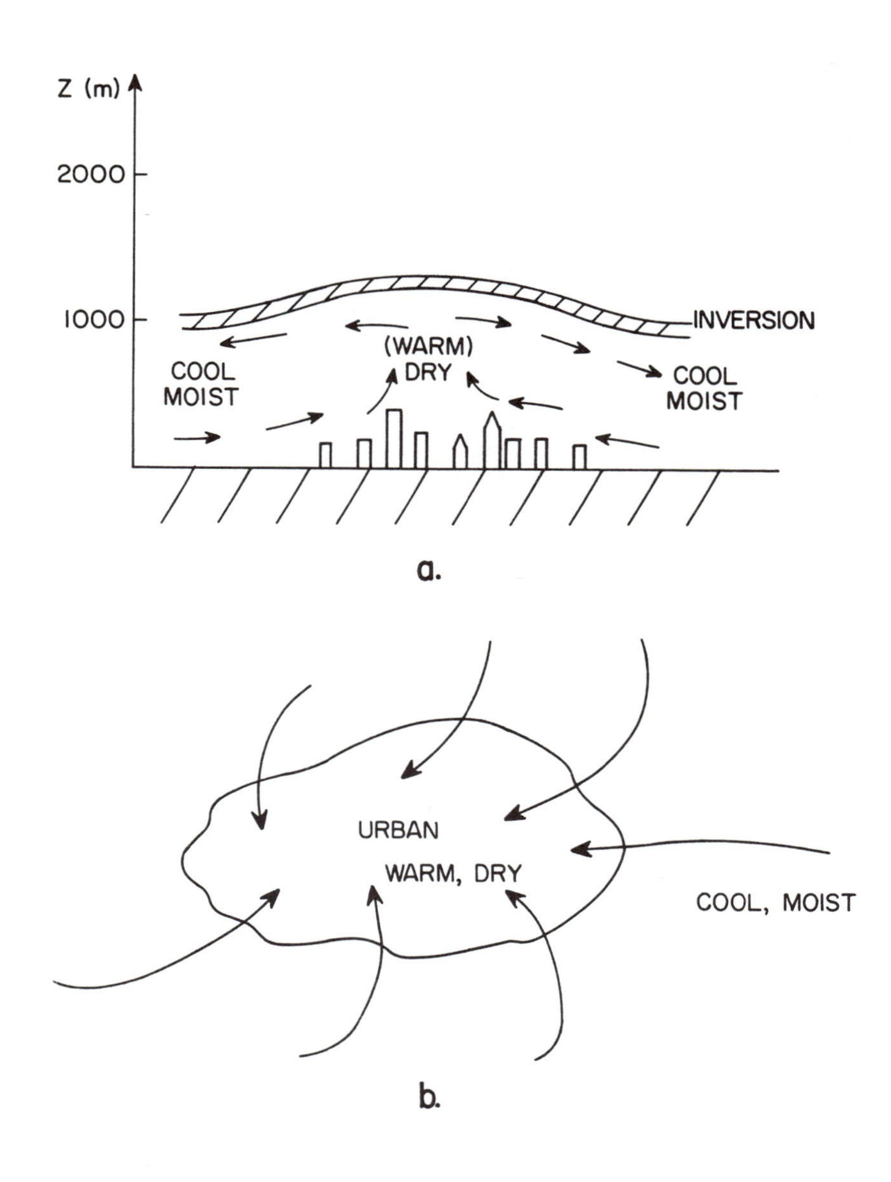

Figure 5.2: *(a) Schematic vertical cross section over a major urban area during the late afternoon in a humid climate region illustrating the effects of the urban heat island. (b) Similar to (a) except a horizontal map of the altered low-level winds by the heat island in the absence of large-scale prevailing flow.*

downwind of the urban area as the heated boundary layer is advected in that direction (Mahrer and Pielke, 1976, Hjelmfelt, 1980).

During the evening hours, heat conduction from the ground in the urban area limits the rate of cooling compared to the countryside. The surface remains warmer and the low-level air is less stable than in the rural areas. Thus the heat island remains stronger over the city throughout the night.

In the next subsections, the observed behavior of clouds and precipitation over and downwind of St. Louis are discussed to see if they are consistent with changes in the urban boundary layer.

5.4.1 Observed cloud morphology and frequency

Clouds over the St. Louis urban area were found to have bases 600 to 700 m higher than rural clouds. This is consistent with the observation that the air over the city is warmer and drier. The exception was clouds downwind of refineries, where it is believed that moisture injections by the refineries caused lower cloud bases. Air motion into the bases of the clouds were stronger which is consistent with the expected more vigorous thermals due to the heat island.

Cloudiness (defined as the percent coverage of clouds over an area) was found to be greater over the urban area in the later afternoon (1600 LDT) consistent with the observed convergence and upward motion due to the heat island.

5.4.2 Clouds and precipitation deduced from radar studies

The first detectable radar echoes is a measure of the initiation of precipitation. Echoes were found to be more frequent over the urban area during the late morning, about 1400 LDT, and after 1930 LDT. This suggests that the heat island-induced convergence field played a major role in creating precipitating cumuli. Moreover, individual cumulus cells over the urban area were found to grow deeper and have slightly longer durations than over the rural areas. Again, this is consistent with stronger convergence over the urban heat island favoring deeper longer lasting precipitating cumuli.

Clouds over the urban area were also found to merge more frequently with cells over the city, grew taller and lasted longer than did merged cells over rural areas. As noted previously, this is consistent with observations and modeling studies which suggest that cloud merger is enhanced by low-level convergence, such as that caused by the urban heat island effect. Because it is generally found that taller and longer lasting cells create more rain and a greater likelihood for hail, these findings are consistent with the hypothesis that the urban heat island enhances convective rain systems over and downwind of the city.

Analysis of cells that contributed to the nocturnal rainfall maxima downwind of St. Louis (Changnon and Huff, 1986) suggested that this urban related anomaly was associated with the enlargement of rain areas from well organized storms that existed upwind of St. Louis and then moved over and downwind of the city, as well as the development of new cells over the urban area. Changnon and Huff (1986) and Braham (1981) speculated that this behavior of the storms may have been due to the injection of drier air into the storms as they passed over the urban area, causing them to weaken prematurely and release stored water downwind of the city. This interpretation, however, is inconsistent with the observation that organized, nocturnal storms normally draw on air that has its origin over a large area 50 or more kilometers away from the storm. This warm, moist air typically glides over the nocturnal low-level inversion so that the nocturnal storms do not readily ingest much surface air. Even if the weaker nocturnal inversion over the city allow the storms to tap the drier urban surface air, it seems that the volume of urban air ingested would be a small fraction of the total volume of air ingested into the storm. It is our opinion that enhanced mesoscale ascent associated with the urban heat island could have intensified the nocturnal storms. The fact that nocturnal storms are typically less severe than afternoon convection means that they could strengthen without exhibiting any increase in severe weather. Further studies are needed (probably with multiple Doppler radar) to determine if the storms contributing to the nocturnal urban rainfall anomaly were actually weakening or strengthening, on average.

In summary, there is considerable evidence indicating that the St. Louis urban area enhances rainfall and possibly the occurrence of severe weather. The actual physical processes responsible for those effects, however, have not been fully identified. Both the glaciation mechanism and urban heat island induced mesoscale changes are leading contenders. Further observational and modeling studies are required to identify the actual causal mechanisms.

One may ask: is it really necessary to identify the actual mechanisms responsible for an urban precipitation anomaly? Can't we be satisfied that the rainfall analysis shows a strong rainfall anomaly downwind of the urban area? **The answer is clearly no!** For one thing we cannot be sure that the statistical analysis of the rainfall records did not produce an urban 'signal' purely by chance. Another reason for establishing a cause and effect is St. Louis, like many major urban areas, is situated in a river valley. Could local physiographic features such as the higher terrain of the valley sidewalls and moisture sources from the low, relatively wet river bottomland or channeling of the moist, low-level jet through the river valley be the primary causal factors in creating a rainfall anomaly? Attempts to isolate contributions to the rainfall anomaly were made during METROMEX using mesoscale models (Vukovich et al., 1976; Hjelmfelt, 1980). These models revealed that there may be important interactions between the local topography and the downwind thermal plume of the heat island. It was concluded by the METROMEX team that these effects were small, at least in the afternoon hours. They could not dismiss the possibility that physiographic effects could have contributed to the nighttime maximums, however. It should be noted that models used at that time could not simultaneously simulate both the mesoscale responses to the physiography and urban heat island, and the response of deep precipitating convection to those forcings. It is possible to perform such computations today, but those computations would require large amounts of computer resources. Nonetheless, such calculations should be performed to help isolate the factors contributing to the inferred urban precipitation anomaly.

A third reason for isolating the causes of the urban precipitation anomaly is that it may become necessary to reduce the rain anomaly and enhanced severe weather occurrences. Without a clear identification of the causal factors, one cannot decide if reductions in emission of gases and aerosols contributing to CCN, or alterations in landuse patterning, is required to reduce the anomaly.

The results of METROMEX apply to urban environments in midlatitude humid areas in which the natural vegetation is a deciduous forest which has been replaced by agriculture. Avissar and Pielke (1989) modeled such an environment where the urban area was assumed to contain 20% built-up areas, 10% bodies of water, 40% agricultural crops, and 30% forests. Shown in Figure 5.3a, for 1400 LST, a substantial modification of the boundary layer over the urban area and the development of low-level convergence into the city were simulated, which is consistent with the interpretation of the METROMEX data. In contrast, when this same heavily irrigated urban area is inserted into an arid or semi-arid environment (such as Denver, Colorado), the impact on the local environment is even more pronounced as shown in Figure 5.3b with the urban area acting as an oasis during the day, rather than a daytime urban heat island. Large differences in boundary layer structure between the urban area and the surrounding desert terrain result in a well-defined local wind circulation.

In desert environments, the urban island effect has been documented. For example, the nighttime summer temperature at Sky Harbor Airport in Phoenix increased an average of 1.1°C every decade from 1948 to 1984 (Balling and Brazel, 1987) apparently as a result of the reduction in the urban area that contains irrigated vegetation and an increase in coverage by buildings, concrete, and asphalt. The demand for air conditioning resulted in an increase of peak electricity demands of 1 to 2% per degree Centigrade (Akbari et al., 1989). McPherson and Woodard (1990)

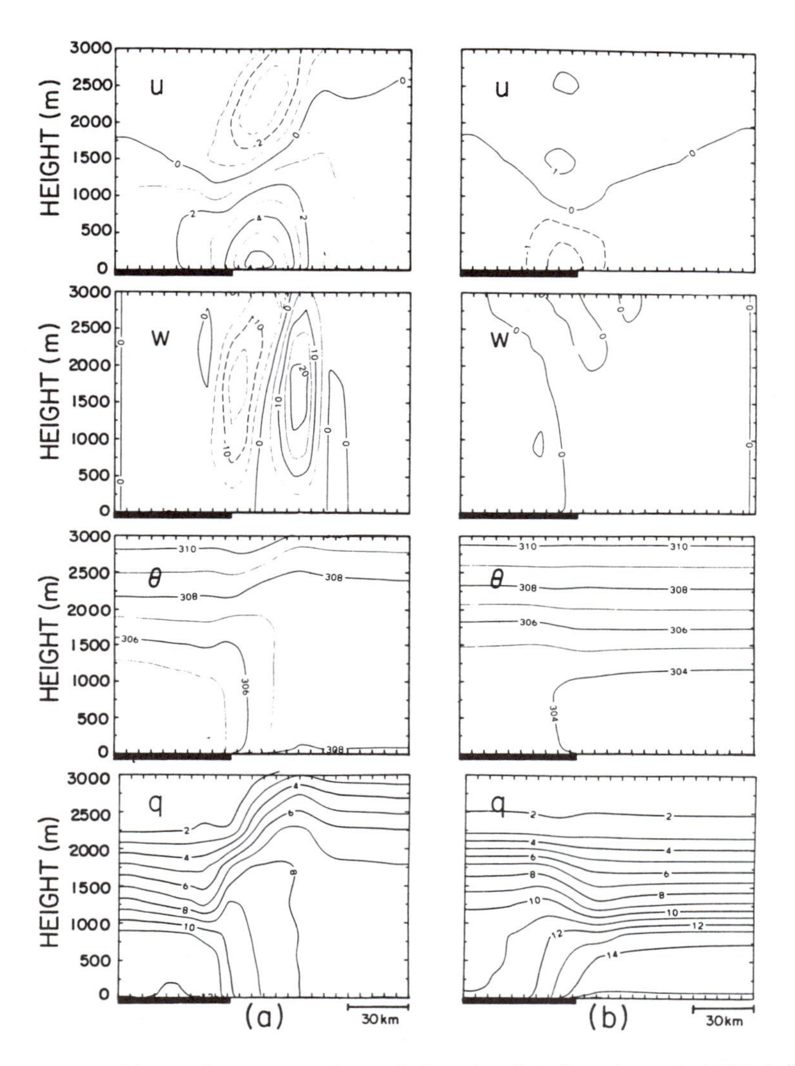

Figure 5.3: *Vertical cross section of the simulated region at 1400 LST for: (i) the horizontal wind component parallel to the domain (u) in m* s^{-1}*, positive from left to right; (ii) the vertical wind component (w) in cm* s^{-1}*, positive upward; (iii) the potential temperature (θ) in K; and (iv) the specific humidity (q) in g* kg^{-1}*, resulting from the contrast of a 60 km wide, heterogeneous land surface region (indicated by the dark underbar) which consists of 20% built-up areas and wastelands, 10% bodies of water, 40% agricultural crops, and 30% forests. The adjacent region is (a) bare and dry, and (b) completely covered by unstressed vegetation.* [From Avissar, R., and R.A. Pielke, 1989: A parameterization of heterogeneous land surfaces for atmospheric numerical models and its impact on regional meteorology. *Mon. Wea. Rev.*, **117**, 2113-2136.]

suggest that the ratio of water and energy costs determine the optimal landscape type which should be used in this environment to minimize their costs. The types of landscaping include *zeroscape*, which is primarily rock-covered ground, *xeriscape*, which utilizes drought-tolerant vegetation such as mesquite, palo verde, and heritage oak, and *mesiscape*, that includes moderate or high water users such as magnolias and ashes. The optimal planting is one that permits shading and cooling by transpiration to minimize air conditioning needs, yet water loss is constrained as much as possible. McPherson and Woodard (1989) estimate that in Tucson, the projected annualized cost of a mature tree is $7.76, while its benefits are $26.18 with $19.20 of this amount resulting from cooling due to transpiration.

The influence of landuse on climate and weather is discussed in more detail in the next chapter. It is clear, however, that the urban effect will vary depending on its geographic location and we need to explore a range of urban environments in even more detail than was achieved during METROMEX.

Finally, Karl and Jones (1989) compared urban and rural temperature records to show that the growth of cities during this century has resulted in a 0.4°C bias in the United States climate record. Although the study has not yet been completed, if this bias exists in the global climate data set, since the urban heating is local, its use to represent a wider geographic record for climate change studies will be misleading. The neglect of the urban bias would suggest the global climate is warming more rapidly than it actually is.

Chapter 6

Other Landuse Changes

6.1 Landscape Changes

The influence of landuse on climate has been considered for at least several hundred years. Thomas Jefferson, for example, made the following comment in Notes on the State of Virginia (1976) in 1861.

> "The eastern and southeastern breezes come on generally in the afternoon. They formerly did not penetrate far above Williamsburg. They are now frequent at Richmond and every now and then reach the mountains... As the lands become more cleared, it is probable they will extend still further westward."

As human population has grown further since the time of Jefferson, the natural distribution of vegetation on earth has been altered, often in a major fashion. In Europe, for example, the hardwood forests of France and Germany have been removed in order to permit large areas of agriculture production. Similarly in the eastern United States, only token traces of the original climax forest cover remain with the current landscape covered by second-growth forests, cropland, and urban developments. In India, forest cover has decreased 3% since 1975, while in the Philippines only 29 million acres remain of an original forest area of 70 million acres. The Amazon rain forest, which originally covered an area of about the size of one half the contiguous United States is being reduced at an average rate of 20,000 km^2 per year (Nobre et al., 1991). In arid and semi-arid lands, large areas of irrigation have replaced the natural scrub and desert vegetation. Much of the Texas panhandle, for example, is now covered with irrigated cropland.

There are potential regional and global climatic impacts associated with these landscape changes. Two of the most important influences are (i) changes in the fraction of solar radiation reflected back to space (i.e., the albedo) and (ii) changes in the fraction of heat which is used to evaporate and transpire water to the atmosphere. A schematic illustration contrasting the difference between bare soil and ground covered with vegetation is shown in Figures 6.1 and 6.2. Note that vegetation provides an

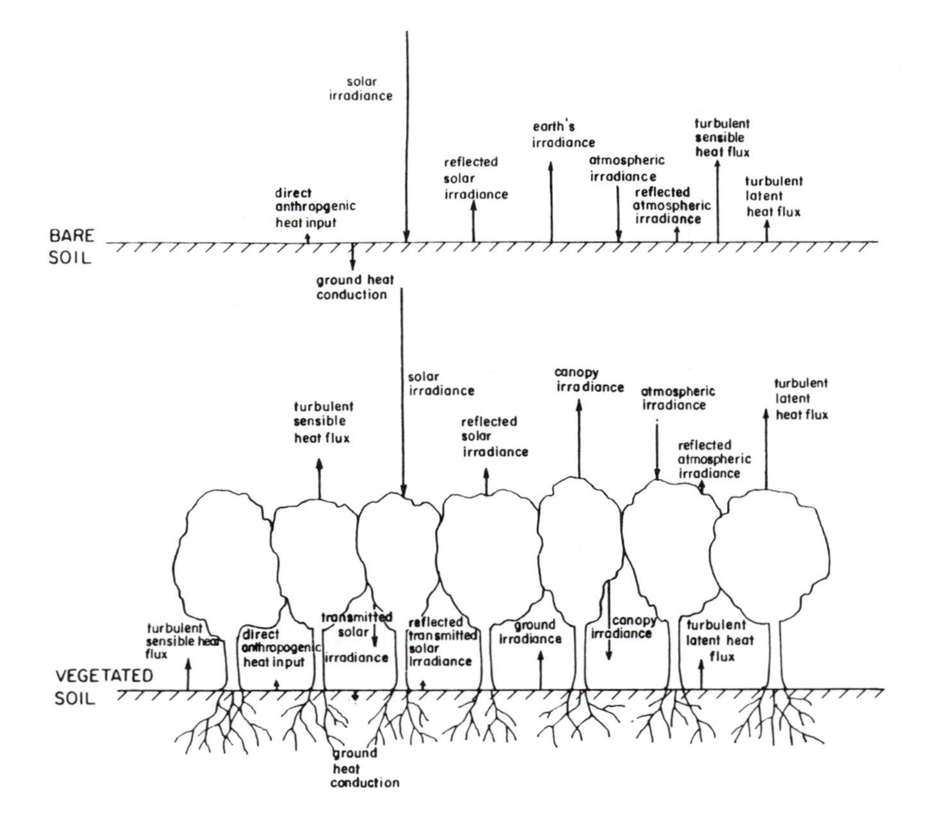

Figure 6.1: *Schematic illustration of the surface heat budget over bare soil and vegetated land. The roughness of the surfaces (and for the vegetation, its displacement height) will influence the magnitude of the heat flux. Dew and frost formation and removal will also influence the heat budget.* [From Pielke, R.A. and R. Avissar, 1990: Influence of landscape structure on local and regional climate. *Landscape Ecology*, **4**, 133-155.]

effective conduit to extract water from within the soil in the root zone and to transfer it to the atmosphere above. Bare soil, in contrast, relies on only diffusion of water through the soil which often becomes limited when the surface crusts over during drying. On a warm day Lee (1978, pg. 117) reports on water losses from vegetation in excess of 5 liters m^{-2}. The leaf area index (LAI), which is the ratio of leaf area per unit ground area, is proposed as the single most useful structural variable for quantifying energy and mass exchanges from a terrestrial ecosystem of the to the atmosphere (Running and Coughlan, 1988; Nemani and Running, 1989; Running, 1990).

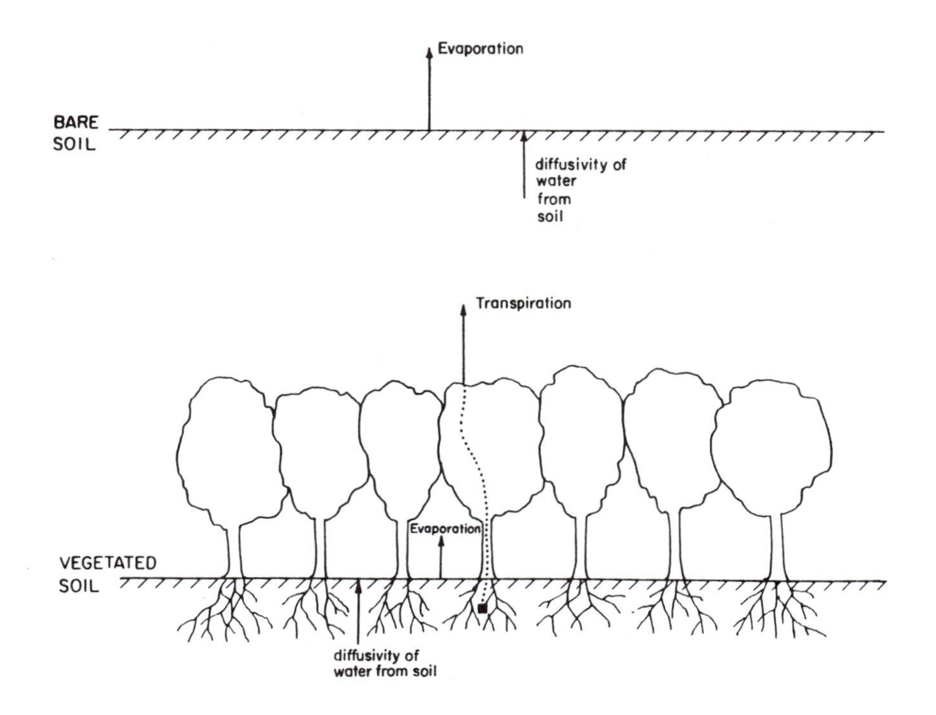

Figure 6.2: *Schematic illustration of the surface moisture budget over bare soil and vegetated land. The roughness of the surface (and for the vegetation, its displacement height) will influence the magnitude of the moisture flux. Dew and frost formation and removal will also influence the moisture budget.* [From Pielke, R.A. and R. Avissar, 1990: Influence of landscape structure on local and regional climate. *Landscape Ecology*, **4**, 133-155.]

6.1.1 Influence on boundary layer structure

The presence of transpiring vegetation substantially alters the atmospheric boundary layer structure. As reported in Segal et al. (1988) the substantial persistence of surface temperature gradients over 10°C, as observed from a satellite, due to the different landscapes (i.e., irrigated cropland adjacent to a natural short grass prairie) was documented over Colorado (Figure 6.3). Aircraft cross sections flown across the boundary of these two land surface types (e.g., see Figures 6.4 and 6.5) document the different boundary layer structure at different heights up to 440 m, with the irrigated area being cooler and substantially more moist. Rabin et al. (1990) used satellite imagery to show that clouds preferentially form first over mesoscale sized areas of harvested wheat in Oklahoma. Pielke and Zeng (1989) used sounding data from that experiment to demonstrate the larger energy for deep cumulus convection over the wetter

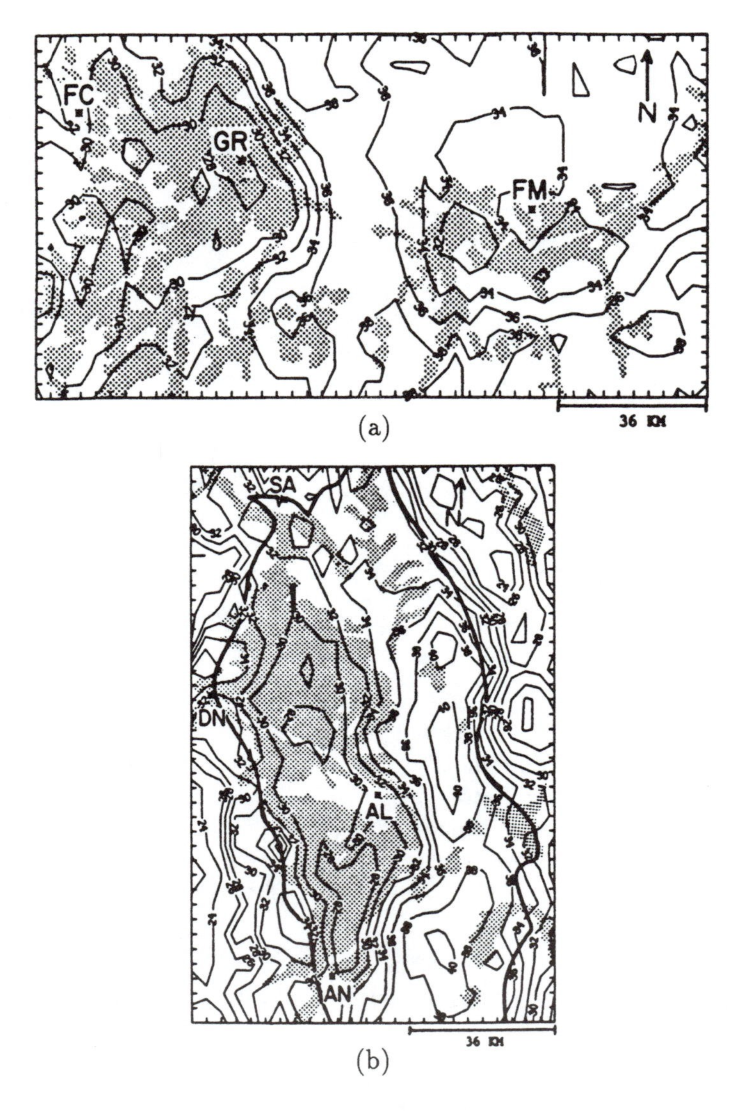

Figure 6.3: *Composite of GOES derived surface temperature at 1300 LST for the period 1 August 1986 to 15 August 1986 (a) for northeast Colorado (FC–Fort Collins: FM–Fort Morgan; GR–Greeley), (b) for the San Luis Valley in Colorado (AL–Alamosa; AN–Antonito; DN–Del Norte; SA–Saguache). The lower valley is outlined by a dark line separating it from significant elevated terrain. Irrigated areas are shaded.* [From Segal, M., R. Avissar, M.C. McCumber, and R.A. Pielke, 1988: Evaluation of vegetation effects on the generation and modification of mesoscale circulations. *J. Atmos. Sci.*, **45**, 2268-2292.]

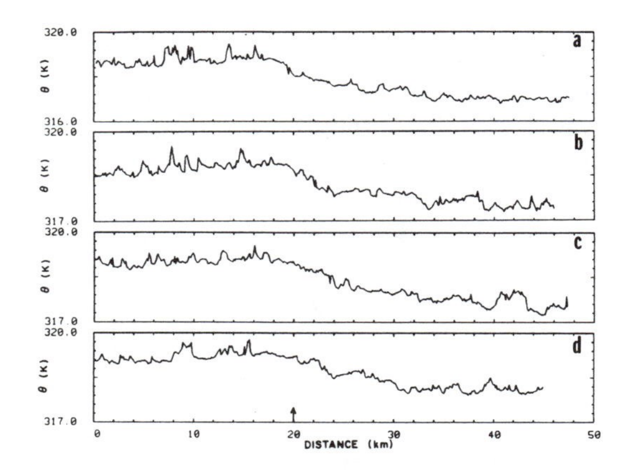

Figure 6.4: *Measured potential temperature for a transect from Briggsdale to Windsor on July 28, 1987 at the altitude of (a) ≈ 140 m, (b) ≈ 240 m, (c) ≈ 345 m, and (d) ≈ 440 m above the ground. The observed crop-dry land boundary is indicated by an arrow.* [From Segal, M., W. Schreiber, G. Kallos, R.A. Pielke, J.R. Garratt, J. Weaver, A. Rodi, and J. Wilson, 1989: The impact of crop areas in northeast Colorado on midsummer mesoscale thermal circulations. *Mon. Wea. Rev.*, **117**, 809-825.]

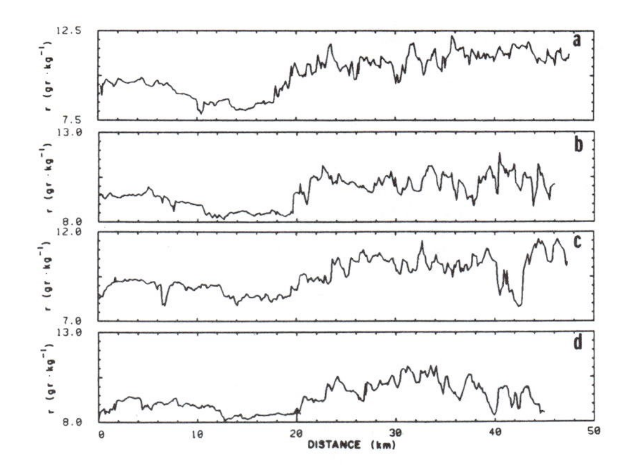

Figure 6.5: *As in Figure 6.4 except for moisture mixing ratio.* [From Segal, M., W. Schreiber, G. Kallos, R.A. Pielke, J.R. Garratt, J. Weaver, A. Rodi, and J. Wilson, 1989: The impact of crop areas in northeast Colorado on midsummer mesoscale thermal circulations. *Mon. Wea. Rev.*, **117**, 809-825.]

surface in response to the larger lower-level values of equivalent potential temperature.

The First International Satellite Land Surface Climatology Project (ISLSCP) Field Experiment (FIFE) was compiled from 1987–1989 in order to provide ground truth data to relate to satellite measurements of land surface irradiances. This study also provides additional evidence of the importance of land surface characteristics on surface layer fluxes of heat and moisture. There has been extensive analyses of these data with a summary of many of the results reported in Sellers and Hall (1992). Among the conclusions of the field study is the change in surface fluxes of heat and moisture by even small slope changes (Smith et al., 1992 and Davis et al., 1992), as well as whether or not the area is grazed (Turner et al., 1992).

6.1.2 Generation of physiographic-forced circulations

Spatial variations in transpiring vegetation generate mesoscale circulations as strong as sea breezes. Ookouchi et al. (1984) showed that gradients of soil moisture can result in local wind circulations as strong as sea breezes along coastal areas. Mahfouf et al. (1987) demonstrate using the two-dimensional model of Nickerson et al. (1986), that soil texture gradients (e.g., sand, loamy sand, loam, clay, peat) influence surface moisture availability and hence the strength of the resultant local winds.

Segal et al. (1988; 1989) and Avissar and Pielke (1989) have shown that sea breeze strength circulations can occur when transpiring vegetation is juxtaposed to dry land. Pinty et al. (1989) have shown that the intensity of these local wind circulations is sensitive to the soil moisture under the vegetation cover. When the underlying soil is sufficiently dry, the stoma on the plants close and little transpiration cooling occurs.

Figure 6.6, reproduced from Segal et al. (1988), presents results for a sea breeze and for the situation of unstressed vegetation juxtaposed to dryland. As evident in the figure, the intensity of the simulated circulations are very similar. Since it is well-known that sea-breeze convergence can generate areas of deep cumulonimbus convection and substantial anvil production (e.g., see Pielke, 1974b), gradients in sensible heat flux resulting from spatial variations in the magnitude of water vapor flux from transpiring vegetation are likely to have the same result.

A major research program in southwest France referred to as HAPEX-MOBILHY[1] was conducted in 1986 in order to evaluate the influence of landscape variation on boundary layer structure and the resultant influence on local wind flows (André et al., 1989; 1990; Mascart et al.,

[1]HAPEX and MOBILHY are acronyms for Hydrologic Atmospheric Pilot EXperiment and MOdélisation du BILan HYdrique.

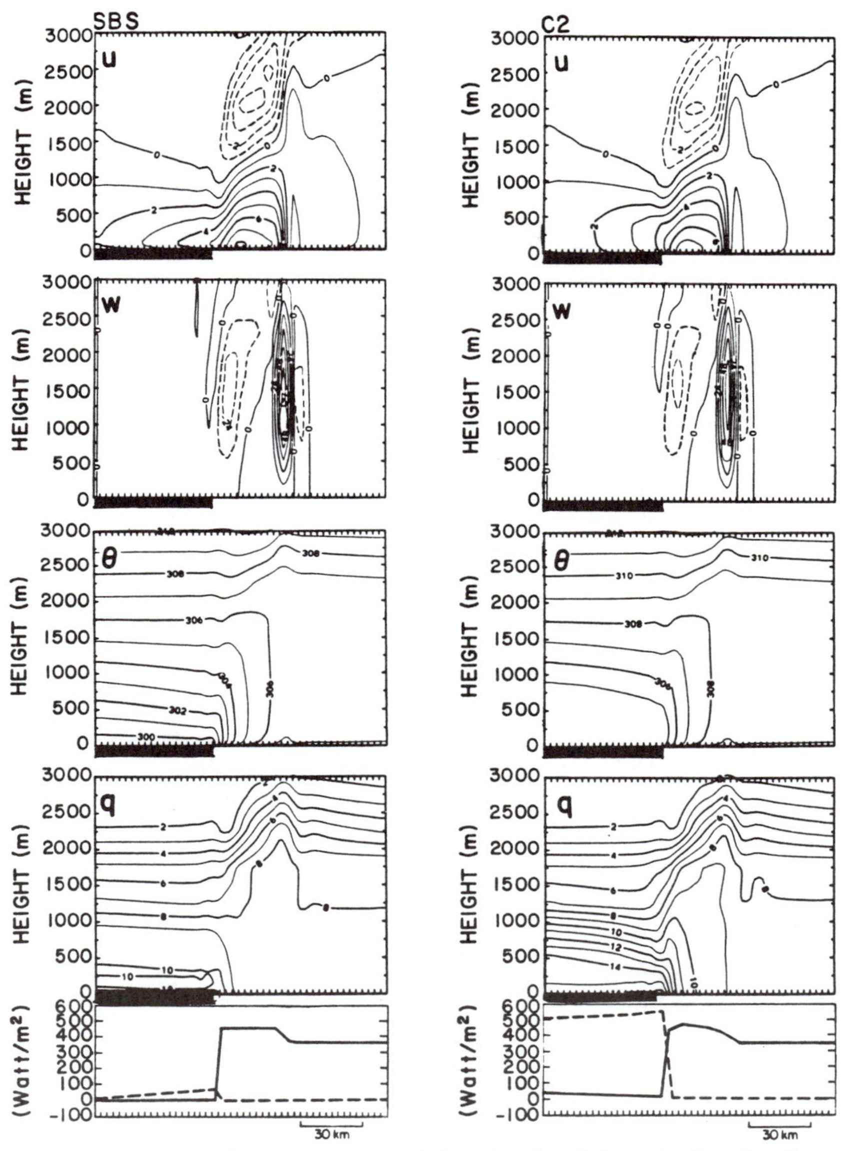

Figure 6.6: *Vertical cross section of the simulated domain for the Cases SBS (where the dark segment indicates the sea) and C2 (where the dark segment indicates the vegetation section) at 1400 LST for (i) u – west-east component of the wind in m* s^{-1} *(dashed contours indicate negative component-easterly), (ii) w – vertical wind component in cm* s^{-1} *(dashed contours indicate negative component-downward vertical velocity), (iii)* θ *– the potential temperature in K, (iv) q – the specific humidity in g* kg^{-1}*, (v) the sensible (solid line) and latent (dashed line) heat fluxes at the surface.* [Adapted from Segal, M., R. Avissar, M.C. McCumber, and R.A. Pielke, 1988: Evaluation of vegetation effects on the generation and modification of mesoscale circulations. *J. Atmos. Sci.*, **45**, 2268-2292.]

1991). In this region an area of corn, which covered about 80% of the soil surface, was adjacent to a pine forest. Observations and model results for a case study day (June 16, 1986) shown in Figures 6.7 and 6.8 document the larger latent heat flux over the corn as a result of its lower stomatal resistance to water vapor flux. André et al. (1989) show, using a two-dimensional version of the Nickerson et al. (1986) model and the three-dimensional model of Bret and Bougeault (1988), that these gradients of surface heat flux result in significant influences on the local air flow. Figure 6.9 shows the three-dimensional model results for surface and near surface air temperature in which the pine forest area (shown by the square) is about 2°C cooler at the surface and the air temperature is 1°C warmer than over the agricultural crop. The smaller atmospheric vertical temperature gradient is attributed to the larger sensible heat fluxes over the forest.

An index, referred to as the normalized difference of the vegetation (NDVI) is being used to assess the geographic distribution of transpiring vegetation. Nemani and Running (1989) have documented statistically that latent heat exchange as measured by NDVI is the dominant explanation for spatial variations in surface radiant temperatures. NDVI is based on the observation that healthy, photosynthetically active vegetation is higher in the near infrared surface radiant temperature than in the visible band, with the differential reflectance in the two bands a measure of the health and spatial density of the plants. NDVI measurements have been shown to be well correlated to LAI values (Spanner et al. 1990). An overview of this index is presented in Karl et al. (1989) and Running (1990). The use of NDVI in the HAPEX-MOBILHY program in France is described in Phulpin et al. (1989). NDVI has been used to demonstrate the large spatial and temporal variation of transpiring vegetation over the Great Plains of the United States (Pielke et al., 1990a; 1991b). Based on model results, these gradients of vegetation characteristics should produce surface heat flux differences during summer days which are large enough to produce well-defined thermally-forced mesoscale circulations. An implication of this result is that thunderstorm evolution over the Great Plains (and for other areas where such gradients exist) during the summer is strongly modulated by this landscape patchiness (Chang and Wetzel, 1991).

In Pielke et al. (1990b), it was demonstrated that vegetation *type* is critical with respect to the evolution of its overlying boundary layer structure. It was also shown that if a spatial gradient in vegetation type exists, a significant mesoscale circulation can result. Using the tall grass/deciduous forest boundary in the eastern High Plains of the U.S. it was illustrated in that paper that a substantial mesoscale circulation is produced as a result of the cooler, more moist boundary layer over the forest. A surprising result is that when the natural grassland is replaced by wheat, the intensity of the mesoscale circulation is increased due to its dif-

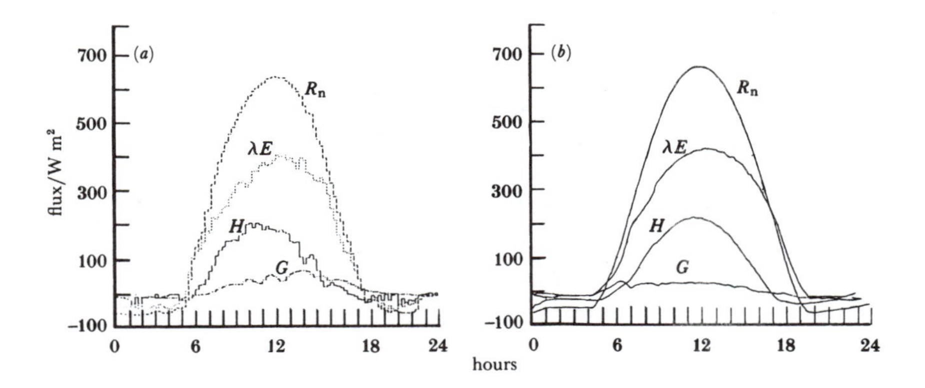

Figure 6.7: *Diurnal variation of surface-energy budget and net radiation (R_n), sensible (H), latent (λE), and ground (G) heat fluxes over corn growing on sandy soil: (a) observed; and (b) modeled. This case corresponds to 16 June 1986 over southwest France.* [From André, J.-C., P. Bougeault, J.-F. Mahfouf, P. Mascart, J. Noilhan, and J.-P. Pinty, 1989: Impact of forests on mesoscale meteorology. *Phil. Trans. Roy. Soc. London*, **B324**, 407-422.]

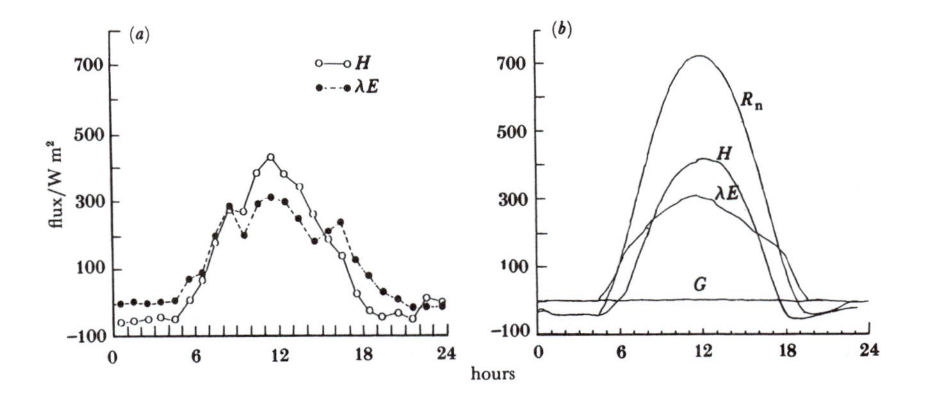

Figure 6.8: *Same as Figure 6.7, but for the 'Landes' pine forest.* [From André, J.-C., P. Bougeault, J.-F. Mahfouf, P. Mascart, J. Noilhan, and J.-P. Pinty, 1989: Impact of forests on mesoscale meteorology. *Phil. Trans. Roy. Soc. London*, **B324**, 407-422.]

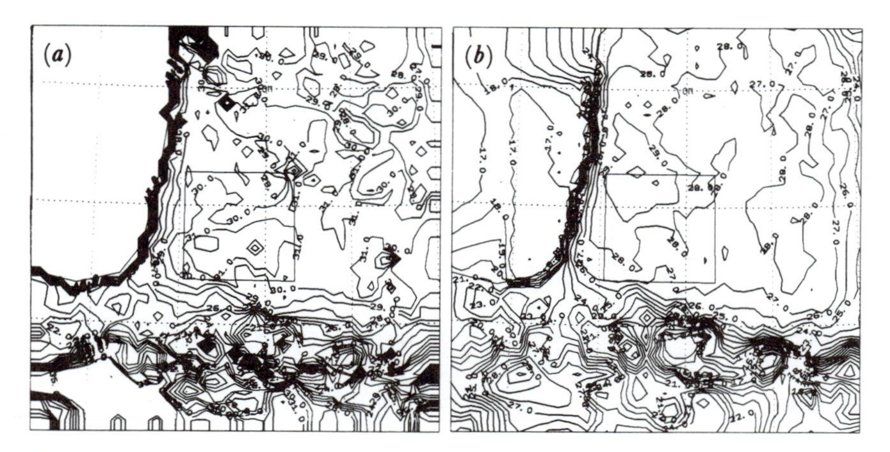

Figure 6.9: *Horizontal variations of (a) surface, and (b) air temperature over southwestern France, as modeled by a mesoscale model for 16 June 1986. Isolines are labeled in Kelvins.* [From André, J.-C., P. Bougeault, J.-F. Mahfouf, P. Mascart, J. Noilhan, and J.-P. Pinty, 1989: Impact of forests on mesoscale meteorology. *Phil. Trans. Roy. Soc. London* **B324**, 407-422.]

ferent canopy conductance and albedo. In Pielke et al. (1990c), satellite imagery of the highest surface irradiance (observed from GOES) during a five-week period during the summer over the eastern Great Plains was used to demonstrate that large surface temperature differences between the deciduous forest and adjacent land (primarily agricultural) actually do exist and are consistent with the surface temperature gradients produced in the model simulations.

Lanicci et al. (1987) have shown that dry soil conditions for northern Mexico and the soil moisture conditions of the southern Great Plains were critical to convective rainfall patterns observed over Texas and Oklahoma in the SESAME IV case study of 1979. This surface forcing was associated with differential surface heating and the generation of large values of convective potential energy as a result of surface evaporation.

Gradients in landscape will also influence atmospheric dispersion, as reported for example, in Pielke et al. (1991a). In the absence of landscape variability, under synoptically quiescent conditions, boundary layer turbulence is the only mechanism available to disperse pollution. With mesoscale circulations that result from landscape variations however, the resultant vertical motion and differential horizontal velocities can enhance the boundary layer dispersion.

The spatial distribution of transpiring vegetation over large areas can best be assessed from satellite. Figure 6.10a and b provide examples of NDVI images over the Great Plains of the United States for two time periods. Among the important implications from this figure are the sub-

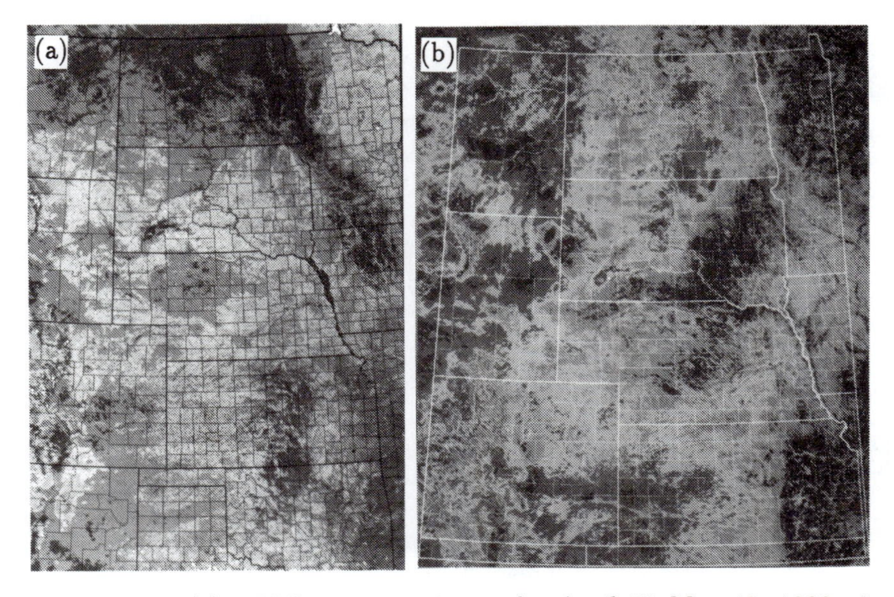

Figure 6.10: *(a) NDVI composite image for April 27–May 10, 1990. 1 km pixel data. From a map prepared by the EROS Data Center, Sioux Falls, SD.* [From Pielke, R.A., G. Dalu, J.S. Snook, T.J. Lee, and T.G.F. Kittel, 1991b: Nonlinear influence of mesoscale landuse on weather and climate. *J. Climate*, **4**, 1053-1069.] *(b) NDVI composite image for the week ending June 6, 1988 covering the northern and central U.S. Great Plains. Photograph courtesy of the EROS Data Center.* [From Pielke, R.A., G. Dalu, J.R. Garratt, T.G.F. Kittel, R.A. Stocker, T.J. Lee, and J.S. Snook, 1990: Influence of mesoscale landuse on weather and climate and its representation for use in large-scale models. *Proceedings, Indo-U.S. Seminar on "The Parameterization of Subgrid-Scale Processes in Dynamical Models of Medium-Range Prediction and Global Climate"*, Pune, India. August 6-10, 1990.]

stantial variations in space of NDVI and the large temporal changes that occurred over a relatively short time period. As regions dry out and the vegetation becomes stressed, the NDVI values drop. Correspondingly, after a rainfall over the natural grasslands, the vegetation is invigorated and transpiration is enhanced.

Pielke et al. (1991b), present model results of the relative contribution of boundary layer turbulence and mesoscale motions to the vertical flux of heat at about 500 m, as a function of the size of a landscape patch (see Figures 12.1 and 12.2. While a prevailing larger-scale wind reduces the magnitude of the mesoscale heat flux, there is a significant contribution to the atmospheric heat budget due to these variations in landscape. The mesoscale heat flux, for example, is deeper than occurs with the turbulent heat flux for a homogeneous surface case. Landscape features much smaller than 10 km, however, can be represented by a weighting of surface fluxes from a one-dimensional model (i.e., for each landscape type for use in a mesoscale resolution grid, as discussed by Avissar and Pielke, 1989, Avissar and Verstraete (1990), and suggested by the analytic results of Dalu et al., 1991; Pielke et al. 1990d). A similar conclusion has also been obtained by Shuttleworth (1988) and André et al. (1990) who proposed a cut-off of 10 km below which the land patches have no apparent organized planetary boundary layer response and spatial scales larger than 10 km which can produce organized planetary boundary layer variations. The Pielke et al. (1991b) result is consistent with the Shuttleworth (1988) and André et al. (1990) conclusion except for zero large-scale flow the cutoff horizontal scale is less than 10 km.

Since natural and man-caused landscape variations exist, and because substantial heat and moisture are transported vertically by mesoscale circulations due to the surface patchiness, a model which is investigating climate change, as well as regional and synoptic numerical prediction tools must consider their impact on weather. When considered globally, the alteration of landscape through human activities such as farming, urbanization, deforestation, etc. could have already had an effect as large as greenhouse gas warming (Pielke and Avissar, 1990), as discussed further in Chapter 12.

6.2 Influence of Irrigation

Definitive evidence of long-term regional climate changes due to irrigation is generally lacking, primarily because of the absence of observational studies. Those that exist include that of Schickendanz (1976) who found rainfall anomalies in the vicinity of irrigated areas in parts of Kansas, Nebraska, and Texas during the months of June, July, and August from 1931–1970. He estimated that there was an increase of summer rainfall associated with the irrigated areas in these three states of 14.5 million acre-feet. Moreover, a hail maximum was found to be associated with the core of the heavily irrigated region (the Lubbock-Plainview area). The work of Beebe (1974) further substantiated this effect as shown in Figure 6.11 where substantially more tornadoes were observed over the irrigated areas of the Texas panhandle than over the adjacent prairie.

Modeling and observational studies presented in Section 6.1.1 and 6.1.2, provide evidence that local areas of irrigation significantly influence boundary layer structure and the development of physiographically-forced circulations. Other case study observational examples of the influence of irrigation include Doran et al. (1992a,b) where differences in sensible heat flux of over 400 W m^{-2} were observed in the middle of the

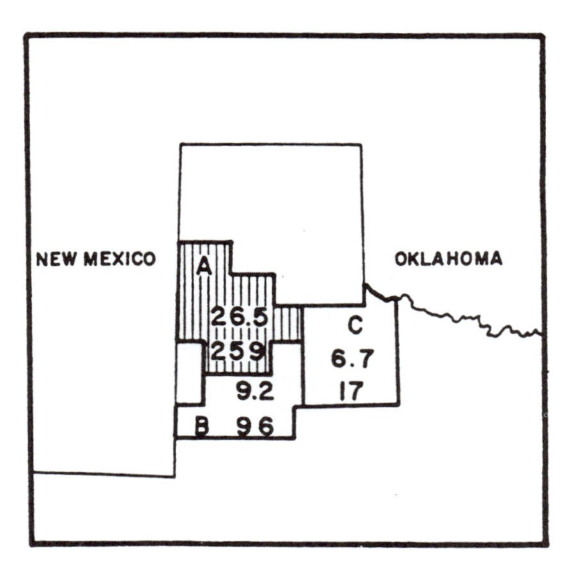

Figure 6.11: *The average number of tornadoes per 1000 mi² (top number) and average number of acre-feet of water irrigated per square mile for 3 regions in the Texas panhandle. Region A included all those counties averaging greater than 200,000 acre-feet/yr. and B and C represent "control" areas adjacent to A but considerably less irrigated.* [From Beebe, R.C., 1974: Large scale irrigation and severe storm enhancement. *Proc., Symposium on Atmospheric Diffusion and Air Pollution of the American Meteorological Society.* Co-sponsored by the World Meteorological Organization. AMS, Boston, September 9-13, Santa Barbara, California, 392-395.]

day in June 1991 in northeast Oregon between steppe and an irrigated wheat field.

Colorado can be used as an example of a region in which irrigation is a dominant agricultural practice. In Colorado, the statewide annual precipitation is 17.1 inches which corresponds to 95 million acre-feet (31 trillion gallons) of water over the 104,247 square miles of the state. (About 85% of the state's streamflow originates as snowmelt.) Approximately 14% of this water exits the state through streamflow. The remainder of the water not in the streamflow eventually is evaporated or transpired to the atmosphere, although some of the water resides for a period of time as ground water. Agriculture consumes 88% of the 1.825 trillion gallons of water used by man in the state each year. Using these data, it can be estimated that of the 26.66 trillion gallons of water emitted to the atmosphere each year, approximately 1.61 trillion gallons of this are due to irrigation. Since much of this water is obtained through tunnel and channel diversions from rivers and would otherwise be unavailable for transfer to the atmosphere within the state, the man-induced moistening of the atmosphere over the state represents about a 6% state-wide average increase over the natural situation. Much of this water, of course, is transpired and evaporated to the atmosphere within local areas of the state so that the local percentage impact is much larger.

Vegetation also directly influences the flux of trace gases to the atmosphere. As discussed in Chapter 12, deforestation of tropical rain forests would contribute to an increase of the carbon dioxide in the atmosphere. To compensate for such anthropogenically-caused increases in the atmospheric concentrations of these gases, Glenn et al. (1992) has proposed planting of seaweed and salt-tolerant terrestrial plants (halophytes) in arid and semi-arid regions as a means to uptake carbon from the atmosphere. The role of halophytes in irrigated agriculture is discussed in O'Leary (1984). The seaweed carbon would be stored as ocean sediment after it dies, while the halophyte debris could be accumulated in the soil. Glenn and associates estimate that approximately 500 tons of carbon per km^2 could be stored. They calculate that about 1.3×10^6 km^{-2} of continental shelf and 1.3×10^6 km^{-2} of salt desert are amenable to this type of agriculture. Several of these vegetation products are also valuable as animal and/or human foods. The terrestrial halophyte, Salicornia bigelovii Torr., for example, has been shown to have oil, protein, and biomass levels which equal or exceed equivalent levels in freshwater oilseed crops such as soybean and sunflower (Glenn et al., 1992). This salt water irrigated vegetation, if the areal coverage is sufficiently large, would also result in major alterations in local weather, similar to that found in fresh water irrigated areas as discussed above. It is already planned by E. Glenn and associates to plant areas of Saudi Arabia and Mexico with halophytes in order to test their hypothesis. The influence of salt-water irrigated plants in enhancing winter rains through enhanced transpiration needs to be investigated, as well as regional effects such as on the Arabian heat low if the coverage by these plantings becomes large enough.

Anthes (1984) and Yan and Anthes (1987) have proposed that spacing of such irrigated areas could result in enhanced convective rainfall, providing an optimal horizontal separation was chosen. The modeling work of Yoshikado (1981) and Xian and Pielke (1991) supports the Anthes hypothesis. For light large-scale flow, in lower latitudes, the optimal spacing is on the order of 50–100 km.

Black and Tarmy (1963) proposed coating large areas in coastal deserts with asphalt in order to enhance the sea breeze circulation and, thus, produce sea-breeze convergence induced rainfall. Unlike the planting of halophytes, however, the hotter asphalt (as contrasted with desert soil) might simply result in a deeper, but still dry boundary layer as drier air from aloft is entrained into the boundary layer. In addition, there is the environmental degradation due to covering natural ground with asphalt. The Gulf War fires in Kuwait in 1991 might, unfortunately, provide an opportunity to test the Black and Tarmy hypothesis since the deposition of the soot blackened areas of the adjacent desert.

6.3 Desertification

Desertification has an opposite effect on climate than irrigation. Desertification was originally defined, by Aubreville (1949) as cited in Odingo (1990), as land degradation in semi-arid and subhumid regions such that arid conditions develop and persist. Vegetation loss due to human activities can, therefore, create deserts. The Rajasthan desert in northwest India is an example of desertification. In that area, in the millennium before the Christian era, a well developed civilization existed. Currently the region is a tropical desert with only archaeological ruins remaining. It has been suggested that overpopulation in the area denuded the vegetation, as the populace used firewood and cleared land for agriculture. While the change in weather could have been part of the natural evolution of climate, major landscape changes by man on the scale of this desert region have occurred elsewhere.

In the Middle East, Neumann and Parpola (1987), for example, have documented that the political, military, and economic decline of Assyria and Babylonia in the twelfth through tenth centuries B.C. coincide with a notable period of warming and drying in the region which started around 1200 B.C. and lasted until about 900 B.C. While natural large-scale fluctuations in climate could have caused this aridity, as implied by their paper, desertification provides another explanation. The return of weather to wetter conditions could have occurred in response to a rejuvenation of vegetation as human stress on the landscape was reduced, and the larger-scale weather patterns which influence the region were not more permanently displaced as apparently has occurred in northwest India.

Charney (1975) proposed a mechanism of desertification over northern Africa in which the removal of vegetation increased the albedo of the land. Since more solar radiation was reflected back into space, the result was increased subsidence in the lower atmosphere in order to compensate for the loss of heat energy. In a later study, Charney et al. (1977) used the Goddard Institute for Space Studies (GISS) model to conclude that local evaporation rates are as important as albedo in influencing rainfall patterns in semi-arid regions, with their relative effects dependent on location. This study found a more complicated interaction between the land surface and the atmosphere than was originally hypothesized in the Charney (1975) paper.

The process of desertification continues today in the Sahel regions of Africa as illustrated in Figure 6.12 in which a portion of Chad is shown. The darker region in the Figure corresponds to an area in which the government controlled grazing while the adjacent areas were extensively

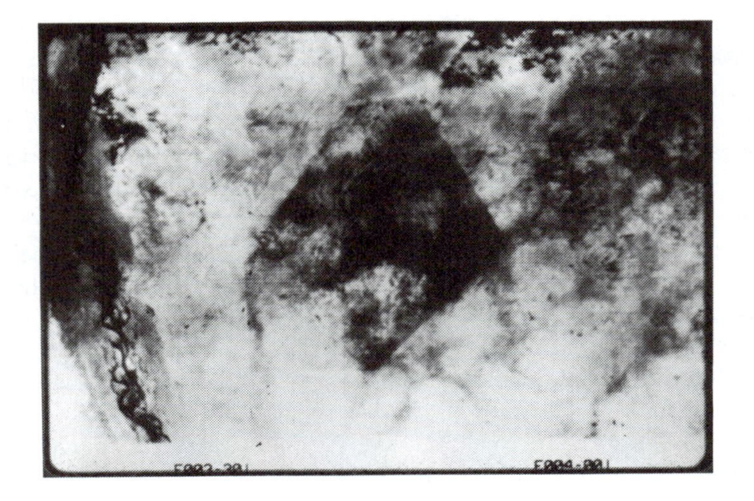

Figure 6.12: *Landsat imagery of Chad.*

utilized by cattle and goats. The significantly higher albedo observed in the overgrazed area is evident in the Figure.

In western Australia, as reported by Lyons (1991; personal communication), the use of fencing to prevent the movement of rabbits into an agricultural area is also evident from satellite imagery. In this region, 130,000 km^2 of native perennial vegetation has been transformed into winter growing annual species. Lyons reports a 30% decline of winter rainfall in this region which he attributes to the change of vegetation types. Presumably, the winter crop is harvested such that subsequent transpiration of water into the atmosphere is lost, whereas the perennial plants would be reaching their peak growth (and transpiration) in the latter part of the winter and in the spring.

Otterman (1974) and Otterman and Tucker (1985) have also used an area of contrasting landuse in the Negev Desert – Sinai Peninsula region to demonstrate how desertification works. Figure 6.13a presents satellite imagery of this region, in which the political boundary between Egypt and Israel is clearly shown. In Israel grazing was controlled such that dark plant debris and limited clumps of living vegetation reduced the albedo of the surface. In the Egyptian area, in contrast, overgrazing by goats

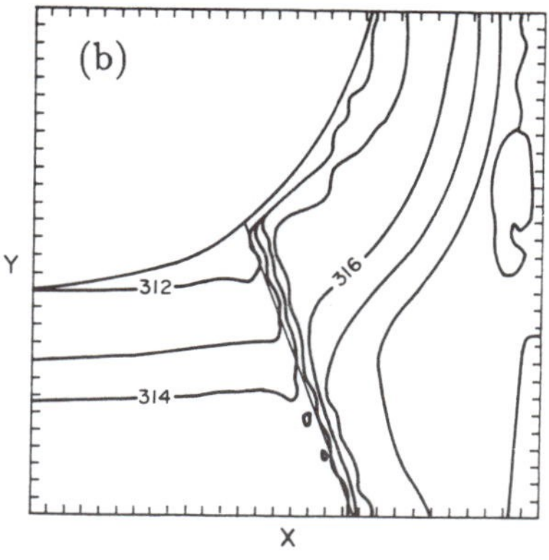

Figure 6.13: *(a) The ERTS-1 image E-1091-07482, taken on 22 October 1972, band MSS-6, showing the denuded high albedo regions of the Sinai and Gaza Strip, in contrast to the darker western Negev.* [Photo courtesy of the EROS Data Center, Sioux Falls, SD.] *(b) Predicted surface temperature at 1300 LST for typical fall meteorological conditions over the same area.* [From Mahrer, Y., and R.A. Pielke, 1978: The meteorological effect of the changes in surface albedo and moisture. *Israel Meteorological Soc. (IMS)*, 55-70.]

permitted few plants to survive so that the soil became more mobile and covered the vegetation debris. The result was a higher albedo from the light bare soil. Figure 6.13b documents using a numerical model simulation that surface ground temperature differences over 5° can result with the area desertified being cooler. Otterman suggests that these cooler temperatures result in less annual precipitation than otherwise would occur due to the reduced surface supplied buoyant energy for showers. Thus it would be less likely for the vegetation to regenerate even if the grazing pressure were removed. Mahrer and Pielke (1978) performed a model simulation of this region to show that the gradient of surface temperatures due to the albedo differences will cause preferential ascent (and thus more likely showers when the atmosphere is favorable for cumulus convection over the Negev side of the boundary). Otterman et al. (1990) suggest that natural vegetation in a semi-arid or arid region could have a more positive impact on rainfall than the tarring of high albedo sandy desert areas as proposed by Black and Tarmey (1963).

6.4 Creation of Man-Made Lakes

Flooding depressions in desert areas has been proposed as a limited solution to locally counteract desertification. In Egypt, for example, the Qattara Depression reaches 133 meters below sea level. By piping water from the Mediterranean Sea, these depressions could be filled. Segal et al. (1983) showed that the new salt water lake would cool the environment in its vicinity and perhaps slightly increase rainfall along its shore when the infrequent large-scale rain system moves across the area. Planting of halophytes (see Section 6.2) along the lake shore could enhance this effect. A side benefit of this project is that electric power generation could be achieved through the movement of the water downhill into the depression.

Similar creation of lakes in desert depressions has been proposed by Enger and Tjernström (1991) in which the Chott region of Algeria and Tunisia could be flooded with Mediterranean or Atlas-Aures mountain water to create a 6000 km^2 lake with a depth of around 10-15 m. Using the model of Tjernström (1987a,b) along with synoptic climatological studies, they estimated that local precipitation could be increased on the land near the new lake by up to 150 mm yr^{-1}.

6.5 Deforestation

Deforestation is another type of landscape change. Deforestation occurs when lumbering and fires remove extensive areas of trees from a region. Such deforestation occurred, for example, in the eastern United States in the 1800s and resulted in an almost total removal of the original uneven aged, climax forest. Only in the Great Smoky Mountain Forest, where about 40% of the virgin trees remain, can one appreciate the large diameter deciduous and evergreen trees and sparse understory vegetation that originally covered the area prior to European settlement. During the last 50 to 75 years, much of the forest has returned as the nutrient rich organic soils in the midlatitude region permitted regrowth from root sprouts, natural seedings, and forestry planting of seedlings. Thus while the deforestation was extensive, a new forest, albeit of a different species composition and age, has evolved in non-agricultural and non-urban areas in the eastern United States.

More recent deforestation has occurred in the Amazon region and the influence of this removal of trees on global climate has been the subject of considerable concern (Dickinson, 1987; Lean and Warrilow, 1989; see Chapter 12 for a discussion). There are local effects of this deforestation as well. Currently, it is estimated that about 50% of rainfall within the Amazon Basin is from local evaporation and transpiration (Salati et al., 1978). Figure 6.14 shows that the removal of trees is not completed as one vast clear-cutting but is accomplished through a series of strips of several kilometers in width and in large clear cut areas. A major question yet to be addressed is whether local wind circulations develop in response to these landscape alterations, such as discussed earlier in this chapter. This could result, for example, in an enhancement of local rainfall and cloudiness.

A major difference between the midlatitude forests and forests in the humid tropics such as the Amazon is the absence of substantial organic material and trace metals in the soils in the tropical environment. The heavy rains in the region leach nutrients to below the root layers with the result that much of the nutrients required for vegetation development and growth are within the living and recently dead plants. A rapid recycling of nutrients occurs as new vegetation generates from the rapidly decomposing plants. In the absence of this rapid recycling, rains would deplete the root zones of the needed plant food. Thus, the removal of trees by lumbering short-circuits the recycling. Even an increase of rainfall due to the patchiness of the timber cutting is unlikely to influence significantly any regeneration of the forest.

Several field studies have been undertaken in the Amazon in order to better understand the influence of this natural vast forest area on climate and weather. In 1985 and 1987, the Amazon Boundary Layer Experiment (ABLE) was completed (Garstang et al., 1990). Among the results it was shown that local wind circulations apparently develop in response to large rivers in the area being adjacent to the forests (Miller et al., 1988). A similar atmospheric response would be expected when large areas are cleared of trees. The Anglo-BRazilian Amazonian Climate Observation Study (ABRACOS) was initiated at the end of 1990 to investigate the influence of clear cutting in the region. Initial results from the field study

Figure 6.14: *Examples of clear-cutting of the tropical forest in two areas of the Amazon.* [Photos provided by Carlos Nobre of the Center for Weather Prediction and Climate Studies – CPTEC, INPE, Brazil.]

(see Figure 6.15) demonstrate that the forest captures more radiative energy than an area which is pasture. During the daytime, the albedo, on average, is about 2.5% larger over the cleared area with the resultant wind speed, temperature, and moisture profiles shown in Figures 6.16, 6.17, and 6.18.

These measurements show that the temperature change is twice as great over the cleared area with even a greater variation in wind speed. Early morning fog or mist is observed in the clearings, while such moisture saturation is almost unknown in the forest.

The influence of deforestation in another tropical area, southern Nigeria, is discussed in Ghuman and Lal (1986). Figure 6.19 is consistent with the results from the Amazon in that the diurnal range of temperature is substantially greater over the cleared ground. This figure also shows that the influence of forest results in somewhat cooler temperatures deeper in the soil (i.e., 50 cm).

6.6 Summary and Conclusions

There is strong observational and modeling evidence that indicate changes in landscape due to irrigation, deforestation, desertification, etc. have had a significant influence on local and regional weather and climate. However, considerable field measurement and theoretical work yet remains to quantify their impact. We do not know in sufficient detail how these influences on weather and climate vary as a function of the large-scale atmospheric flow, the time of year, or latitude. Among the major needs are field experiments at representative sites around the world in order to characterize in detail the weather and climate over natural areas, as contrasted with locations in the same region which have been substantially modified by man.

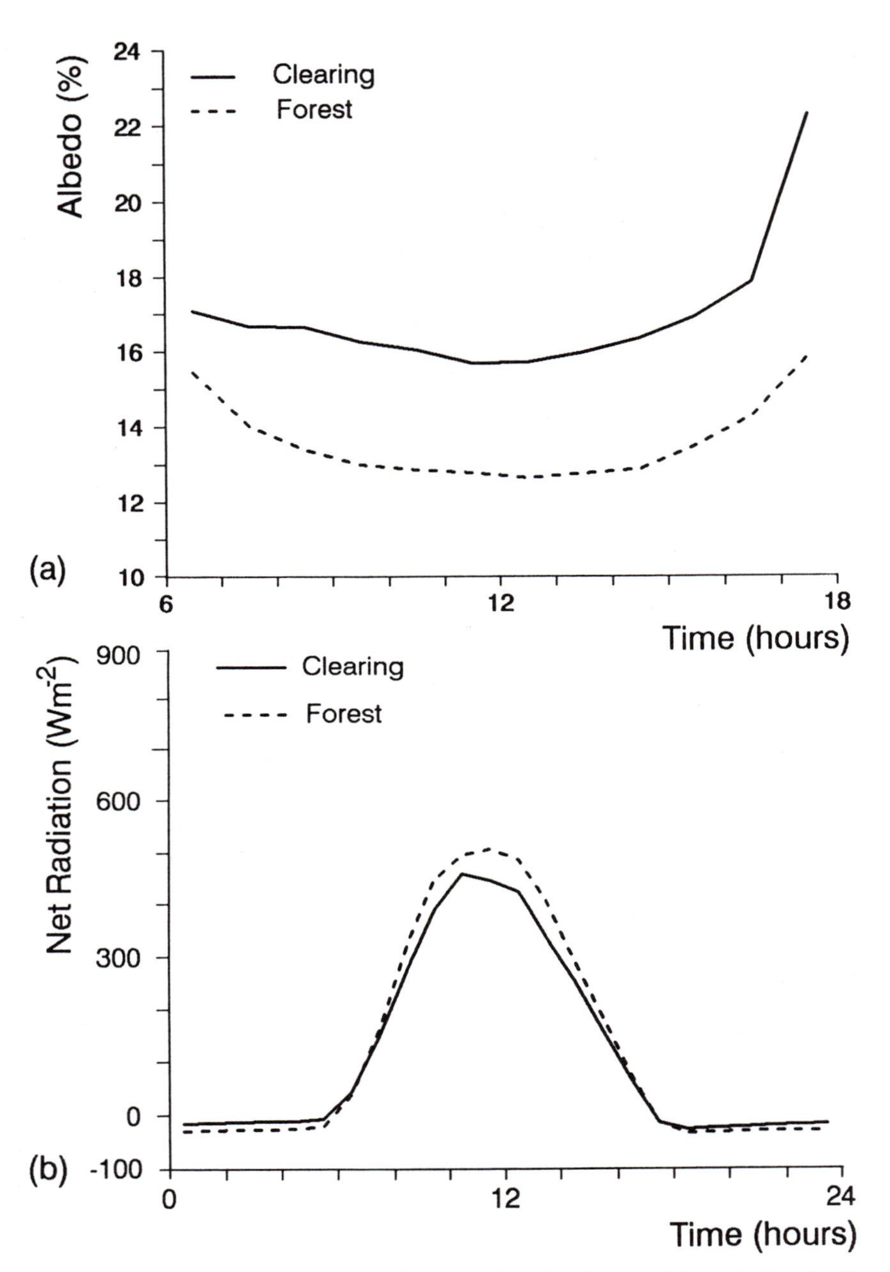

Figure 6.15: *(a) Mean hourly albedo at the clearing and forest sites in the Amazon for the study period from 12 October to 10 December 1990; and (b) mean hourly net all-wave radiation at the clearing and forest sites for the same study period.* [From Bastable, H.G., W.J. Shuttleworth, R.L.G. Dallarosa, G. Fisch, and C.A. Nobre, 1992: Observations of climate, albedo and surface radiation over cleared and undisturbed Amazonian forest. *Int. J. Climatol.*, (submitted).]

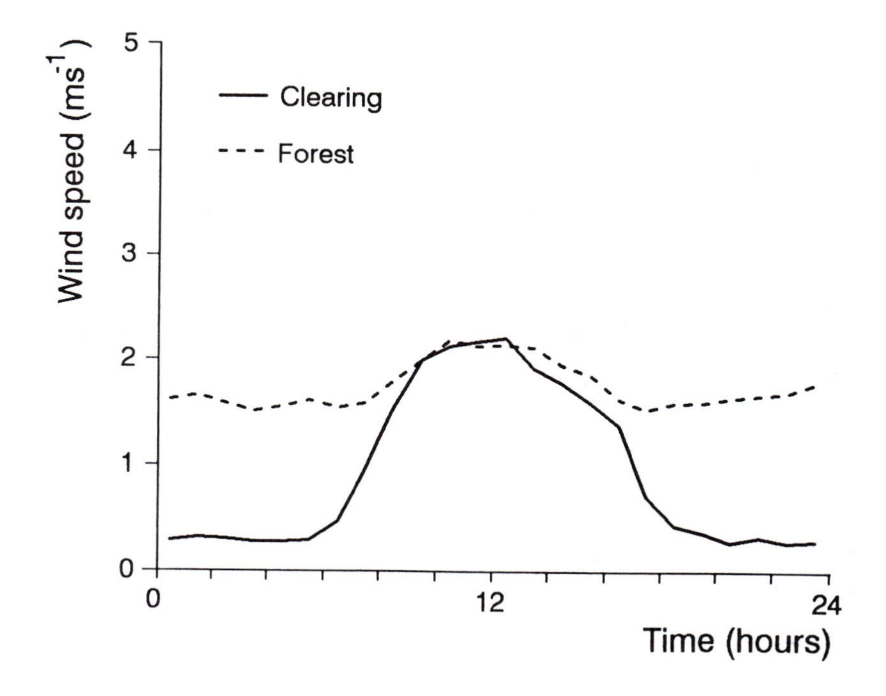

Figure 6.16: *Mean hourly wind speed at clearing and forest sites in the Amazon for the period 12 October to 10 December 1990.* [From Bastable, H.G., W.J. Shuttleworth, R.L.G. Dallarosa, G. Fisch, and C.A. Nobre, 1992: Observations of climate, albedo and surface radiation over cleared and undisturbed Amazonian forest. *Int. J. Climatol.*, (submitted).]

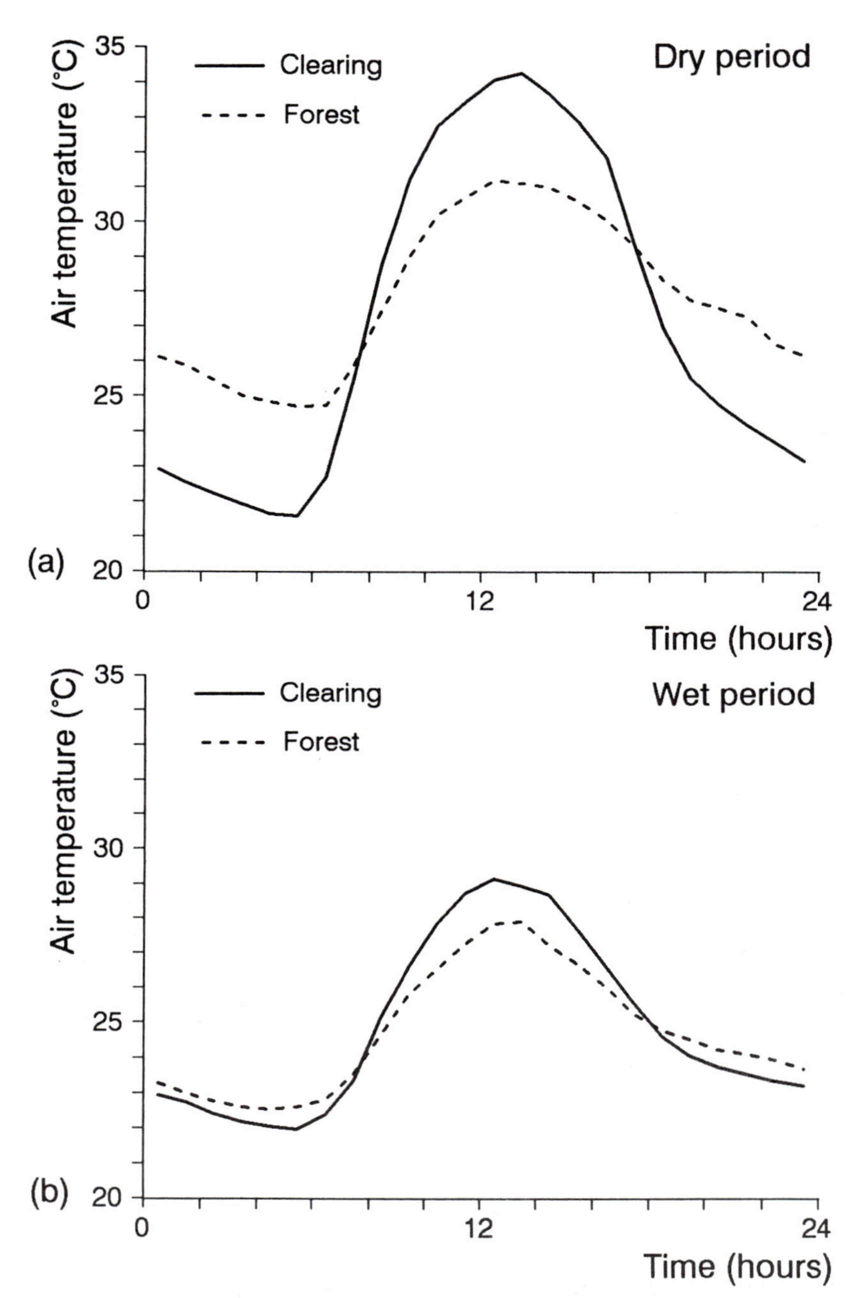

Figure 6.17: *Mean hourly air temperature at screen height at the clearing and forest sites in the Amazon for (a) the dry period from 12 to 21 October, and (b) the wet period from 1 to 10 December 1990.* [From Bastable, H.G., W.J. Shuttleworth, R.L.G. Dallarosa, G. Fisch, and C.A. Nobre, 1992: Observations of climate, albedo and surface radiation over cleared and undisturbed Amazonian forest. *Int. J. Climatol.*, (submitted).]

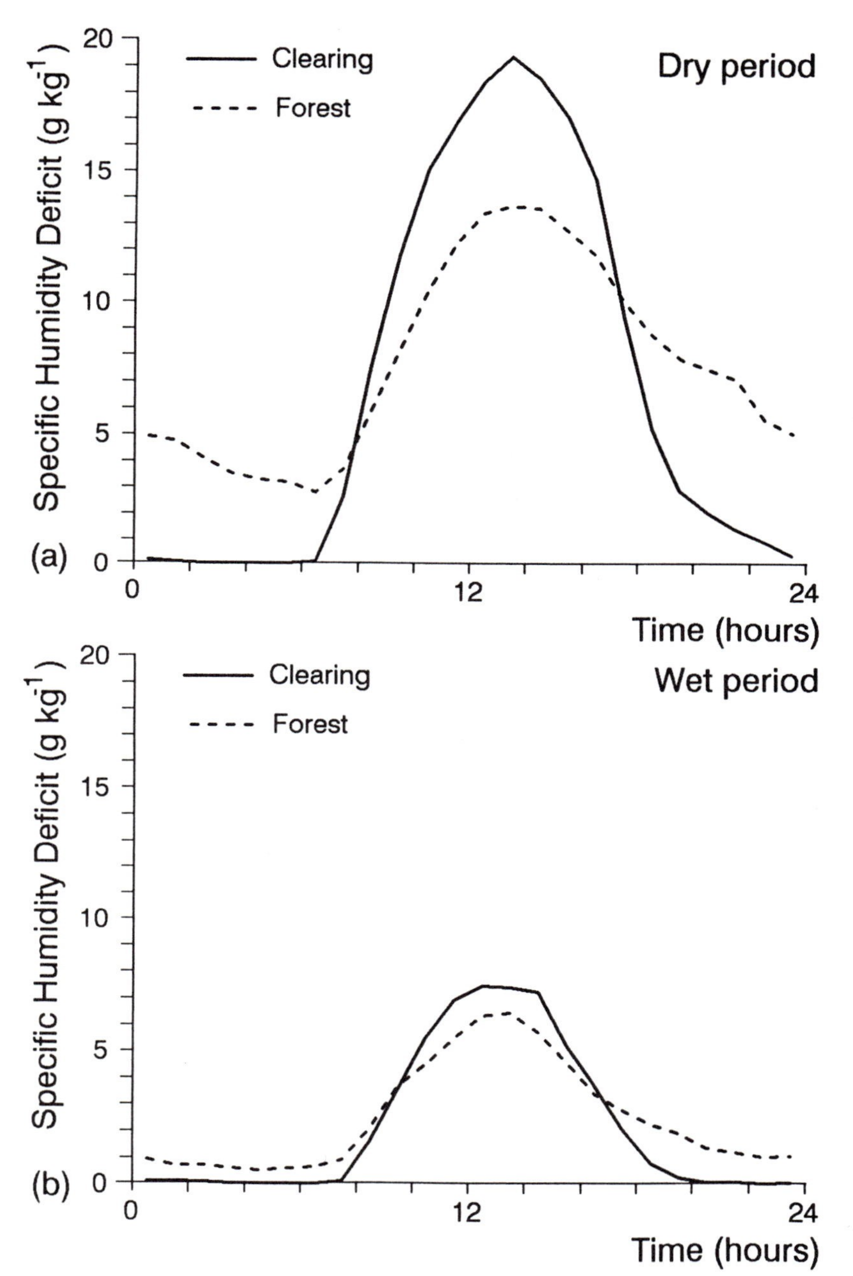

Figure 6.18: *Mean hourly specific humidity deficit at screen height at the clearing and forest sites in the Amazon for (a) the dry period from 12 to 21 October, and (b) the wet period from 1 to 10 December 1990.* [From Bastable, H.G., W.J. Shuttleworth, R.L.G. Dallarosa, G. Fisch, and C.A. Nobre, 1992: Observations of climate, albedo and surface radiation over cleared and undisturbed Amazonian forest. *Int. J. Climatol.*, (submitted).]

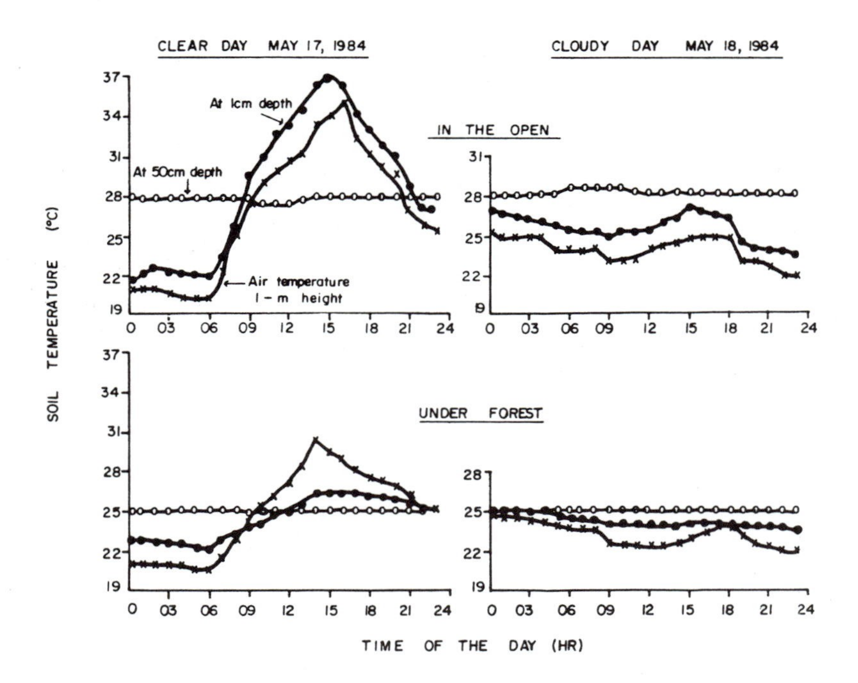

Figure 6.19: *Diurnal fluctuations in soil and air temperature under forest and on cleared land.* [From Ghuman, B.S. and R. Lal, 1986: Effects of deforestation on soil properties and microclimate of a high rain forest in southern Nigeria. In *The Geophysiology of Amazonia*, (R.E. Dickinson, Ed.), John Wiley and Sons, New York, 225-244.]

Chapter 7

Concluding Remarks Regarding Human Impacts on Regional Weather and Climate

We have seen that there is considerable evidence suggesting that anthropogenic activity, either in the form of constructing major urban areas, changing natural landscape to agricultural and grazing areas, or emission of particles and gases, has contributed to changes in weather and climate on the regional scale. In most cases the evidence has not adequately documented the physical linkage between human-caused influences and the meteorological and climatological response. The statistical analyses could have created an inferred rain anomaly purely by chance, or other physiographic factors associated with siting of cities and specific industries and placing of irrigated areas could be responsible for the inferred anomalies. It is curious that the scientific community has accepted the results obtained in studies such as METROMEX as being valid, yet question the validity of cloud seeding induced changes in rainfall inferred from well-designed, randomized cloud seeding experiments.

The answer to this paradox lies in human psychology. As an example, Dr. Stan Changnon described a conversation he had with Dr. John Tukey, one of the world's leading statisticians, following a meeting of the Weather Modification Advisory Board about ten years ago. Stan asked John why the statisticians had been very critical of attempts to prove planned weather modification of clouds and rainfall was successful, yet were not so critical of inadvertent weather modification (i.e., the cities are not randomized). John looked up at Stan and said, "Well, Stan, in the end it is just a lot more believable that a big city can cause clouds, rain, and hail than it is that a small amount of seeding material can." In other words, no matter how objective we attempt to be, a certain amount of subjectivity is involved in accepting the results of any scientific study.

In summary, while there is considerable evidence supporting the hypothesis that human activity is inadvertently modifying weather and climate on the regional scale, much more research is required to pinpoint the causes of inferred human related weather anomalies and to strengthen the statistical inferences and physical understanding.

Unfortunately, United States federal funding of inadvertent weather modification fell sharply in the 1980s along with that of deliberate weather modification. For some incomprehensible reason, funding of inadvertent weather modification has apparently been tied to funding of advertent weather modification. To some extent funding for inadvertent weather modification fell through the cracks of the funding agencies. It does not fit into the mission-oriented agencies that have supported planned weather modification, so little support came from those agencies. It does fit within the mission of the Environmental Protection Agency (EPA), but the research program there, with respect to inadvertent local and regional weather and climate changes, was seriously weakened under the Reagan administration. Because inadvertent modification and planned weather modification were placed within the same program in the National Science Foundation, cuts in the planned weather modification program also cut the already meager program in inadvertent weather modification. Again, the lack of a lead agency or a funded, coordinated program in weather modification appears to be limiting progress in furthering our quantitative understanding of human impacts on weather and climate on the local and regional scale.

Part III

Human Impacts on Global Climate

In Part III we examine the scientific basis of the hypothesis that human activity is resulting in changes in global climate. Included in our discussion are the impacts of anthropogenic emissions of aerosol and gases, and changes in the earth's biosphere on the global atmospheric/ocean system. We begin by presenting a discussion of the basic physical concepts that are fundamental to understanding human impact on global climate. Some of the principles and concepts that we have already described in Parts I and II are also relevant to global change, particularly since the global atmospheric responses to human activity are the sum total of local and regional human impacts.

Chapter 8

Fundamental Principles Important to Understanding Global Climate Change

8.1 Atmospheric Radiation

The principal source of energy driving the atmospheric/ocean system comes from the sun. In order to understand the potential for human impact on global climate, we must have at least a general understanding of how the sun's energy is distributed through the atmosphere/earth/ocean system. The sun emits most of its radiant energy over wavelengths ranging from 0.2 μm to 1.8 μm, with a peak in intensity at 0.470 μm. As shown in Figure 8.1, the spectrum of energy emitted by the sun closely approximates the energy spectrum emitted by a perfect absorber/emitter substance (or what we call a blackbody) having a temperature of 5900 K. On the other hand the spectrum of energy emitted by the earth emits radiant energy corresponding to a blackbody at the approximate temperature of 250 K, and thereby exhibits a peak in spectral density at wavelengths between 10 and 15 μm. Thus the combination of radiation energy emitted by the sun as received by the earth at its orbital distance and the earth span from less than 0.2 μm to greater than 50 μm. The spectrum of energy emitted by the sun and the earth, however, exhibit very little overlap. We therefore refer to the energy emitted by the sun as *shortwave radiation* and the energy emitted by the earth and its atmosphere as *longwave radiation.*

As atmospheric radiation passes through the atmosphere it interacts with the gases and particulates. Some of the radiant energy is absorbed, some scattered, and some unaffected. The amount of energy absorbed and scattered varies with the wavelength of the radiant energy, the type of gases, and the size and chemical composition of the particles. The absorbed energy is converted to heat in the atmosphere and that en-

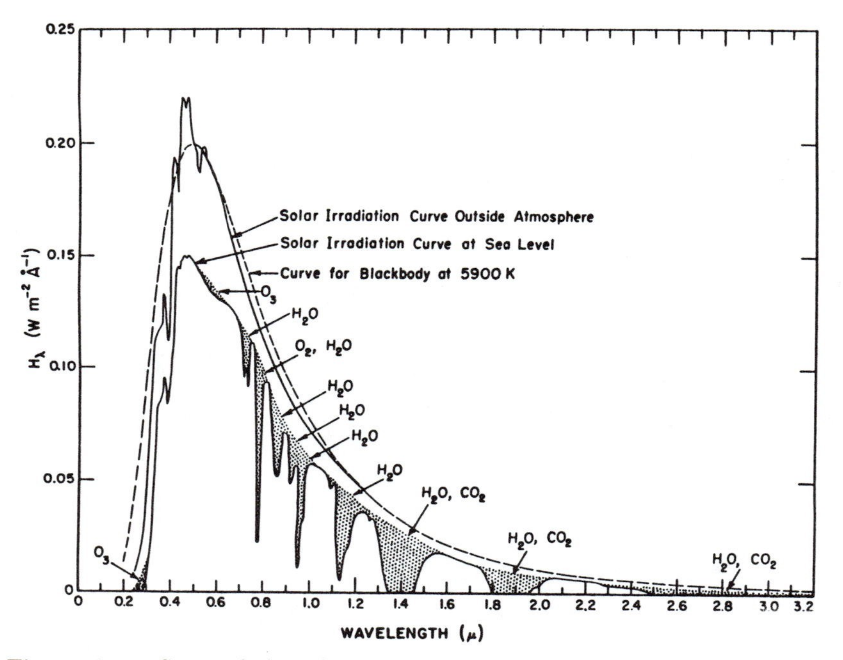

Figure 8.1: *Spectral distribution curves related to the sun; shaded areas indicate absorption at sea level, due to the atmospheric constituents shown.* [From Gast, P.R., A.S. Jursa, J. Castelli, S. Basu, and J. Aarons, 1965: Solar electromagnetic radiation. In *Handbook of Geophysics and Space Environments*, S.L. Valley, Ed., pp. 16-1–16-38. McGraw-Hill, New York. ©McGraw-Hill. Reprinted by permission.]

ergy is reradiated at a wavelength and intensity that corresponds to the temperature and composition of the absorbing gases or particulates. The amount of energy that is scattered, however, is simply redirected in space (or reflected) with no change in intensity. The amount of energy scattered and the directions in which it is scattered depends on the wavelength of the radiation as well as the types of gases and size and composition of particles.

8.1.1 Absorption and scattering by gases

As shown in Figure 8.1, the energy spectrum incident at the top of the atmosphere is significantly attenuated by the time it reaches the earth's surface at sea level. It is not attenuated uniformly across the spectrum, however, but instead major energy losses occur over rather narrow bands, called *absorption bands*. In a cloud-free atmosphere the primary absorbers of shortwave radiation are ozone and water vapor. Airborne particles, or aerosols, in typical non-polluted air contribute less to ab-

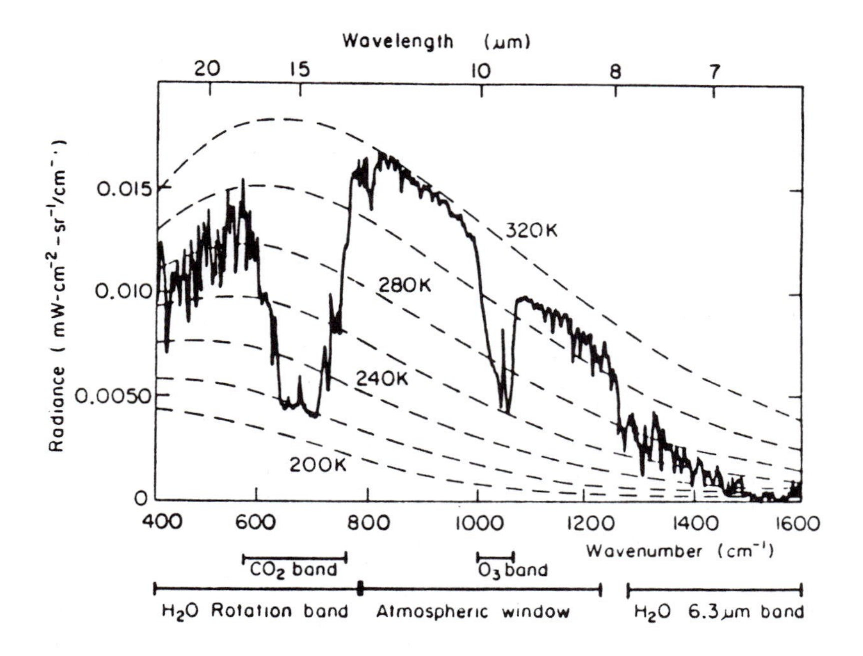

Figure 8.2: *Atmospheric spectrum obtained with a scanning interferometer on board the Nimbus 4 satellite. The interferometer viewed the earth vertically as the satellite was passing over the North African desert.* [After Hänel, R.A., B.J. Conrath, V.G. Kunde, C. Prabhakara, I. Revah, V.V. Salomonson, and G. Wolford, 1972: The Nimbus 4 infrared spectroscopy experiment, 1. Calibrated thermal emission spectra. *J. Geophys. Res.,* **11,** 2629-2641. From Paltridge, G.W., and C.M.R. Platt, 1976: *Radiative processes in meteorology and climatology.* Developments in Atmospheric Science, 5. Elsevier Science Publishers, New York. ©American Geophysical Union.]

sorption. At wavelengths less than 0.3 μm, oxygen and nitrogen absorb nearly all the incoming solar radiation in the upper atmosphere. Between 0.3 and 0.8 μm, however, little gaseous absorption occurs. Only weak absorption by ozone takes place over this spectral range. At wavelengths, less than 0.8 μm Rayleigh scattering by oxygen and nitrogen molecules also depletes the radiant energy. At longer wavelengths absorption in various water vapor bands is quite pronounced while weak absorption by carbon dioxide and ozone also occurs.

Absorption of longwave radiation occurs in a series of bands. The principal natural absorbers of longwave radiation are water vapor, carbon dioxide, and ozone. Figure 8.2 illustrates the infrared spectrum obtained by a scanning interferometer looking downward from a satellite over a desert. Strong absorption by CO_2 at a wavelength band centered at 14.7 μm is shown by the emission of radiance at a temperature of 220 K which is representative of stratospheric temperatures. The stratosphere contributes mainly to the peak of this band. Water vapor absorption bands at 1.4, 1.9, 2.7, and 6.3 μm, and greater than 20 μm, cause emissions corresponding to midtropospheric temperatures. Little absorption is evident in the region called the '*atmospheric window*' between 8 and 14 μm. Here the satellite observed radiance corresponds to the surface temperature of the desert except for a slight depression in magnitude due to departures of the emittance of sand from its block body value of unity. A distinct ozone absorption band is evident at about 9.6 μm in the middle of the window. Although not very evident in the figure, weak continuous absorption also occurs across the window believed to be due to the presence of clusters or dimers of water vapor molecules.

In summary the principal natural gaseous absorbers of terrestrial radiation are carbon dioxide and water vapor. These are the so-called principal *'greenhouse'* gases. We will discuss the greenhouse effect next. Other strong absorbers of terrestrial radiation are methane, nitrous oxide, and chlorofluorocarbons (CFC's). Currently the concentrations of these gases are so small that their contributions to infrared absorption cannot be easily detected by satellite. The concentrations of CFC's, for example, are six orders of magnitude (a million times) smaller than carbon dioxide. The concern about CFC's is that they are increasing in concentration in the atmosphere at a rate of about 4% per year. Per molecule they are 20,000 times more effective at absorbing infrared radiation than carbon dioxide.

8.1.2 Absorption and scattering by aerosols

The radiative properties of aerosol particles is a complicated function of their chemistry, shape, and size-spectra. Moreover, if the aerosol particles are hygroscopic, their radiative properties change with the relative humidity of the air. At relative humidities greater than 70%, the hygroscopic particles (called haze) take on water vapor molecules and swell in size, thus changing their radiative properties not only because of size effects but also because of changes in their complex indices of refraction as the water-solution/particle mixture changes in relative amounts.

Dry aerosol particles in high concentrations such as over major polluted urban areas and over deserts can cause substantial absorption and scattering of solar radiation. In some polluted boundary layers, aerosol absorption has been estimated to result in heating rates on the order of a few tens to several degrees per hour (Braslau and Dave, 1975; Welch and Zdunkowski, 1976). In the Saharan dust layer, aerosol absorption has been calculated to produce a heating of 1–2° C/day.

The effects of aerosols on infrared radiation transfer is less, being mainly limited to the atmospheric window where gaseous absorption is weak (Ackerman et al., 1976; Welch and Zdunkowski, 1976; Carlson and

Benjamin, 1980). Some evidence of aerosol effects on longwave radiation heating/cooling have been detected over very deep Saharan dust layers and very polluted air masses (Welch and Zdunkowski, 1976; Carlson and Benjamin, 1980; Saito, 1981). There is also some evidence that the swelling of haze particles at high relative humidities enhances their effectiveness in absorbing infrared radiation.

8.1.3 Absorption and scattering by clouds

Like aerosol particles, cloud droplets, raindrops, and ice particles (hydrometeors) interact with radiation in complex ways. The amount of absorption and scattering depends on the wavelength of the incident radiation and on the size of the hydrometeors, and the hydrometeor phase and shape. Numerous small, liquid cloud droplets are strong reflectors of solar radiation, while raindrops, though much fewer in number, are strong absorbers of solar radiation, but small contributors to scattering. Likewise, numerous small ice crystals can be strong reflectors of solar radiation while being weak absorbers. The amount of energy reflected by ice crystals depends on the ice crystal habit (i.e., dendrites vs. needles or columns) and on the preferred fall orientation of the ice crystals relative to the direction of incident radiation.

Important to the radiative properties of clouds is their *liquid water path*. The liquid water path is the cumulative condensed water (or frozen water) along the direction of the incident radiation in a cloud. If this radiative energy becomes sufficiently scattered such that it is isotropic, the liquid water path becomes equal to the remaining depth of the cloud. A shallow cloud, such as some stratocumulus clouds, may have a relatively small liquid water content through a depth of a kilometer or so. As a result much of the sun's energy can pass through the cloud without being appreciably attenuated. It will still appear relatively bright beneath the clouds and the disk of the sun may still be visible. The same is true for high thin cirrus clouds which often cause only weak attenuation of solar radiation. In contrast deep cumulonimbus clouds contain appreciable amounts of condensed liquid water through their depths. As a result these clouds often appear very dark, even black, because solar radiation is nearly completely attenuated.

Except for very deep, wet clouds, absorption of solar radiation by cloud droplets and ice crystals is small. By contrast, absorption of longwave radiation by clouds is quite large, while they reflect longwave radiation rather poorly. As much as 90% of incident longwave radiation can be absorbed in less than 50 meters in moderately wet clouds. Clouds such as deep convective clouds and thick stratus are such effective absorbers of longwave radiation that they are often viewed as black bodies, or perfect absorbers and emitters of terrestrial radiation. A cumulonimbus cloud, for example, behaves like a blackbody after longwave radiation has pene-

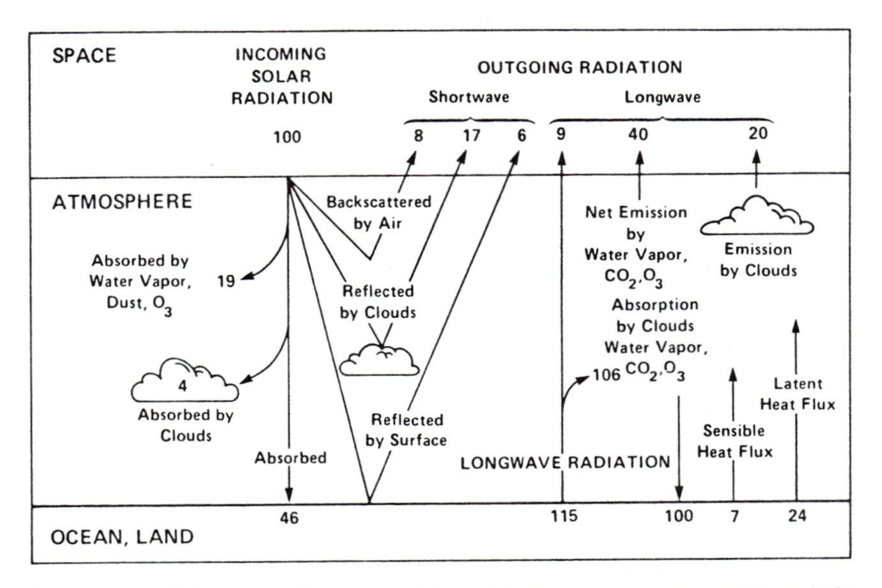

Figure 8.3: *Schematic diagram of the global average components of the Earth's energy balance.* [Adapted from MacCracken, M.C., 1985: Carbon dioxide and climate change: Background and overview. In *The Potential Climatic Effects of Increasing Carbon Dioxide.* MacCracken, M.C., and F.M. Luther, eds., U.S. Department of Energy, Washington, DC, and the University of California, Lawrence Livermore National Laboratory, (DOE/ER-0237), 381 pp. ©U.S. Government.]

trated a distance of only 12 meters. By contrast, thin cirrus clouds must be greater than several kilometers in depth before they behave as black bodies, which is generally greater than their depths.

We noted previously that in a cloud-free atmosphere, little gaseous absorption takes place between 8 and 14 μm, or the 'atmospheric window'. In a cloudy atmosphere, on the other hand, there are no spectral regions where gaseous absorption of longwave radiation is small. Clouds effectively slam the atmospheric window shut, thereby limiting the amount of radiation emitted to space and increasing the amount of longwave radiation re-emitted downward towards the ground.

8.1.4 Global energy balance and the greenhouse effect

To understand the role of radiation on global climate change, consider the globally-averaged energy budget as illustrated schematically in Figure 8.3. Of the 340 units of solar energy (in watts m^{-2}) entering the atmosphere, 65 units are absorbed by water vapor, dust, and ozone, 14 units are absorbed by clouds, 27 units are scattered by air molecules, 58 units are reflected by clouds, 20 units are reflected by the earth's surface back into space, and 156 units are absorbed at the earth's surface. The earth's albedo as viewed from space is 0.31, which means that 31% of the solar radiation incident on the earth is reflected back to space.

At the earth's surface, 390 units of longwave radiant energy are emitted with 31 units escaping directly to space through the atmospheric window. A total of 359 longwave radiation units are absorbed by water vapor, carbon dioxide, ozone, and clouds. Of that amount, 340 units are re-emitted and absorbed at ground, while 136 units are re-emitted to space by water vapor, carbon dioxide, and ozone, and 68 units are emitted to space at the tops of clouds. In addition, 24 units of heat energy are emitted from the ground as sensible heat and 82 units are emitted as latent heat. The climate of earth will remain constant as long as the apportionment of energy contributions to the global budget remain the same. Any systematic change in one or more components can lead to an imbalance in the global budget and lead to warming or cooling of globally-averaged temperatures. For example, if all existing greenhouse gases were removed from the atmosphere, the amount of longwave energy emitted to space would be greatly enhanced. This would result in an average surface temperature of the earth that is 30° C cooler than it is today. That is why the warming caused by absorption of longwave radiation by water vapor, CO_2, etc. is called the *'greenhouse effect'*.[1]

Let us consider some of the possible changes in the global energy budget that can occur naturally.

8.1.5 Changes in solar luminosity and orbital parameters

The total energy output from the sun cannot be viewed as being a constant. There is evidence suggesting that early in earth history the sun's output was 25% less than it is today. In recent history solar luminosity has been observed to decline from 1980 to 1986 and to increase since 1986 (Willson and Hudson, 1988; Willson et al., 1986). The current solar irradiance on a plane perpendicular to the incident energy at the top of the atmosphere is about 1365 watts m^{-2} (Barkström et al., 1990). Attempts have been made to relate changes in solar luminosity to observed solar parameters such as sunspot activity, solar diameter, and umbral-penumbral ratio (Wigley et al., 1986; Pecker and Runcorn, 1990; Friis-Christensen and Lassen, 1991). The recent data suggest that solar

[1]The term 'greenhouse' is somewhat of a misnomer as discussed by Bohren (1989). Actual greenhouses work primarily through their influence in preventing the turbulent removal of heat from the building as a result of the glass panes. The emissivity of the glass panes to infrared radiation is generally insignificant. The glass panes permit sunlight to enter and heat the interior of the greenhouse.

luminosity is positively correlated to sunspot number with an 11-year cycle having an amplitude of 0.1 watts m^{-2} at the top of the atmosphere (Willson and Hudson, 1988). Estimates of global mean surface temperature changes due to observed changes in sunspot number are generally quite small, being on the order of less than 0.1° C and as a result undetectable (Wigley, 1988). Nonetheless, convincing statistical studies suggest a correlation between sunspot number and meteorological parameters such as temperatures at the 30 mb level in the Arctic (Labitzke, 1987). The implication of those findings to surface temperatures remains unknown, however. There is recent evidence (Kerr, 1991, Friis-Christensen and Lassen, 1991) that the *length* of the solar cycle correlates very closely to the global average surface temperature. It still remains a mystery, however, as to how such small changes in solar irradiance caused by variations in sunspot number and the length of the solar cycle could have a significant climatic impact. Some scientists speculate that some unknown indirect amplification mechanism must exist.

There have been a number of attempts to use indirect measures of solar activity to identify possible climatic impacts. Wigley (Labitzke, 1987), for example, attempted to infer changes in solar luminosity associated with variations in C^{14} concentrations in tree rings. Because the production rate of C^{14} in the atmosphere is related to the output of energetic particles from the sun or solar wind, variations in C^{14} concentrations may be indicative of variations in solar irradiance. Historical records, for example, show a correlation between positive C^{14} anomaly and sunspot minima. According to Wigley and Kelly (1990) the C^{14} concentration peaked during the Little Ice Age in the 17th century and during earlier intervals of glacial advance. Assuming that the C^{14}-climate link is real, Wigley (1988) estimated the change in solar irradiance associated with observed changes in C^{14} concentrations and inferred changes in surface temperatures over the last several hundred years. He estimated that changes in solar irradiance associated with C^{14} anomalies would translate into changes in global mean temperature of a magnitude of 0.1–0.3° C.

The net output of energy from the sun is not only important to climate variability, but the distribution of the sun's energy on the earth/atmospheric system is also important in controlling global temperatures and whether or not earth may be moving into an ice age. Variations in the pattern of energy reaching the earth's surface, in turn, are caused by slow changes in the geometry of the earth's orbit around the sun and in the earth's axis of rotation. The fundamental theory predicting the onset of ice ages in response to changes in orbital parameters is attributed to M. Milankovitch (see Imbrie and Imbrie, 1979; Berger, 1982). As shown in Figure 8.4 the important orbital parameters are (a) the eccentricity of the orbit, (b) the axial tilt which affects the distribution of sunlight, and (c) the precession of the equinoxes. The amount of radiation reaching polar latitudes in summer is important to the onset of ice ages, since it is the

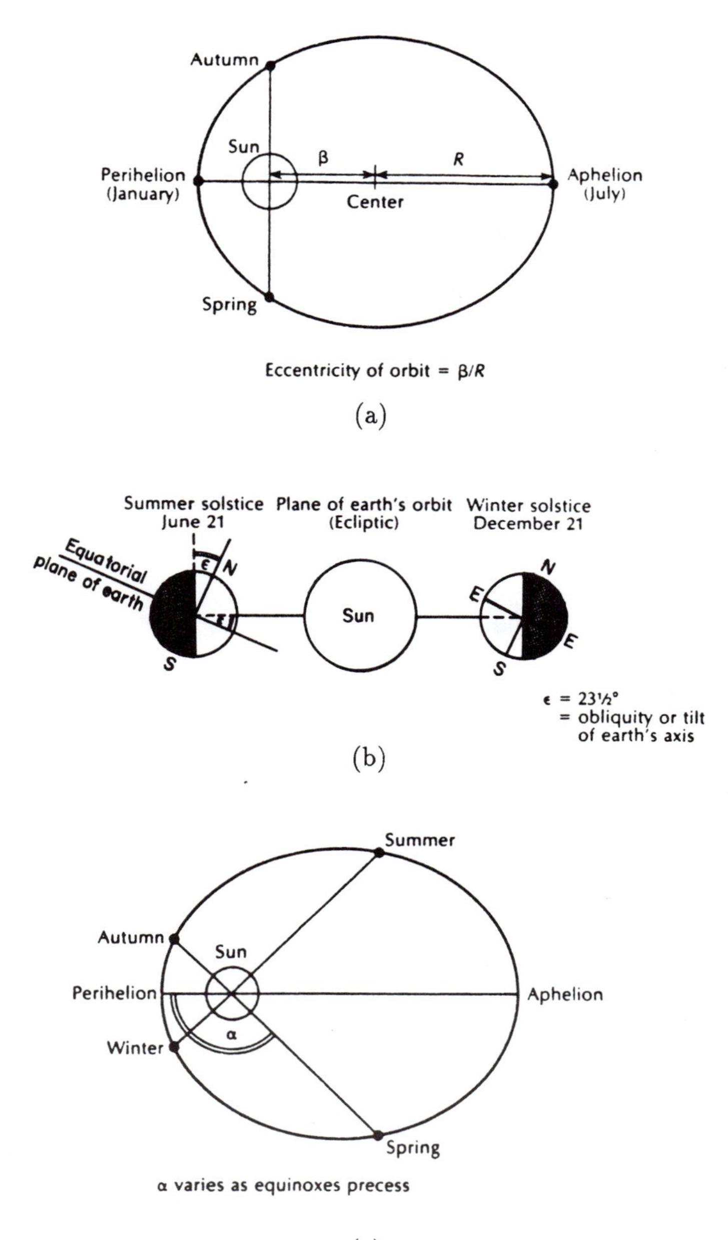

Figure 8.4: *Important components of earth sun geometry. Important orbital parameters: (a) eccentricity of orbit, (b) axial tilt, and (c) precession of the equinoxes.* [From Griffiths, J.F., and D.M. Driscoll, 1982: *Survey of climatology.* Charles E. Merrill Publishing Co., Columbus, Ohio.]

amount of summer melting of glaciers which largely determines whether glaciers are growing or receding. Thus a decrease in axial tilt causes a decrease in summer radiation. Likewise an increase in earth-sun distance in any season causes a decrease in radiation during that season and the strength of those effects varies systematically with latitude. The effect on the net solar radiation received at the surface in response to the 41,000 year oscillation in the earth's tilt is large at the poles and small at the equator. On the other hand, the influence of the 22,000 year precession cycle is small at the poles and large near the equator.

The Milankovitch theory predicts that earth will gradually be moving into an ice age over the next 5000 years. It remains to be seen if the impact of anthropogenic activities on climate can alter the effect of the orbital forcing on climate trends.

8.1.6 Natural variations in aerosols and dust

We have seen that aerosol particles can significantly attenuate solar radiation. A major source of natural dust and aerosol particles in the upper troposphere and stratosphere are volcanos. A major volcanic eruption such as Tambora in 1815 and Pinatubo in 1991 can spew large quantities of dust, aerosol, and gases into the lower stratosphere where they can reside for months to several years. The gases such as sulfur dioxide are converted to sulfate aerosols through photochemical processes. The high-level aerosol particles reflect some of the incoming radiation, thus increasing planetary albedo, and absorb solar radiation in the stratosphere, thus reducing the amount of energy reaching the earth's surface. Therefore, a single large volcanic eruption can reduce surface temperatures by several tenths of a degree for several years (Hansen et al., 1978; 1988; Robuck 1978; 1979; 1981; 1984a).

Because of the thermal inertia associated with the oceans, a two to three year period of reduced solar heating can impact average surface temperatures for several decades. Periods of active volcanism can have a substantial counteracting effect to global warming scenarios. In fact, a period of global warming between 1920 and 1940 has been attributed to very low volcanic activity (Robuck, 1979). Because large volcanic eruptions occur rather frequently and they cannot be predicted with any confidence, they represent a major source of uncertainty in predicting climatic trends.

Another potentially major source of dust and aerosol in the atmosphere can result from a collision between the earth and a large meteor or comet. It is hypothesized that major meteor impacts sent up a cloud of dust into the atmosphere that was so dense it caused such dramatic cooling that major life forms such as dinosaurs were destroyed (see review by Simon, 1981). Recent support for the hypothesis was provided by the Alverez team who found high concentrations of iridium, as well as other rare metals on earth that are abundant in extraterrestrial material in the geologic strata corresponding to the Cretaceous-Tertiary geological boundary. Thus large meteor collisions provide another source of uncertainty in climate prediction, although the frequency of such events is so low that they are generally thought to be a minor factor in predicting climate change over the next several hundred years.

8.1.7 Surface properties

As shown in Figure 8.3, the earth's radiation balance can be altered by variations of surface properties. The net albedo of the surface of the earth is determined by the percent coverage of ocean versus land, the amount of glacial coverage, and properties of the land surface such as the amount of desert versus forested lands. The surface properties introduce a major non-linearity into the system, since if summer temperatures are cooler at high latitudes, glaciers advance, which in turn increases net surface albedo contributing to cooler temperatures and so forth. Likewise, as the glaciers advance, less water is available for the oceans and the percent cover of ocean surface is decreased, also altering the net albedo. More importantly, lower sea levels may block certain ocean currents from transporting warm sea water into high latitudes, enhancing sea ice coverage, which again affects the net surface albedo.

Vegetation can also respond to changes in surface temperature and rainfall causing another complicated feedback through changing global albedo. All these complicated feedbacks must be included in any climate change model. Also the local albedo of the surface can influence cloud cover, thereby modifying the radiation balance as the cloud cover is increased or decreased or becomes thinner or thicker. Moreover, the effect of the surface albedo on the planetary albedo will depend on this cloud coverage since the surface would be shaded when it is cloudy.

Let us now examine some of the basic concepts and philosophy of modeling global change.

8.2 Modeling Global Change and the Scientific Method

In order to estimate natural or human-induced changes on global weather and climate, quantitative models of the earth/atmosphere/ocean system are required. A model represents a mathematical description of the earth/atmospheric/ocean system. They can vary greatly in complexity from those that can be solved using paper and pencil to very sophisticated computer models requiring days or even weeks of computer time on the fastest supercomputers. Climate models solve a set of mathematical equations that describe the interactions among the atmosphere, oceans, snow and ice, the biosphere including plants and animals, and land surfaces including surface hydrology. Figure 8.5 illustrates schematically some of the components important to climate models.

The following are some examples of climate modeling approaches.

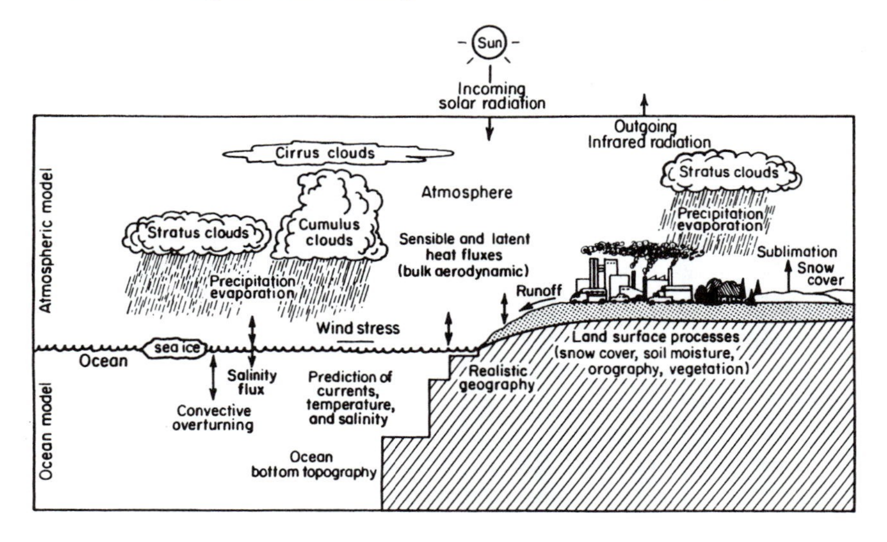

Figure 8.5: *Schematic illustration of some of the processes simulated in global climate models.* [Adapted from Building an Advanced Climate Model. Program Plan for the Computer Hardware, Advanced Mathematics, and Model Physics (CHAMMP) Climate Modeling Program. DOE/ER-0479T, U.S. Dept. of Energy, December 1990.]

8.2.1 Energy balance models (EBMs)

Energy balance models do not distinguish among the various components of a climate system, but instead determine the effective radiative temperature of the planet, assuming there is a balance between the solar radiation absorbed by the earth and the longwave radiation emitted by the earth to space. Important parameters for these models are the solar irradiance, the albedo of the planet including the effects of clouds, aerosols, and ice caps, the amount of infrared radiation absorbed by the atmosphere (i.e., the greenhouse effect), and the distribution of zonally-averaged temperature in the north-south direction. In the more comprehensive EBMs, the ice caps (and as a result albedo) change in response to the equilibrium temperature and the north-south (meridional) temperature profiles are determined from a parameterization of the meridional transport of heat by atmospheric and ocean circulations.

Parameters in these models are typically calibrated for the current climate and current equilibrium climate. Attempts to apply such models to an altered climate requires that the calibration of parameters is invariant for a a different radiative balance. Unfortunately, this condition is by no means guaranteed.

8.2.2 Radiative-convective models (RCMs)

Radiative-convective models are one-dimensional models that determine the vertical distribution of globally-averaged temperatures for the atmosphere and the underlying surface. A RCM responds to prescribed atmospheric composition and surface albedo. It includes sub-modules that describe solar and terrestrial radiative transfer, heat exchanges between the earth's surface and the atmosphere, the vertical transfer of heat and water vapor in the atmosphere, and the vertical distribution of clouds having varying radiative properties (see reviews by Ramanathan and Coakley, 1978; Schlesinger, 1983). Some RCMs have been coupled to one-dimensional mixed layer models of the ocean and include heat exchanges with deeper ocean layers.

The predicted responses of RCMs to changes in atmospheric composition are very sensitive to the prescription or parameterization of the radiative effects of clouds and to the parameterization of heat exchanges with the underlying ocean, particularly the rate of heat exchange between the ocean-mixed layer and deep ocean.

8.2.3 Zonally-averaged models (ZAMs)

Zonally-averaged models are two-dimensional models capable of simulating vertical and meridional variations in surface and atmospheric properties averaged around latitude circles. Their advantage over RCMs and EBMs is that they can respond to prescribed or modeled latitudinal variations in ice cover, land and ocean surface properties, and cloud distributions. By numerically integrating the equations of motion, heat, and water vapor transports on a two-dimensional, vertical-meridional grid, vertical and meridional transports by large-scale circulations such as the Hadley Cell can be explicitly simulated. Meridional transports by extratropical and tropical cyclones and vertical transports by convective clouds must be parameterized, however. Some ZAMs are coupled with zonally-averaged models of the ocean mixed layer. Like RCMs, they are very sensitive to parameterizations of cloud distributions (vertically and latitudinally) and to heat exchanges with the ocean, particularly the deep ocean.

Both ZAMs and RCMs share a common problem when responding to reductions in solar insolation. That is, when the insolation is reduced below a certain threshold, both types of models jump from one stable climate consisting of a partially ice-covered earth to another stable climate, consisting of a totally frozen earth (Gerard et al., 1990). Moreover, once they are 'frozen' into the colder stable mode, it takes an unrealistically large increase in insolation to revert back to the partially ice-covered regime. This model artifact is sometimes referred to as the *'albedo-ice catastrophe'*. This is because they do not model vertical and meridional heat transports properly. In the case of RCMs, vertical heat exchange is modeled by using only simple linear eddy diffusivity that responds to the globally-averaged surface temperature. Likewise, the global ice albedo is parameterized to be simply related to the global-average surface temperatures. In the case of ZAMs, meridional heat transport by large-scale eddies, such as extra-tropical cyclones, is parameterized by a linear eddy viscosity. Both these modeling approaches miss important nonlinear interactions in the earth/atmosphere/ocean system that produce a much more gradual response to changes in insolation and other alterations in the net radiative budget.

8.2.4 General circulation models (GCMs)

General circulation models are fully three-dimensional global models that attempt to simulate climate and climate change using numerical weather prediction techniques (i.e., finite-difference, or spectral, or hybrid spectral and finite-difference, or finite element techniques). Therefore GCMs explicitly simulate atmospheric, and in some cases oceanic, circulations that contribute to meridional, zonal, and vertical transports of heat, water vapor, and other properties. The highest resolution GCMs have horizontal grid meshes down to about 2 degrees latitude and longitude (~222 km) and can thus capture some of the features of extra-tropical cyclones and tropical cyclones, although rather crudely (e.g., Broccoli and Manabe, 1990). Since cyclones are major contributors to meridional and vertical heat transports, GCMs should be less susceptible to the 'albedo-ice catastrophe'. They may not be totally immune to this problem, however, since only the highest resolution GCMs exhibit sufficient non-linearity in the simulations to prevent the simulated catastrophe from occurring.

GCMs are not numerical weather prediction (NWP) models, even though they use computational procedures similar to forecast models. Global NWP models are initialized with real data, have higher resolution, and only exhibit predictability of the largest-scale atmospheric wave structures for periods of a week or two. GCMs, however, do not have sufficient horizontal resolution to properly represent the dynamics of important weather systems such as extratropical and tropical cyclones. These weather systems play an important role in meridional, and vertical heat and moisture transports, as well as in the global hydrological cycle. To some extent, the inability of GCMs to represent these important weather systems properly is offset by calibration of some of the model parameterizations, such as horizontal eddy viscosity, large-scale precipitation cumulus convection, and gravity wave drag. Thus, GCMs are able to represent certain climate features of the general circulation quite well (Houghton et al., 1990; Randall, 1992). It is possible, though, that they may be getting the right results for the wrong reasons, which could impact their ability to simulate climate change properly.

GCMs are typically initialized from a previous run or an observed atmospheric structure and must simulate the evolution of the atmospheric general circulation for periods of decades or even centuries; far beyond the range of predictability of any 'realization' of atmospheric structure. Instead the aim of general circulation modeling is to represent the statistical properties (or climate) of the longer wave structure of the atmosphere and to 'simulate' climate change scenarios. We underscore the word 'simulate' because they represent simulations of climate change scenarios under the constraint 'that all other things are the same'. To be a true forecast of climate change, GCMs should consider the effects of all the important physical processes responsible for climate change, including changes in solar irradiance and orbital parameters, volcanism and other sources of dust, greenhouse effects, global ocean dynamics, etc. Moreover, the climate at a particular time should be predicted from an ensemble of realizations in which the model initial conditions (e.g., snow/ice cover, soil moisture, ocean circulations and associated sea surface temperature anomalies and atmospheric circulations) are perturbed a small amount about some realistic initial state, so that a true ensemble-averaged climate can be predicted. Alternatively, one can integrate the climate model for very long integration times (e.g., 1000 years), assuming constant external forcing. Otherwise one cannot be certain that the atmospheric structure obtained in a single realization of a period of time in the future is but only one of an entire spectrum of atmospheric states that could be present at a later time. We need to perform these experiments in order to determine if there is as much variability in long-term climate as a result of nonlinearities in the system (i.e., chaos) as occurs on shorter scales of weather. Due to the enormous amount of computer time required for a single realization, however, only single GCM runs are used for simulation of climate change scenarios.

Like all of the climate models, GCMs are sensitive to the depiction of clouds and to the representation of the effects of the oceans. Some GCMs are being formulated to explicitly simulate cloud water (e.g., Sundqvist, 1978; 1981), but their limited vertical resolution (usually 15 to 20 levels) and coarse horizontal resolution prevents them from adequately depicting thin clouds and mesoscale cloud formation processes. Most of the early simulations of climate change with GCMs were performed with fixed cloud distributions derived from observed cloud cover data. This, of course, does not allow any feedback between changing atmospheric/ocean temperatures and motions, and cloud distributions. Most scientists view the cloud radiative feedback problem to be an 'Achilles Heel' of GCMs.

The representation of the effects of the ocean on the general circulation is likewise quite crude. Some of the earliest GCMs contained fixed, prescribed sea surface temperatures and sea ice distributions. Others contained what is essentially a *swamp* — a stagnant, wet surface that absorbs heat and releases heat and water vapor (e.g., Manabe and Wetherald, 1975; 1980). In the swamp model, the ocean surface temperatures are calculated from a balance of net shortwave and longwave radiation, evaporation, and sensible heat transfer between the ocean surface and the air. Later GCMs included a slab ocean 50 m or so thick or of variable depth that allowed heat storage in the summer and release in the winter seasons (e.g., Manabe and Stouffer, 1979; 1980). These models do not include the effects of ocean currents, but represent seasonal responses of the upper ocean. The neglect of ocean circulations on heat transport, however, results in tropical oceans being too warm and polar regions too cold. This is because warmer water near the surface transports heat poleward, and cold, dense water near the poles, sinks and travels toward the equator well below the surface. Coupled ocean GCMs are only coming on line, but still contain rather crude representations of ocean circulations (e.g., Stouffer et al., 1989 and Washington and Meehl, 1989).

The representation of vegetation and other biophysical feedbacks (e.g., dimethylsulfide or DMS – see Section 11.3.3) in GCMs has been even more crude than the characterization of ocean effects. The earlier GCMs used a water reservoir (a 'bucket') to represent the land surface hydrology, while even current models use only a single vegetation type to represent vast geographic areas.

One can readily see that the most sophisticated, computationally-intense, climate models are still only first-order approximations to the earth/atmosphere/ocean/biosphere system. What are often called 'predictions' of climate are in fact 'numerical experiments' in which the response of a model to an imposed set of changes in model parameters is examined. Such 'numerical experiments' only represent one component of the modern scientific method illustrated in Figure 8.6. Other components to the computer-age scientific method include observation and analysis of field data, such as climatic records, and postulation of hypotheses which can be tested through physical and numerical experiments. Because it is nearly impossible to perform physical experiments on the global scale, climatologists have had to rely heavily on numerical experiments to obtain insight into the behavior of the earth system. Nonetheless, the credibility of those models used for numerical experimentation must be continuously questioned and tested. We must always ask how well a model represents current or past climates. Can the model simulate previous cases of climatic change and account for the natural variability of climate? How credible are the various component physical/chemical/biological modules or parameterizations that are the building blocks of the model? We must answer questions such as these before we feel confident that we have arrived at a true understanding of the factors controlling climate and climate change.

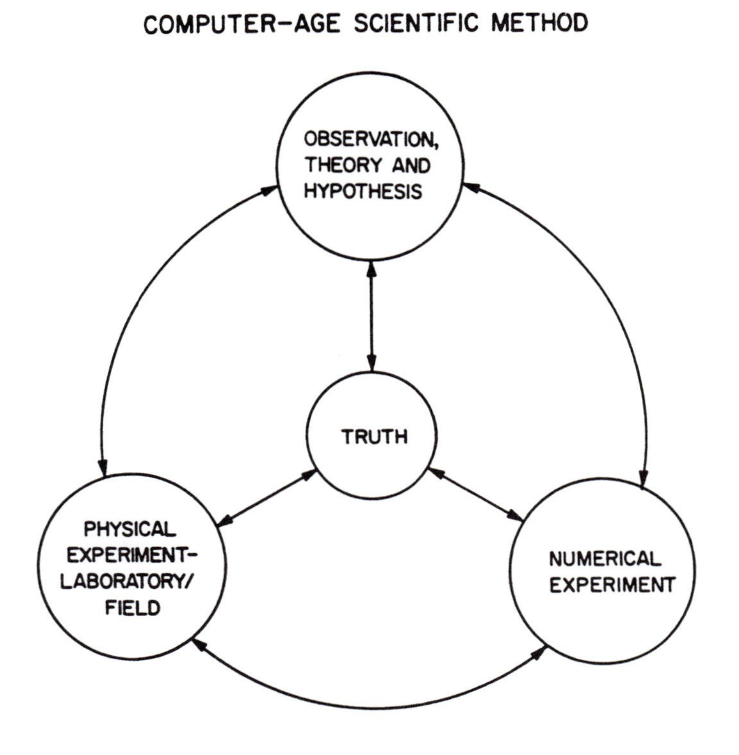

Figure 8.6: *Flow diagram of the computer-age scientific method.* [From Cotton, W.R., 1990: *Storms.* Geophysical Science Series, Vol. 1. ASTeR Press, Fort Collins, CO, 158 pp.]

Chapter 9

Climatic Effects of Anthropogenic Aerosol

9.1 Introduction

In Chapter 8 we showed that aerosol particles, particularly those emitted by volcanic eruptions, can cause significant anomalies in atmospheric radiation and global climate. Likewise, in Chapter 4, Sections 4.1 and 4.4, we examined evidence suggesting that human production of aerosol particles has local and regional impacts on clouds, precipitation, and atmospheric temperature. In this chapter we examine the evidence indicating potential impacts of anthropogenic aerosol on global climate.

Estimating the effects of aerosol on climate is particularly challenging. We noted in Chapter 8 that the radiative response to aerosol particles vary with size and chemical composition of the particles relative to the wavelength of the incident radiation. Moreover, because most aerosol particles are heterogeneous in structure, some components of the particles are very absorbing while others reflecting. As a consequence only gross estimates of the radiative properties can be made. Aerosol particles also have limited lifetimes in the atmosphere. Particles greater than a few micrometers may survive for only a few days, while particles on the order of 0.1 μm and less may reside in the lower troposphere for several weeks. This results in pronounced regional and hemispheric variations in aerosol concentrations. Unfortunately, there have been few systematic observations of aerosol, their size-spectra, and chemical composition. It is, therefore, usually necessary to resort to a variety of less-than-direct measurements to infer changes in aerosol concentrations and their radiative effects.

As discussed in Chapter 4, aerosol particles can have direct effects on atmospheric radiation as well as indirect effects through their impact on cloud microstructure. We shall examine their direct and indirect effects separately below.

9.2 Direct Aerosol Effects

It is well known that aerosol particles in polluted urban areas deplete direct solar radiation by about 15%, sometimes more in winter and less in summer (Landsburg, 1970). Less well known, however, is how far the pollution-caused aerosol extend from the urban areas. Recently, Schwartz (1989) summarized measurements of the concentrations of sulphate (or sulphur) aerosol at remote locations in both the northern and southern hemispheres. These particles form from sulphur dioxide emitted naturally largely by decay of plant and animal matter, by wildland fires, by volcanos, and by anthropogenic activity. The particles form primarily from the in situ oxidation of sulphur dioxide either as a primary gas or as an intermediate stage of oxidation. Two mechanisms for aerosol formation from sulphur dioxide are: (1) dissolving in cloud droplets to create sulfurous acid, which oxides further to form sulphuric acid aerosol particles, and (2) photochemical oxidation to form sulphate particles. Although sulphate particles are not the only human-caused aerosol, they are certainly the most prolific, so identification of their distribution is important to understanding the role of human-caused aerosol on climate.

Anthropogenic sulphur dioxide emissions have increased to their pre/-sent level almost entirely within the last 100 years (Cullis and Hirschler, 1980) and, furthermore, as summarized by Schwartz (1988), the bulk of those emissions are in the northern hemisphere. Aerosol sulphate is quite common at remote sites in both the northern and southern hemispheres, with northern hemispheric concentrations substantially exceeding those in the southern hemisphere. At several remote sites the observed high aerosol sulphate concentrations have been attributed to transport from regions of industrial activity over 1000 kilometers away (Prahm et al., 1976; Wolff, 1986). There is also evidence that sulphate concentrations have increased substantially over the last century in polar ice at northern hemispheric sites (Barrie et al., 1985; Neftel et al., 1985a; Mayewski et al., 1986) but there is no such increase in Antarctica (Delmas and Boutron, 1980; Herron, 1982). While it is difficult to make accurate estimates of the concentration of sulphate aerosol, and its spatial and temporal variability, there is little doubt that it is increasing, particularly within a few thousand kilometers of industrial regions. As an example, Figure 9.1 shows the drastic reduction in visibility that has occurred in the eastern United States since 1948. Much of this visibility degradation is attributable to sulfate particles.

An indirect measure of aerosol concentrations is the electrical conductivity of the air. The electrical conductivity, in turn, is controlled by the concentration and mobility of ions and small, charged aerosol particles. In general, increases in the concentration of aerosol particles decreases conductivity because the more highly concentrated aerosol particles collect small ions and charged small aerosol particles, thus immobilizing the

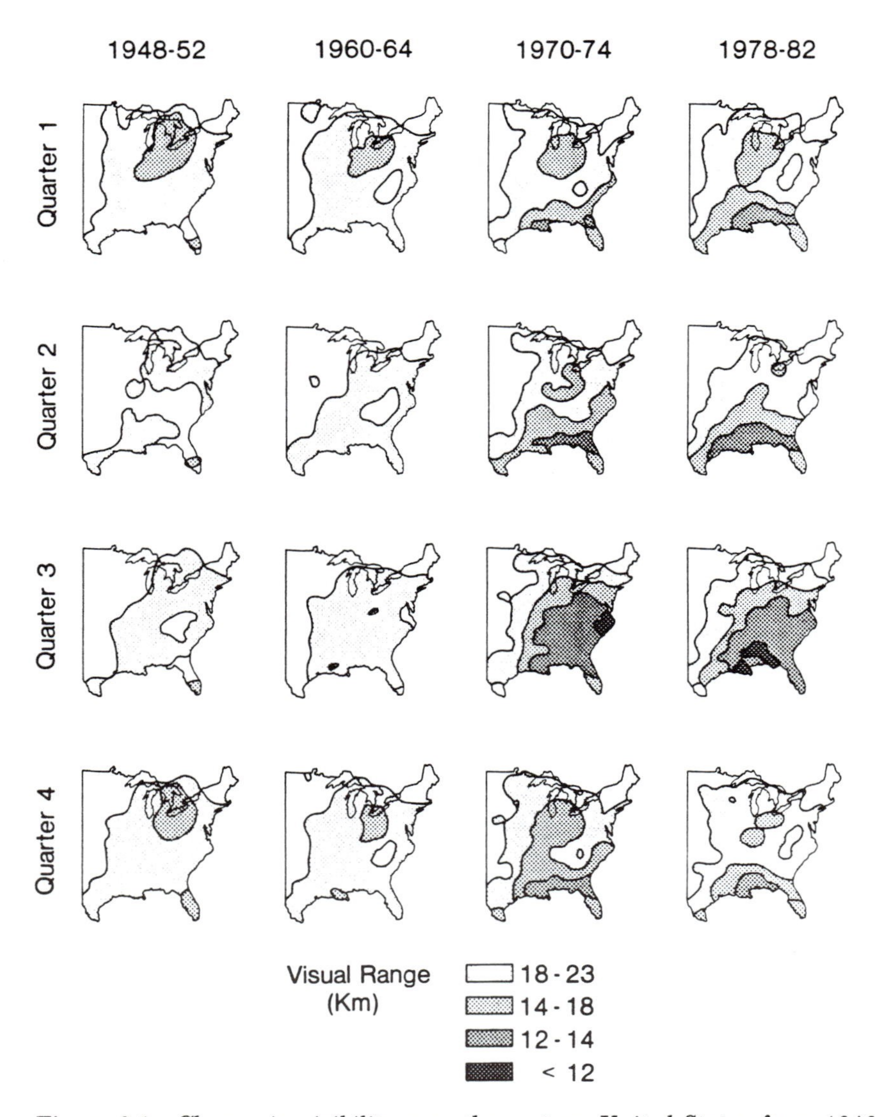

Figure 9.1: *Change in visibility over the eastern United States from 1948 to 1982.* [From Malm, W.C., 1989: Atmospheric haze: Its sources and effects on visibility in rural areas of the continental United States. *Environ. Monitoring Assessment,* **12,** 203-225. Reprinted by permission of Kluwer Academic Publishers.]

charge. An exception is radioactive contamination which makes the atmosphere more conductive. Extensive conductivity measurements were made during the first half of this century, but unfortunately the practice has not been continued in the latter part of the century. Cobb and Wells (1970) summarized the results of a few such measurements that were collected from 1907 to 1970. In general, they suggested a 20% decrease in conductivity took place in the North Atlantic over this period. This corresponds to roughly a doubling of small aerosol particle concentrations over the area. Limited data in the southern Pacific show no such trends.

In general, there is compelling evidence that anthropogenic activity is increasing the concentration of tropospheric aerosol, particularly in the northern hemisphere. Let us now examine the potential impacts of those increased aerosol concentrations on climate. There is considerable evidence that naturally occurring aerosol particles can affect climate and impact the global circulation of the atmosphere (Hansen et al., 1980; Randall et al., 1984; Tanre et al., 1984; Coakley and Cess, 1985; Ramaswamy, 1988; Hansen et al., 1988). These assessments are generally made with some type of climate or general circulation model. The responses of those models to naturally occurring aerosols, varies with the concentration and type of aerosol, the characteristics of the underlying surface (i.e., the surface albedo), the solar zenith angle, cloud cover, and the way the models treat ocean responses. The latter effect is particularly important since, if the ocean temperatures are fixed, the oceans serve as an infinite heat reservoir, and no long-term global temperature responses can be expected (Coakley et al., 1987). Furthermore, unless sea surface temperatures can vary, important feedbacks such as variations of the flux of moisture from the ocean cannot occur (e.g., Ramanathan, 1981).

In general, absorption of solar radiation by aerosols reduce solar heating at the surface while it heats the layer of air in which the aerosols reside. The impact of aerosols on the surface, however, varies with the albedo of the underlying surface. If a surface has a relatively low albedo such as over the ocean, a given aerosol may increase the surface albedo, while over the higher albedo deserts, it may decrease it. In general, the impact of aerosol on albedo dominates over absorption at most latitudes, but in higher latitudes over snow- or ice-covered surfaces, aerosol absorption can dominate (Charlson et al., 1992). Likewise, the atmospheric response to aerosol heating varies widely depending on the height and depth of the aerosol layer, and on the basic stability of the layer. For example, the stabilization caused by aerosol heating over a moist moderately stable layer may shut down deep convection and precipitation which could have important climatic implications through the hydrological cycle. In other regions where there is not sufficient moisture or instability to support deep convection, aerosol impacts would be less.

Taking into account the geographic distribution of cloud cover and surface albedo, Charlson et al. (1992) estimate that the radiative cooling due to sulfate aerosols in the atmosphere is comparable, but opposite in sign to the radiative heating of greenhouse gases for an ~25% increase in CO_2 for the period since industrial activity became significant. These are just rough estimates, however, so that the total climatic impact of anthropogenic aerosols remain unknown.

9.3 Aerosol Impacts on Clouds — The Twomey Effect

Clouds, we have seen, are good reflectors of solar radiation and therefore contribute significantly to the net albedo of the earth system. We thus ask, how might aerosol particles originating through anthropogenic activity influence the radiative properties of clouds and thereby affect climate?

First of all, there are indications that in urban areas aerosols make clouds 'dirty' and thereby decrease the albedo of the cloud aerosol layer and increase the absorptance of the clouds (Kondrat'yev et al., 1981). This effect appears to be quite localized; being restricted to over and immediately downwind of major urban areas, particularly cities emitting large quantities of black soot particles. Kondrat'yev et al. noted that the water samples collected from the clouds they sampled were actually dark in color.

A potentially more important impact of aerosol on clouds and climate is that they can serve as a source of cloud condensation nuclei (CCN) and thereby alter the concentration of cloud droplets. Twomey (1974) first pointed out that increasing pollution results in greater CCN concentrations and greater numbers of cloud droplets, which, in turn, increase the reflectance of clouds. Subsequently, Twomey (1977) showed that this effect was most influential for optically thin clouds; clouds having shallow depths or little column integrated liquid water content. Optically thicker clouds, he argued, are already very bright, and are therefore susceptible to increased absorption by the presence of dirty aerosol. In Twomey's words: "*an increase in global pollution could, at the same time, make thin clouds brighter and thick clouds darker, the crossover in behavior occurring at a cloud thickness which depends on the ratio of absorption to the cube root of drop (nucleus) concentration. The sign of the net global effect, warming or cooling, therefore involves both the distribution of cloud thickness and the relative magnitude of the rate of increase of cloud-nucleating particles vis-a-vis particulate absorption.*" Subsequently, Twomey et al. (1984) presented observational and theoretical evidence indicating that the absorption effect of aerosols is small and the enhanced albedo effect plays a dominate role on global climate. They argued that the enhanced cloud albedo has a magnitude comparable to that of greenhouse warming (see Chapter 11) and acts to cool the atmosphere. More recently, Kaufman et al. (1991) concluded that although coal and oil emit 120 times as many CO_2 molecules as SO_2 molecules, each SO_2 molecule is 50-1100 times as effective in cooling the atmosphere than each CO_2 molecule is in warming it. This is by virtue of the SO_2 molecules' contribution to CCN production and enhanced cloud albedo.

Twomey suggests that if the CCN concentration in the cleaner parts of the atmosphere, such as the oceanic regions, were raised to continental atmospheric values, about 10% more energy would be reflected to space by relatively thin cloud layers. He also points out that an increase in cloud reflectivity by 10% is of greater consequence than a similar increase in global cloudiness. This is because while an increase in cloudiness reduces the incoming solar radiation, it also reduces the outgoing infrared radiation. Thus both cooling and heating effects occur when global cloudiness increases. In contrast, an increase in cloud reflectance due to enhanced CCN concentration does not appreciably affect infrared radiation but does reflect more incoming solar radiation which results in a net cooling effect.

Let us now examine the evidence supporting or refuting the Twomey hypothesis. The major focus of these studies is on the world's oceans where shallow stratocumulus clouds reside. These shallow clouds, which are believed to be most susceptible to the Twomey effect, cover over 34% of the world's oceans. Therefore, any consistent trend towards increasing CCN concentration over the oceans has the potential for increased atmospheric albedo and global cooling. Unfortunately, there have not been long-term systematic measurements of CCN concentration anywhere on earth.

We must therefore resort to indirect methods of assessing whether global pollution is affecting climate. In general, oceanic CCN concentrations are low; being on the order of 50 to 100 cm^{-3}. Over continental areas CCN concentrations range from 500 to 1000 cm^{-3} with some heavily polluted regions reaching several thousand per cubic centimeter. The main source of natural CCN over the oceans is believed to be dimethylsulphide (DMS) which is excreted by plankton and then liberated into the atmosphere where it is oxidized, probably photochemically, to form non-seasalt-sulphate aerosol (Bigg et al., 1984; Charlson et al., 1987). Charlson et al. (1987) speculate that because the largest flux of DMS comes from the tropical and equatorial oceans, the rate of production of DMS is temperature-dependent so that any increase in ocean temperature, such as is expected in a greenhouse warming scenario (see Chapter 11), would enhance CCN and cloud droplet concentrations, and this increases cloud albedo, which would create a counteracting cooling effect.

This hypothesized negative feedback is argued to be but just one example of the Gaia hypothesis (Lovelock, 1972) described more fully in Chapter 12. The DMS climate hypothesis has a number of yet to be proven hypothesized components: (1) phytoplankton production of DMS varies significantly with temperature; (2) DMS is the dominate source of CCN over the remote oceans; (3) changes in DMS fluxes produce corresponding changes in CCN; (4) changes in CCN produce corresponding changes in cloud droplet concentrations, particularly at upper levels of clouds; (5) changes in cloud droplet concentrations produce corresponding changes in cloud albedo over a large fraction of the population of oceanic clouds; and (6) the resultant changes in cloud albedo will significantly impact global mean temperature. It is interesting that this multiple chain hypothesis resembles a number of the multiple chain hypotheses that we found so difficult to prove correct for purposeful weather modification discussed in Part I (i.e., the dynamic seeding hypothesis or the STORMFURY hurricane modification hypothesis). Moreover, there are parallels in the motivations that encouraged scientists to propose such multilevel hypotheses. The STORMFURY hurricane modification program was *based on faith* in the concept that hurricanes could be weakened by cloud seeding even though we had (and still have) only a rudimentary understanding of the behavior of such storms. Likewise the DMS-albedo hypothesis is *based on faith* that the Gaia hypothesis works, such that the earth's biosphere will respond to human-caused climatic changes by producing negative feedbacks that will preserve earth climate as we know it today. The DMS-albedo hypothesis provides a potentially observable mechanism in support of the broader Gaia hypothesis. *In many ways, however, the Gaia hypothesis resembles more a religion than science!*

One benefit of the DMS-albedo hypothesis is that it has generated considerable interest and debate on the CCN-albedo concept (e.g., Meszaros, 1988; Schwartz, 1988; Slingo, 1988; Henderson-Sellers and McGuffie, 1989; Wigley, 1989).

Schwartz (1988), for example, argued that one can test part of the DMS-albedo hypothesis by showing that anthropogenic emissions of sulphur dioxide have caused a similar change in cloud albedo. It is well documented that emissions of sulphur dioxide and sulphates have increased substantially over the last century and are about double estimates of the DMS flux. Because those emissions are predominately in the northern hemisphere, he argues one should be able to detect hemispheric differences in cloud albedo and in average temperatures caused by increased CCN/cloud droplet concentrations. He then demonstrates that clouds are neither brighter in the northern hemisphere nor is the northern hemisphere substantially cooler. Slingo (1988) points out, however, that the cloud contribution to total albedo is controlled not only by cloud brightness, but also by cloud amount. There is also evidence that cloud coverage over the oceans has increased by 3% in the northern hemisphere since 1930 (Warren et al., 1988; see Figure 9.2), but these results are quite uncertain and the causes of any systematic change in cloud cover are unknown. Furthermore, we have noted that cloud reflectance is influenced both by droplet size and by column integrated liquid water.

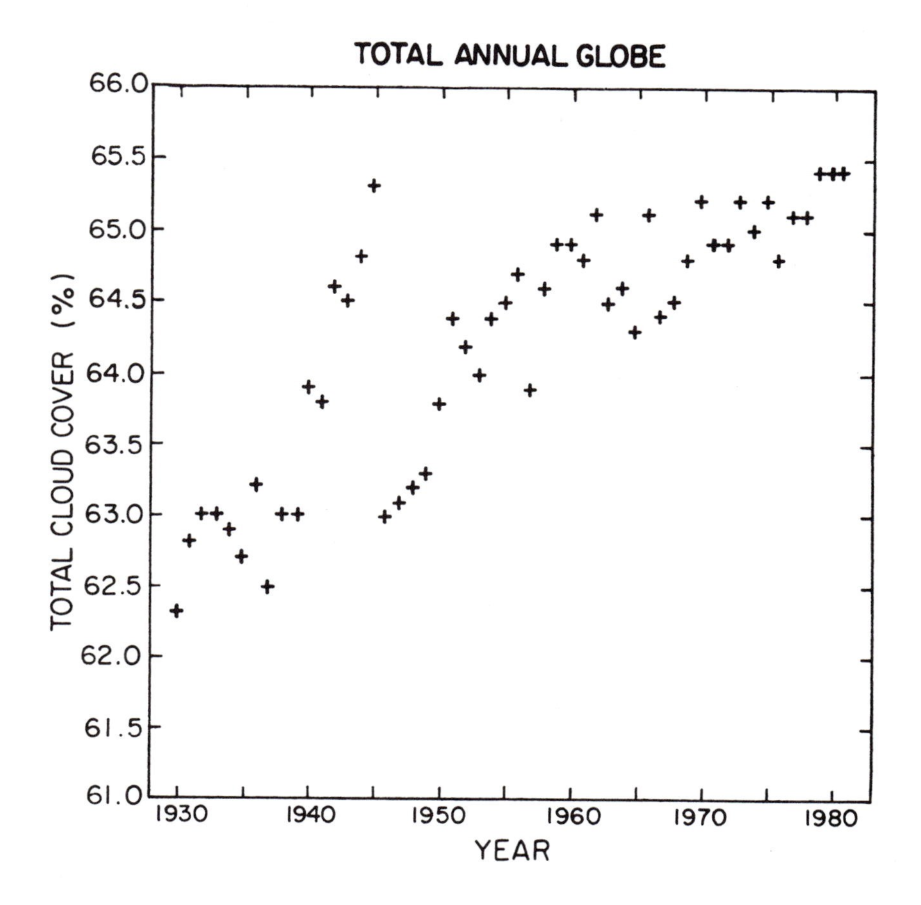

Figure 9.2: *Total annual global cloud coverage in percent over the oceans.* [From Warren, S.G., C.J. Hahn, J. London, R.M. Chervin, and R.L. Jenne, 1988: Global distribution of total cloud cover and cloud type amounts over the ocean. Prepared for the United States Department of Energy, Washington, DC and the National Center for Atmospheric Research, Boulder, CO.]

Slingo points out that differences in cloud amounts and liquid water contents could easily mask the effect of differences in droplet concentrations, thus giving a misleading signal in the hemispheric albedo. Furthermore, he notes that changes in cloud amounts and liquid water contents, as well as other quantities, over the period of the temperature record could have combined to minimize inter-hemispheric temperature differences. Thus Schwartz's conclusion, that the lack of a discernible response of mean global or hemispheric albedo, or temperature, to anthropogenic sulphur dioxide emissions indicates that control of these properties is too complex to be governed by simply CCN-albedo interactions, is not justified. Nonetheless, it points out the complexity of the problem and the need to obtain more direct measurements of CCN, cloud droplet concentrations, and cloud albedo.

Ship trails as seen from satellite have also been used to examine potential interactions between CCN and albedo. In fact Porch et al. (1990) referred to them as the possible *rosetta stone* connecting changes in aerosol over the oceans and cloud albedo effects on climate. As discussed in Section 4.1, ship track trails appear as a line of enhanced brightness in satellite imagery. The prevailing hypothesis is that the ship trails appear brighter because the effluent from the ships is rich in CCN particles. We showed that the ship track clouds contain higher droplet concentrations, smaller droplet sizes, and higher liquid water content than surrounding clouds (Radke et al., 1989). The observed higher liquid water content does not at first appear to be consistent with the CCN-albedo hypothesis, but Albrecht (1989) proposed that the higher droplet concentration in ship track clouds would reduce the rate of drizzle formation, resulting in wetter, brighter clouds.

The CCN-albedo hypothesis is strengthened by introducing the impact on drizzle. This is because cloud reflectance is increased both by increased droplet concentrations and by increased column integrated liquid water. Furthermore, an altered drizzle process could affect the lifetime, and hence, cloud coverage of some clouds such as small cumuli and thin stratocumuli. We have seen in Part I and II, however, that it is difficult to access the extent to which the drizzle formation process is susceptible to changes in CCN. We know, for example, that clouds with high liquid water content will form precipitation-sized particles regardless of moderate changes in CCN. Thus, for example, a stratocumulus cloud having a peak liquid water content on the order of 0.4 g m^{-3} may produce drizzle if the cloud droplet concentrations (activated CCN) are 50 cm^{-3} or 100 cm^{-3}. On the other hand, the drizzle formation process in a stratocumulus cloud with peak liquid water contents of 0.2 g m^{-3} may be very susceptible to 30% increases in CCN concentrations. We have also seen in Chapter 4 that effluent from paper pulp mills, though high in CCN concentrations, appear to result in enhanced precipitation downwind. It has been hypothesized that the large and ultra-giant aerosol particles emitted by paper mills serve as coalescence embryos and initiate precipitation-sized particles. Thus the susceptibility of the drizzle process in marine stratocumulus clouds to anthropogenic emissions of CCN may depend on the presence or absence of large and ultra-giant aerosol particles in the subcloud layer. Over the open sea, the dominant large and ultra-giant aerosol particles are sea-salt particles. While these particles contribute only 10% or less to the total CCN population, they represent major contributors to the large end of the aerosol size spectrum. Their concentration in the marine boundary layer varies with wind speed, being in greater numbers with stronger winds. We have noted on a number of occasions while living in the maritime tropics and subtropics, that shallow, coastal cumuli produce vigorous showers on very windy days when you can almost taste the sea-salt in the air, while on less windy days similar looking clouds do not rain. *We therefore hypothesize* that the susceptibility of the drizzle process in marine stratocumuli to anthropogenic CCN will depend on wind speed, being less susceptible at higher wind speeds than lower.

Returning now to our ship track trail *rosetta stone*, we find there is another complication in interpreting the observed brighter clouds in terms of the CCN-albedo hypothesis. Porch et al. (1990) hypothesized that the ship track trails may be brighter because the heat and moisture emissions from the ships invigorate the air motions in the clouds, thereby increasing their depths and vertically integrated liquid water content. The observed higher liquid water content in the ship track clouds is consistent with that hypothesis. Porch et al. showed that the ship track clouds are not only characterized by greater brightness, but also clear bands on the edges of the cloud tracks (see Figure 4.2). They speculated and provided some modeling evidence that the heat and moisture fluxes from the ship effluent excited a dynamic mode of instability which, in some marine stratocumulus environments, leads to sustained, enhanced upward and downward cloud circulations in the wake of the ship. Clearly, more observational and modeling studies are needed to determine if ship track trails represent a 'rosetta stone' with respect to the CCN-cloud albedo hypothesis or a 'cement overcoat' suitable for burial at sea.

In summary, the CCN-cloud albedo hypothesis is physically consistent and has the potential of causing a climatic impact opposite in sign to the greenhouse gas hypothesis. Charlson et al. (1992) made a rough calculation that a 15% global mean increase in droplet concentration in marine stratus and stratocumulus clouds, results in a radiative cooling effect comparable (and opposite in sign) to current estimates of greenhouse warming. A number of questions must be answered before estimates like these can be placed on a sound scientific basis: (1) Are CCN concentrations increasing globally and particularly over the pristine oceanic areas where shallow stratocumulus clouds are prevalent? (2) Are cloud droplet concentrations in shallow marine stratocumuli changing with time? (3) What clouds are susceptible to changes in albedo due to CCN concentration changes? (4) Is the areal coverage of susceptible clouds great enough to cause a significant change in global albedo and global mean temperature? The answer to (3) and, hence, (4) presents a major challenge to theoreticians and modelers. Since the drizzle formation process is potentially involved, it will require the use of models that are capable of realistically simulating the drizzle formation process and its sensitivity to both CCN concentrations and giant and ultra-giant aerosol particles. It will also require very complicated three-dimensional radiative-transfer calculations in which radiation interacts with the complicated microstructure and geometrical structure of clouds.

Currently, cloud radiation models are unable to explain the anomalously low values of cloud reflectances that are observed by satellites (Fritz, 1954; 1958; Robinson, 1958; Drummond and Hickey, 1971; Rozenberg et al., 1974; Reynolds et al., 1975; Herman, 1977; Stephens et al., 1978; Twomey and Cocks, 1982; Wiscombe et al., 1984). This could be interpreted that more radiation is absorbed in the interior of clouds than predicted because they are dirty, or because of greater absorption rates by drizzle, or because clouds are lumpy, three-dimensional structures in which multiple scattering and leaking of radiation through the sides of cloud hummocks transmit more radiation in the downward hemisphere than current models predict. In a recent review Stephens and Tsay (1990) argued that aerosol and large drop effects are small, and that the most likely candidate for creating the so-called *absorption anomaly* is that owing to their three-dimensional structure, clouds are more porous than thought and, as a result, the phenomena should be called the *reflection darkening anomaly.* Stephens and Tsay also pointed out that the anomaly is greatest in the near infrared part of the spectrum, while evidence that such a discrepancy between theory and observations exist over the visible part of the spectrum is less conclusive. Nonetheless, cloud-radiation models must be able to demonstrate that they can simulate observed cloud albedos realistically before we can have confidence that they can realistically simulate changes in cloud albedo due to anthropogenic sources of aerosol.

Chapter 10

Nuclear Winter

10.1 Introduction

The nuclear winter hypothesis in simplest terms contends that a large-scale nuclear war would generate large amounts of smoke and dust in the atmosphere which would attenuate solar radiation and cause so much cooling of land areas that winter-like conditions would prevail in the summer months and major crop failures would occur. The rudiments of the hypothesis were first presented by Crutzen and Birks (1982) who calculated the amounts of smoke that would be lofted into the atmosphere by fires following large-scale nuclear warfare resulting in obscuration of solar radiation of the northern hemisphere for several weeks or more. They then speculated on the possible climatic responses to the resultant reduced surface temperatures. This paper was followed by Turco et al. (1983) or the so-called TTAPS paper in which a one-dimensional, globally- and annually-averaged, radiative-convective model was used to calculate surface temperatures following massive injections of smoke into the atmosphere. They calculated that the smoke and dust emitted by fires following a massive nuclear exchange would cool land surface temperatures in midsummer below freezing in the northern hemisphere, with surface temperatures falling as much as ~35° C. the term '*nuclear winter*' was established to describe this modeled cooling.

These papers stimulated intense interest in the topic in both the scientific community and in the political arena. As a result a series of national and international committees were established to assess the potential environmental consequences of large-scale nuclear warfare (e.g., Harwell and Hutchinson, 1985; U.S. National Research Council, 1985; Pittock et al., 1986). National and cooperative international research programs were also established.

In the United States, research programs on nuclear winter were established by the Department of Energy, the Defense Nuclear Agency, and to a lesser degree the National Science Foundation and the National Oceanic and Atmospheric Administration. Funding for nuclear winter research was mandated by Congress but no specific funds were allocated. As a result funding was achieved largely by internal reprogramming within the various agencies. In some cases, this meant that researchers in or supported by the agencies were polled to see if they were doing research relevant to nuclear winter and if so, their programs were totalled to identify monies in the agencies going to the nuclear winter program. In other words, the scientists did what they have always done, with the exception, perhaps, of attending a few meetings. In other cases, the agencies re-directed scientists to work on nuclear winter research, sometimes with less than enthusiastic participation. In other cases, funding was reprogrammed from other areas making the program very vulnerable to outside attack once the political support for nuclear winter waned. Still, a few scientists and agencies jumped into nuclear winter research with enthusiasm.

Some scientists, such as Carl Sagan, used the nuclear winter hypothesis as a vehicle for proclaiming anti-nuclear war and anti-war policies. As noted by Dyson (1988), this has put professional scientists in an awkward position. Singer (1989) has also briefly discussed the political aspects of the nuclear winter issue. On one hand, it is the responsibility of scientists to critically examine a new and exciting theory, and in many cases, prove it wrong. This is how science works. Every new theory has to be defended by its proponents against intense and often bitter criticism and scrutiny. This is what keeps science honest! The rare theory that withstands the onslaught of criticism is strengthened and improved by it.

On the other hand, nuclear winter as a political statement makes us want to believe it, because few scientists favor the destructive powers of nuclear warfare, especially if it means the world-wide destruction of society. This is an example of a genuine conflict between the demands of science and the demands of humanity.

Coincident with the Peristroika policy in the Soviet Union, interest in the science of nuclear winter plummeted so that by 1990 no national research program in nuclear winter remained. Nuclear winter experienced a 'rise and fall' in a time span of only eight years. Recently in the Persian Gulf War, Carl Sagan again tried to renew interest in the nuclear winter concept, as applied to the burning of the Kuwaiti oil fields in 1991, but evidence suggests that the meteorological and air quality effects were mainly confined to regional impacts.

In this chapter we set aside our humanitarian concerns and focus on the scientific question: is the nuclear winter hypothesis a viable hypothesis and what is the evidence supporting or refuting it?

10.2 The Nuclear Winter Hypothesis — Its Scientific Basis

The nuclear winter hypothesis is another example of a long, multiple-chain hypothesis. It is composed of the following elements.

- Large-scale nuclear warfare involving a total yield in excess of 100 megatons will occur over much of the northern hemisphere.
- In addition to the death and destruction due to the direct consequences of the bombing, area-wide fires will ensue for a period of several days, and the smoke will be injected into the upper troposphere and lower stratosphere.
- Much of the injected smoke will consist of submicron-sized soot or elemental carbon particles which are excellent absorbers of solar radiation and owing to their small size and low fall velocities, have long residence times in the atmosphere.
- The smoke that escapes early scavenging by clouds and precipitation elements will disperse throughout the northern hemisphere creating a nearly uniform pall over much of the hemisphere.
- The widely dispersed smoke will have sufficiently large optical depths to cause significant absorption of solar radiation with little absorption of terrestrial radiation (assumes that the smoke layer is cloud-free and particles are small and dry) so that widespread lowering of surface temperatures occurs.
- The thermal inertia or heat storage in the oceans will cause only local (coastal?) modification of the air masses thus not damping the widespread cooling effect substantially.
- A threshold amount of smoke-induced cooling will trigger a major climatic shift that will take years or centuries for recovery.
- The smoke-induced cooling will result in major crop losses and it will compound food losses due to disrupted production and distribution systems associated with the direct effects of warfare, leading to widespread famine and death.

Let us now examine each of these subhypotheses, respectively.

10.2.1 The war scenarios

In order to estimate the maximum possible extent of widespread nuclear warfare, one must first estimate the total nuclear arsenal in the United States, Soviet Union, and their allies. Much of this information is classified but educated estimates can be made. Of the total arsenals one must next estimate what fraction of those weapons would be functional and how many would be made inoperative in a first strike (before the defender can respond). Turco et al. (1990) estimate that the combined United States and Soviet arsenals comprise about 25,000 warheads carrying roughly 10,000 megatons (MT). Crutzen and Birks (1982) used two scenarios, a 5,750 MT detonation and a 10,000 MT detonation. The original TTAPS study spanned a range of 100 to 25,000 MT with a baseline scenario of 5,000 MT. The United States Nuclear Regulatory Commission (NRC) (1985) report considered a 6,500 MT war as a baseline scenario. This is the level of bombing that is most often used in the model calculations. Obviously the actual amount of bombing that could occur is a matter of conjecture and we hope that none of these scenarios would ever become reality.

10.2.2 Smoke production

Estimating the amount of smoke produced in any particular bombing scenario is particularly complicated since one must speculate on the nature of targets (i.e., are they urban centers, or rural areas where missile silos, military camps, etc. are located, or a mix of urban-industrial and rural land), the nature of the fuels in each hypothetical target, the magnitude and type of bomb used at each target, and the local weather conditions affecting fire behavior.

In an urban area, for example, smoke production will vary depending on whether the fuel is open to the air or buried beneath piles of rubble. Likewise the size of soot particles and their optical blackness depends on whether the fire burns quickly and at high temperatures or smolders for a long time. Turco et al. (1990) cite references to sources of how fuel loading estimates are made. They conclude that the uncertainty in estimating combustible loading for a particular bombing scenario is probably less than 50%.

Taking factors such as these into account, various scientists and groups of scientists have made rough estimates of the amount of soot that is produced for given bombing scenarios. For example, the U.S. NRC (1985) estimated that a baseline 6,500 MT war would produce 360 Tg (10^{12}g) of smoke containing 65-70 Tg of strongly absorbing soot particles. Penner (1986) estimated the total amount of smoke production for a large-scale nuclear war could range from 24 to 400 Tg, with soot production being between 10 and 225 Tg. It must be recognized that a large uncertainty exists with these probable values.

Once these bulk estimates of smoke production are made, then one must estimate how this smoke is distributed vertically and horizontally and how much smoke is removed by sedimentation and scavenging.

10.2.3 Vertical distribution of smoke

Estimating the vertical distribution of smoke from fires is important in determining how long the smoke will remain in the atmosphere. If most of the smoke remains below 2-3 km above the ground such as occurred in the Kuwaiti oil fires of 1991, turbulent mixing, cloud and precipitation particle scavenging, and sedimentation of particles will result in the removal of the smoke from the atmosphere in a week or so. On the other hand, if much of the smoke rises to the upper troposphere, where turbulence is less common, clouds are low in water content and less able to scavenge soot particles, and precipitation rates are small, the smoke may remain in the atmosphere for periods of several weeks to a month or so. Finally, if a large amount of smoke makes its way into the very stable, lower stratosphere where clouds are uncommon, the smoke could remain in the atmosphere for periods of many months to as much as a year or more, similar to the residence time associated with volcanic aerosols ejected into the stratosphere.

The vertical distribution of smoke produced by a given fire depends on the fuel loading and resultant burn, and on many meteorological conditions including the likelihood that wet, deep convection will participate in the vertical transport of the smoke. Typically wildland fires have modest fuel loadings and as a result most of the smoke from them remains below 3 or 4 kilometers above ground level (Small et al., 1989). Urban fires can be far more concentrated and if there is sufficient moisture and conditional instability in the environment, numerical simulation of urban fire plumes with two- and three-dimensional models suggests that wet deep convection can be triggered by the heat from the fires (Banta, 1985; Cotton, 1985; Penner, 1986; Pittock et al., 1986; Tripoli, 1986). If the heat output is great enough and the atmosphere has sufficient conditional instability, then the models suggest that vigorous convective storms can be triggered that can transport smoke, debris, and water substance into the upper troposphere and lower stratosphere. Stratospheric injections of smoke from burning urban areas, however, is probably relatively little. Turco et al. (1990) estimate as much as 10% of the smoke will reach the stratosphere but this is probably on the high side. First of all, the intense burning period following a bombing is likely to last only a few hours and only a few targets will have sufficiently concentrated fuels and sufficiently unstable atmosphere to support stratospheric penetrating convection.

Even intense firestorms such as occurred over Dresden, Germany during World War II are likely to transport only a small proportion of the smoke produced into the stratosphere. Firestorms are intense fires that produce vigorous whirling winds on the scale of tens of kilometers. The strong winds associated with firestorms ventilate the fires, thereby strengthening them. Tripoli's (1986) simulation of such an urban firestorm revealed, however, that the cyclostrophic reduction in pressure associated with the rapidly rotating storm created vertical pressure gradients that weakened the storm's updrafts, and, as a result, the smoke plume was detrained at lower levels. Moreover, the weaker updrafts increase the time that scavenging processes will operate in the rising updrafts, and increase the efficiency of precipitation processes and wet removal of smoke. We will address the scavenging issue more directly in the next section. In summary, it appears that the bulk of the smoke from wide-spread nuclear warfare will be deposited below the middle troposphere, probably in the range of 4 to 6 kilometers (Small et al., 1989).

10.2.4 Scavenging and sedimentation of smoke

Another uncertainty is the amount of smoke that is emitted by fires and then transported aloft by dry or wet convection which is removed by scavenging and sedimentation. In the case of cloud-free convection, the main removal mechanism is by slow settling of the particles. This process is enhanced by coagulation among numerous, small, slowly settling particles to form larger, faster falling particles. Coagulation occurs very rapidly in the highly concentrated regions close to the fires, but as turbulent diffusion lowers the concentration appreciably, coagulation takes place very slowly, requiring many days to weeks to create numerous, faster falling particles that settle to the earth's surface. Only the very large smoke particles (greater than tens of micrometers) will settle out of the plume in a few hours to days in the absence of clouds.

Clouds greatly enhance the removal of smoke particles. Some of the smoke particles, such as those from wood products, are likely to be hygroscopic and serve as cloud condensation nuclei (CCN). Others, such as from petroleum products, are unlikely to be activated as CCN. Those particles may after a time coagulate with hygroscopic particles and thereby become wetted as the hygroscopic component of the mixed aerosol particle takes on water vapor. Estimates of the amount of smoke that is removed by activation of sooty CCN or what we call *nucleation scavenging* varies widely from less than 10% to more than 90%. In our opinion, for wildland fires and most residential fires it is probably closer to 90%, while in industrial petro-chemical fires it may be as low as 10-20%. These are just educated guesses, however.

Just because such particles are embedded in cloud droplets does not mean they will be removed from the atmosphere. Many cloud droplets remain small and will eventually evaporate releasing the embedded particles back into the atmosphere. During their brief residence in the cloud droplets, the aerosol particles may partially dissolve, undergo wet chemical reactions, and coagulate with other embedded particles. Thus while the particles may be released back into the atmosphere when the droplets evaporate, they probably have changed both in size and chemical composition. Some of the droplets containing embedded aerosol will participate in the formation of raindrops. Most of those aerosol particles will rain out to the ground and no longer play a role in radiation attenuation. The fraction of aerosol actually rained out is proportional to the precipitation efficiency of the clouds. Precipitation efficiencies vary widely among cloud types ranging from 10% to more than 90%. Fire-triggered clouds having very intense updrafts such as in some model calculations may exhibit precipitation efficiencies that are very low, similar to supercell thunderstorms (~10%). Whereas if cloud updrafts are weaker, the longer time available in the updrafts favors higher precipitation efficiencies, probably closer to 70-80%. Because the bulk of the fire-producing clouds are probably in the weaker updraft categories, precipitation efficiencies are probably on the higher side; greater than 50%.

In addition to nucleation scavenging, some remaining particles are directly scavenged by cloud droplets and raindrops. The small aerosol particles that are most likely to attenuate sunlight appreciably are susceptible to scavenging by Brownian diffusion and phoretic scavenging (see Pruppacher and Klett, 1978; Cotton and Anthes, 1989). These processes are relatively slow and probably account for less than 10% of the total removal of small particles. Again, once embedded in droplets, their ultimate fate is determined by the precipitation efficiency of the clouds. Larger aerosol particles are readily scavenged by hydrodynamic capture (see Pruppacher and Klett, 1978) by raindrops. The very large (greater than a few tens of micrometers) non-hygroscopic smoke particles will be capable of colliding with cloud droplets or raindrops thereby becoming large drops and then being rained out as they participate in the precipitation process. Such large particles probably contribute to the black rains seen in major urban fires following nuclear bombardment (National Academy of Science, 1975; MacCracken and Chang, 1975; Whitten et al., 1975), and such black rains were also seen in Basra, Iran, downwind of the oil fires in Kuwait according to news reports. TTAPS originally estimated that 25 to 50% of the bulk smoke mass in a large-scale war scenario would be scavenged immediately and subsequently Turco et al. (1990) downgraded this estimate to 10 to 25%. In our opinion, the original TTAPS estimate is probably an underestimate, with 50-60% being more likely.

10.2.5 Water injection and mesoscale responses

As seen in the above discussion, major nuclear-triggered fires would not only produce the vertical transport of smoke but also the vertical transport of water substance. The source of the water substance transported upward over the fires arises directly from the combustion process and from the low-level convergence of air feeding the fires. Of the fuel that is burned, particularly wood fuels, roughly 10% of the burned mass is water vapor, whereas the amount of submicron-sized smoke that can effectively absorb sunlight comprise less than 1% of the smoke mass (e.g., Cotton, 1985). The combustion process, however, is not the major source of water transported aloft by fires. Feeding the fires and the rising hot gases is low-level convergence of ambient air. In the summer months (which we shall show is the crucial period for major climatic consequences) surface mixing ratios of water vapor range from 15-20 g/kg in the semi-tropics and tropical coastal areas, to as low as 8-10 g/kg in the continental interiors. Even in the driest desert regions, in summer, surface vapor mixing ratios rarely go below 5 g/kg. As the air converging in the fires rises in the buoyant updrafts, it mixes with the moist, fire exhaust-product air as well as drier ambient air at higher levels. Moreover, the rising air column cools nearly adiabatically raising the relative humidity to over 100% where a cloud forms. Eventually some moisture is lost by being precipitated out of the cloud and as a consequence of mixing with drier, high-level air. Nonetheless, the air that is detrained out of the rising fire plumes in the middle and upper troposphere is considerably more moist than the surrounding environmental air.

Ambient water vapor mixing ratios at heights of 8 to 10 km in the atmosphere are typically less than 1 g/kg and so also are saturation values, so that only small additions of water vapor are needed to produce persistent clouds (see discussion in Section 4.2 on contrails).

Turco et al. (1990) conclude that the total mass of water injected over a hemisphere would perturb ambient water vapor concentrations by at most a few percent and therefore have little effect. They argue that the clouds formed by the fire injections would eventually be warmed by smoke injections and thereby be dissipated. We must remember, however, that the smoke and water substance are mutually injected into the atmosphere and with the exception of precipitation removal, they will remain co-located in the atmosphere for an extensive period. Moreover, water vapor in the upper atmosphere is a trace gas that requires only small additions to become saturated to form clouds. An example are jet contrails, which as we saw in Chapter 4, can lead to persistent cloud cover. The moisture released by jet contrails are many orders of magnitudes less than the expected bulk moisture amounts injected into the middle and upper troposphere by nuclear bomb produced fire plumes. As a result we anticipate that associated with the widespread smoke release would be persistent cirrus and altostratus cloud decks which will substantially alter the radiation balance. We examine its potential consequences next.

Several investigators have begun preliminary investigations of the mesoscale responses to smoke and moisture emissions by intense urban fires (Golding et al., 1986; Giorgi and Visconti, 1989; Cotton et al., 1992). Depending on the amount of moisture emitted with the smoke, middle and upper tropospheric clouds formed in their simulations over broad areas (patches greater than 50 or more kilometers) either in the smoke core or at the peripheries of the smoke plumes.

Cotton et al. (1992), for example, examined the mesoscale responses to just a dry smoke plume introduced into the upper troposphere. The emitted smoke amounts were based on the amounts of smoke detrained from a single simulated smoke plume using a three-dimensional cloud model (e.g., Tripoli, 1986), but no moisture was added to the atmosphere by the fires. They found that the model responded to solar heating of the smoke plume by causing lofting of the plume and a sea breeze-like circulation in the upper troposphere at the boundaries of the plume. This sea breeze-like circulation caused the formation of a cirrus-like ice cloud in the rising air entering the base of the smoke plume whenever the ambient relative humidity exceeded 70% relative to water. Thus, in spite of the fact that the heated air in the smoke plume lowered the relative humidity, the ascent of unheated air by the sea breeze-like, solenoidal circulation triggered ice and liquid cloud formation. Moreover the ice cloud persisted owing to the lower saturation vapor pressures with respect to ice relative to liquid water.

They calculated that the persistent ice cloud absorbed the upwelling terrestrial radiation, causing a greenhouse warming effect, particularly at night. Thus while a simulation with a dry smoke cloud (i.e., no cloud was allowed to form in the rising air column) produced average surface temperatures 6 K cooler than the no-smoke simulation after 24 hours, the case with an ice cloud was only 1.4 K cooler than the no-smoke control experiment. This illustrates the potential moderating influence of cloud formation associated with ascent of a smoke plume in the middle and upper troposphere. Because the smoke plumes that rise into the middle and upper troposphere would bring with them large quantities of water substance (relative to what naturally resides at those levels, but not necessarily in absolute values relative to surface moisture values), the moisture and mesoscale circulations associated with the smoke plumes would likely produce extensive stratiform cloud cover which could persist for periods of weeks or more in direct association with the smoke plume. The moderating influence of the fire-induced cloud would delay the onset of strong surface cooling during the phase that the smoke cloud is most concentrated. If the fire-induced cloud is sufficiently optically thick, it could reflect so much solar radiation that it would create a positive feedback to the smoke-induced cooling. Eventually much of the liquid and ice cloud will precipitate out of the atmosphere (faster than the sedimentation of the smoke), but still unanswered is will the persistent cloud be present long enough to allow diffusion of the smoke concentrations to such small magnitudes that only small surface cooling will occur?

In the absence of any cloud responses Haberle et al. (1985), for example, predicted that the solar heated smoke would be lofted into the stratosphere where it can reside for periods of months or more. We shall show that general circulation models simulated similar responses in the absence of any cloud feedbacks. Such behavior of cloud lofting to higher levels, however, did not occur with the Kuwaiti oil fire plumes, even in clear skies, as a result of the strong thermodynamic stability in the midtroposphere. The presence of water clouds provides an additional interaction with the smoke. Clearly, an assessment of cloud feedbacks is important to understanding the ultimate climatic responses to large-scale nuclear warfare. We shall see that a better understanding of cloud feedbacks is important to evaluating any potential impacts of human behavior on climate.

10.2.6 Other mesoscale responses

Atmospheric responses to smoke and water-substance injections as a consequence to nuclear warfare on scales of a few hundred kilometers to a few thousand kilometers (the mesoscale) is important in determining the ultimate fate of the the smoke and soot. Global climate models begin with a uniform layer of smoke introduced into the middle and upper troposphere and lower stratosphere. The actual smoke injections initially would be in discrete plumes originating from the fires over individual targets. Will the plumes remain in relatively narrow corridors and not become widely distributed as is assumed in the larger-scale models? Pielke and Uliasz (1992) have shown that dispersion is substantially enhanced due to spatial variations in surface heating as a result of temporal and spatial variations of turbulence and the generation of coherent mesoscale circulations. A similar response would be expected to occur associated with mesoscale patches of heating within the atmosphere. Even without mesoscale forcing, the results of McNider et al. (1988) suggest that vertical shear of the horizontal wind in the free atmosphere will significantly enhance dispersion from what is normally simulated in larger-scale models. If they do merge and diffuse into a relatively uniform pall, will this dispersion take so long that the smoke concentrations have diminished below levels for any significant solar heating responses?

It is possible that the consequences of such large-scale warfare would result in only mesoscale and regional responses with no longer-term global effects. One can speculate on a number of other potential mesoscale responses to smoke and water substance emissions. For example, patchy surface heating related to attenuation of solar energy by the smoke and cloud plumes can generate low-level, sea breeze-like solenoidal circulations that can trigger deep convective clouds and rainfall. In fact the triggered clouds could not only produce severe weather but penetrate into the elevated plume, scavenging the smoke plumes that have spawned them.

The patchy surface heating can modify natural physiographically-driven circulations (i.e., coastal sea breezes, mountain slope flows, circulations driven by differential surface heating such as described in Part II), strengthening them in some cases, and weakening in others. The patchy surface heating can also alter weak synoptic scale fronts causing those beneath the smoke/cloud cover to penetrate further southward during the daytime and possibly less southward at night.

These are just a few examples of potential mesoscale and regional scale responses to smoke and water-substance injections by large-scale nuclear warfare.

10.2.7 Global climatic responses

The Acute Phase

The estimates of the global climatic responses to smoke injections during large-scale nuclear warfare are normally considered in two stages: an *acute stage* lasting a month or so and a *chronic stage* lasting several months or more. The global models used to calculate potential climatic impacts normally begin with the smoke being uniformly distributed through the northern hemisphere with a specified vertical distribution, and a concentration that varies with the particular war scenario. The first model calculations were done by TTAPS with a one-dimensional radiative-convective model, followed by the application of longitudinally-averaged two-dimensional models (e.g., Cess et al., 1985), and a number of GCMs of varying complexity.

The original TTAPS one-dimensional model calculations suggested that maximum summertime decreases in surface temperatures would be as large as 35° C. The summer months are believed to experience the greatest cooling because too little radiation is present in the winter months to be appreciably diminished in intensity by the presence of the smoke. In addition, during summer the reduction of photosynthetically active radiation by smoke for a sufficiently long period of time below the threshold required by vegetation to sustain their metabolic processes can result in the death of the plants. Sagan (1983) speculated that if the smoke concentrations exceeded a critical level, catastrophic changes in climate would occur. This speculation has not been supported by the more sophisticated model calculations. As more realistic physics was added to the models, the magnitude of predicted cooling diminished appreciably from the original TTAPS estimates. Some of the more important improvements included the allowance for smoke to be transported vertically and horizontally, the inclusion of vertical heat transports from ocean surfaces, and better estimates of scavenging and removal. Vertical transport of the heated smoke, for example, created a lofted stable plume of smoke that formed a very stable layer of air aloft much like the natural stratosphere but with its base much lower at ~5 km (Malone et al., 1986). The strength of that simulated response is only realistic if middle and high-level clouds do not form in association with the lofted layer.

The inclusion of vertical heat transports from the ocean surface plays a major role in moderating the strength of surface cooling particularly in coastal areas (Schneider and Thompson, 1988). For the NRC baseline smoke injection scenario, they calculated maximum summertime, northern hemisphere, land surface temperature changes of 5-15° C, or less than half the original TTAPS estimates. Some short-term sporadic cooling events were still evident in their single realization. Because of the moderation in their simulated responses to the introduction of the smoke, they referred to the phenomena as *nuclear fall* rather than 'nuclear winter'. Figure 10.1 illustrates that the amplitude of predicted temperature changes for baseline war scenarios has diminished appreciably over the seven short years of nuclear winter research.

Several of the models simulate dramatic changes in precipitation in tropical latitudes during the acute phase (Gahn et al., 1988). Sharp drops in precipitation are associated with weakening of major tropical circulation features such as the Asian monsoon. The actual magnitudes of the reduction in rainfall should not be taken too seriously, however, because rainfall predictions with GCMs are not particularly reliable.

The Chronic Phase

The chronic phase or effects of smoke injections by large-scale nuclear warfare on time scales greater than a month are particularly challenging to simulate. Reliable simulations require a coupled ocean model that can simulate changes in sea surface temperatures and associated vertical heat fluxes (e.g., Robock, 1984), as well as changes in meridional and zonal heat transports by ocean circulations. Interactions with sea ice formation and melting, as well as changes in albedo of snow and ice fields caused by soot fallout should also be considered (e.g., Warren and Wiscombe, 1985; Ledley and Thompson, 1986; Vogelmann et al., 1988). Moreover, cloud feedbacks are also important. Not only are the direct responses of clouds important, such as discussed above, but so are more subtle cloud interactions related to changes in land and sea surface temperatures, and changes in important general circulation patterns (i.e., monsoonal circulations, large-scale ridge/trough patterns). Fundamental to simulating long-term responses is the prediction of the horizontal and vertical distribution of smoke and its concentrations, and the removal of smoke by scavenging and sedimentation.

Long-term survival of smoke in the atmosphere requires that the smoke be injected into the upper troposphere and lower stratosphere where residence times can be on the order of several months to as long as a year or more, respectively. The creation of a very stable, upper troposphere in the heated, lofted smoke layer which Schneider and Thompson (1988) called a *smokeosphere* would greatly extend the survival of smoke into a chronic phase.

Overall, the models used thus far for simulating the longer-term effects of smoke injections by large-scale nuclear warfare are too crude to be considered reliable. We must, therefore, wait for a new generation of GCMs to be implemented to examine those potential consequences quantitatively.

10.2.8 Biological effects

A comprehensive assessment of the ecological and agricultural effects of nuclear war has been done by the SCOPE team (Harwell and Hutchinson, 1985). They concluded that agricultural systems, in particular, are very sensitive to even small changes in temperature, photosynthetically active radiation, and rainfall. They conclude that they are so sensitive that many of the unresolved climatic issues we have discussed above are less relevant, since even lower estimates of many effects (i.e., on temperatures, rainfall, and sunshine) could be devastating to agricultural production and thereby to human populations on regional and global scales.

In our opinion the above conclusion represents a rather naïve perception of what is the level of uncertainty of estimates of climatic responses to large-scale nuclear war. There is little question that the level of uncertainty is very great indeed with respect to longer-term chronic impacts, which are important to biological effects. At this point one cannot be certain that the potential anomalies in temperature, precipitation, and sunshine triggered by smoke and soot released by large-scale nuclear war, for example, fall outside the envelope of expected *natural variability* of those parameters that agricultural systems must cope with and have coped with in this century. There is little question that just a naturally poor growing season following a major war would compound the already stressed food distribution system due to disruption of transportation and limited availability of fuels, fertilizers, etc. needed for agricultural productivity. Nuclear fall effects in such a poor growing season would further compound the problem. On the other hand, if a good growing season followed a devastating war, the possibly minor smoke-induced climatic effects would have little impact (although agriculture and the remainder of the biosphere, including people, would have to cope with the accumulation of a variety of harmful radionuclides on the earth's surface).

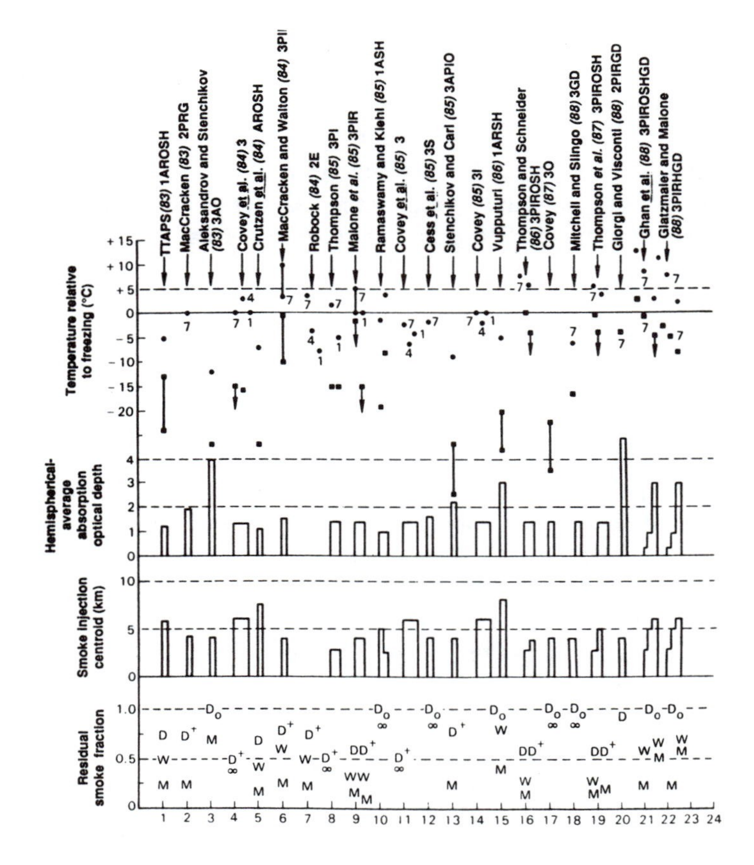

Figure 10.1:

Figure 10.1: *Summary of nuclear winter climate model calculations. Data shown for the following: (i) (•) Average land temperatures (coastal plus inland) in regions beneath widespread smoke layers for the coldest 1- to 2-week period in the simulation (some of the references report only temperature changes; the absolute temperatures have been deduced by subtracting the computed average temperature decrease from the temperature offset given below for each season); the month of the simulation is indicated by a numeral; for the 1-D radiative and convective models, the average land temperature decreases are taken as one-half of the "all-land" temperature decreases to account for the effect of ocean moderation; annual average solar insolation also applies in these cases. (ii) (■) Minimum land temperatures beneath smoke during the acute phase of nuclear winter simulations (again, where necessary, absolute temperatures were obtained by subtracting decreases from offsets); these temperatures are averaged over at least 1 day. (iii) Hemispheric average absorption optical depth of the smoke injection. (iv) Height centroid of the smoke mass injection. (v) Residual smoke fractions at several times in each simulation. The selected calculations roughly correspond to recommended "baseline" smoke injection scenarios; less severe and more severe cases have been investigated, but not as frequently as baseline cases. The studies have been ordered from left to right roughly in chronological sequence and are numbered along the bottom of the figure. For a given study, several cases may be illustrated (for example, the Ramaswamy and Kiehl (1985) results are shown for two smoke-injection profiles.) The data are organized vertically for each simulation. The temperature offset used: 0° C, winter; 13° C annual, fall, and spring; 25° C, summer; and 35° C for cases calculated by Lawrence Livermore National Laboratories. The model treatments: 1, 2, and 3 indicate dimensions; A, annual solar insolation; P, patchy smoke injection; I, interactive transport; R, removal by precipitation, O, optical properties evolve, S, scattering included; H, infrared-active smoke; E, energy balance; G, ground heat capacity; D, diurnal variation; and M, mesoscale (48 hours). Smoke removal: D, after 1 day (prompt removal); D_o, arbitrary initial injection; W, after 1 week; M, after 1 month; $^+$, assumption implicit in smoke scenario adopted; and ∞, no smoke removal after injection.* [From Turco, R.P., O.B. Toon, T.P. Ackerman, J.B. Pollack, C. Sagan, 1990: Climate and smoke: An appraisal of nuclear winter. *Science*, **247**, 166-176. ©American Association for the Advancement of Science.]

Even during the acute phase, the uncertainty of smoke-induced impacts on weather and climate are so great that SCOPE's conclusion is a bit of an overstatement. In our opinion, the GCM experiments by Schneider and Thompson (1988) represent the most useful simulations for examining potential impacts of smoke-induced changes on agriculture and ecosystems. They emphasized the importance of considering geographical and weather variability rather than just time and spatial averages of temperature and other parameters. They noted, for example, that if temperatures fell below a critical threshold for only a few hours or so, crop production could be severely impacted (i.e., subfreezing temperatures for wheat, or temperatures as cool as 15° C during the flowering phase of rice). They examined temperature extremes simulated with the National Center for Atmospheric Research (NCAR) Community Climate Model (CCM) by determining the coldest temperatures reached during a 30-day July control simulation and a smoke-perturbed, baseline war scenario. They found that the regions of subfreezing temperatures that are normally confined to polar latitudes and high mountains, expanded in the smoke-perturbed simulation. They also found considerable regional variability in the simulated responses to smoke. The smoke-perturbed case, for example, exhibited more frequent cooler temperatures over the midwestern United States, but no subfreezing temperatures. They reported some probability of subfreezing temperatures over the Ukraine and some temperatures over China that were low enough to impact rice growing in the smoke perturbed cases.

Let us ignore, for the moment, the uncertainties in the GCM simulation of the climatic response to smoke perturbations due to the specification of the height and concentration of smoke, the dispersion of the smoke, the impact of clouds on scavenging of smoke and interference with smoke-induced radiative anomalies, and the fact that Schneider and Thompson's simulation did not include a diurnal cycle. Instead let us focus on the *natural variability* question again. It is important to ask where do the smoke-induced perturbations stand relative to the *natural variability* that can be expected in July over a given region? Again if the war-produced smoke is to have any climatic or biological impacts of significance, it should produce anomalies greater than the expected *natural variability* over a region. This cannot be addressed from a single GCM realization. To actually define both natural and smoke-perturbed regional variability, a GCM must be run over a number of realizations each of which is initialized by a slightly perturbed initial state. A measure of the validity of the model would be to determine how well the model represents actually observed *natural variability.* Then one could examine the variability of the smoke-perturbed simulations relative to the *natural variability* to determine if the potential smoke-induced changes are substantially different from what could be expected naturally.

As it is, these single realization GCM runs represent only a first step sensitivity experiment showing plausible physical responses to smoke. They should not be taken too literally with respect to biological impacts.

10.3 Summary of the Status of the Nuclear Winter Hypothesis

The nuclear winter hypothesis has been examined mainly with the use of models of varying complexities. There have been some attempts to examine the hypothesis relative to analogs such as meteor impacts and volcanic emissions, but both of these processes involve the deposition of large amounts of smoke and debris in the stratosphere where the expected residence times are considerably longer than in the troposphere. Measurements of smoke emissions from natural wildland fires and from industrial fires have also been made, but these are much smaller than anticipated for nuclear winter with the bulk of the smoke being confined to lower levels in the atmosphere than expected from large urban firestorms. As a result of the almost total reliance on models which have numerous shortcomings, the nuclear winter hypothesis is a long way from being proven scientifically viable. Hopefully, we will never have the opportunity to test the hypothesis experimentally.

One virtue of the hypothesis is that it has triggered many refinements in climate models, particularly in terms of the introduction of aerosol physics and refined radiation physics. It has also brought to the forefront a realization of the many potential ecological consequences of climate change and nuclear war.

Chapter 11

The Greenhouse Gas Theory

11.1 Introduction

The hypothesis that has received much attention from the news media, the general population, politicians, and the scientific community in recent years is that anthropogenic emissions of greenhouse gases such as carbon dioxide are causing a rise in global average surface temperatures. The level of public awareness in the greenhouse concept far exceeds that of cloud seeding at its peak era of publicity. We will see there are many aspects of the greenhouse hypothesis that bear resemblance to cloud seeding during its heyday. Some scientists are clamoring that greenhouse warming is definitely occurring now and that we should get on with the business of curtailing production of greenhouse gases. The challenge is to prove that greenhouse warming produces a signal that can be discerned from the *natural variability* of climate change; a difficult challenge since quality data records are available for only a short time relative to the time scales of natural climatic variability. The number of scientists that are willing to challenge the evidence are fewer than for cloud seeding, probably because most see that only good can come from attempts to reduce greenhouse gas emissions, whereas seeding clouds is viewed to be another example of human intervention in the natural way of things.

In this chapter we will review the basic concepts and examine the evidence that greenhouse gas concentrations are increasing, the theoretical/modeling calculations suggesting the potential impact of the increasing concentration of greenhouse gases and the uncertainties in those estimates, the evidence that the global mean temperature is increasing as well as other 'fingerprints' of climatic change, and the indications for potential regional impacts of global climate change.

11.2 Basic Concepts and Greenhouse Gas Concentrations

The basic greenhouse concept was probably first proposed by Fourier (1827). Later, Arrhenius (1896) calculated the effects of varying CO_2 concentrations on surface temperatures and even included the impact of water vapor absorption in his calculations. He estimated that a 2.5- to 3-fold increase in CO_2 would increase temperatures in Arctic regions by 8 to 9° C. His intent was to examine the impact of greenhouse gases on ice ages and glaciation and was not particularly concerned about human sources. As early as 1938, Callendar (1938) became concerned about the impacts of anthropogenic emissions of CO_2 on global temperatures. Therefore, the basic concept that anthropogenic sources of CO_2 and other greenhouse gases can warm the atmosphere is not new, but recent evidence suggesting that global average temperatures are rising has caused a great deal of concern about potential human impact on climate, and resulted in major research activities in global climate change.

As noted in Chapter 8, in the absence of so-called greenhouse gases, the average surface temperature of the earth would be over 30° C cooler than it is today. This is because the greenhouse gases absorb upwelling terrestrial radiation and radiate the absorbed gases, both upward and downward, causing a net gain in energy (reduced loss) at the earth's surface. Without those greenhouse gases, a larger fraction of the upwelling terrestrial radiation would escape through the atmospheric window to space. The major greenhouse gas is water vapor which varies naturally in space and time due the earth's hydrological cycle (e.g., see Randall and Tjemkes, 1991). The second most important greenhouse gas is carbon dioxide (CO_2). In contrast to water vapor, it is rather uniformly distributed throughout the troposphere.

11.2.1 Carbon dioxide

As noted earlier, concern about the potential impact of industrial and other anthropogenic emissions of CO_2 on the net radiation budget began early in this century. It was not until 1958 that regular measurements of CO_2 were taken at a few locations (e.g., Keeling et al., 1982), so that we have only a period of 33 years for direct assessment of the changing concentration of CO_2. An indirect assessment of CO_2 concentrations can also be made by examining the concentrations of CO_2 and other gases trapped in air-tight bubbles in glacial ice (Raynaud and Barnola, 1985). Estimates of pre-industrial concentrations using the trapped bubble technique range from 265 to 290 ppm (Neftel et al., 1985b), compared with 1958 concentrations of about 315 ppm or a 10 to 20% rise. Some of that rise can be attributed to deforestation and natural variations, but much of it is probably caused by burning of carboniferous fuels such as coal, gas, and oil. Evidence that CO_2 has undergone substantial natural variations can be seen in Figure 11.1. Over the past 20,000 years, CO_2 has

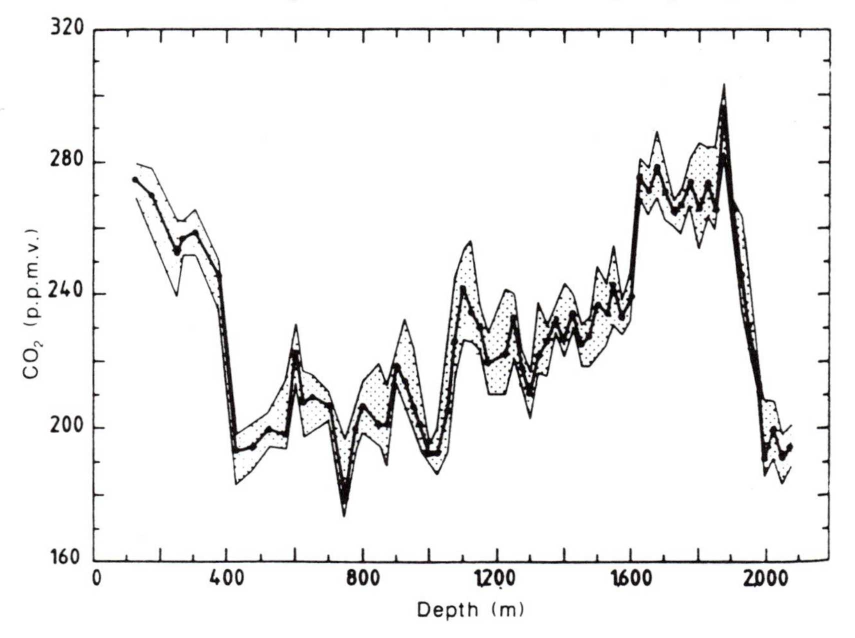

Figure 11.1: *CO_2 concentrations (p.p.m.v.) plotted against depth in the Vostok ice core. The 'best estimates' of the CO_2 concentrations are indicated by dots and the envelope shown has been plotted taking into account the different uncertainty sources.* [From Barnola, J.M., D. Raynaud, Y.S. Korotkevich and C. Lorius, 1987: Vostok ice core provides 160,000-year record of atmospheric CO_2. Reprinted by permission from *Nature*, **329**, 408-414. ©MacMillan Magazines Limited.]

been estimated to vary by as much as 100 ppm, having a strong correlation with polar temperatures. This does not mean that CO_2 variations caused these temperature changes through the greenhouse effect. It is quite probable that the CO_2 changes were linked to large-scale changes in the ocean circulations (which altered polar temperatures) and to associated changes in biological, chemical, and physical processes regulating the CO_2 budget. The CO_2 variations probably played a minor feedback role in determining glacial and inter-glacial changes (Broccoli and Manabe, 1987).

Since 1958, direct measurements suggest that CO_2 has increased steadily (see Figure 11.2) from 315 ppm to about 350 ppm by 1988 or by $\sim$1% per year. The variation of 5 ppm within a year is due to CO_2 uptake and release associated with photosynthesis, primarily in the northern hemisphere. After 1980 the fossil fuel source of CO_2 has remained

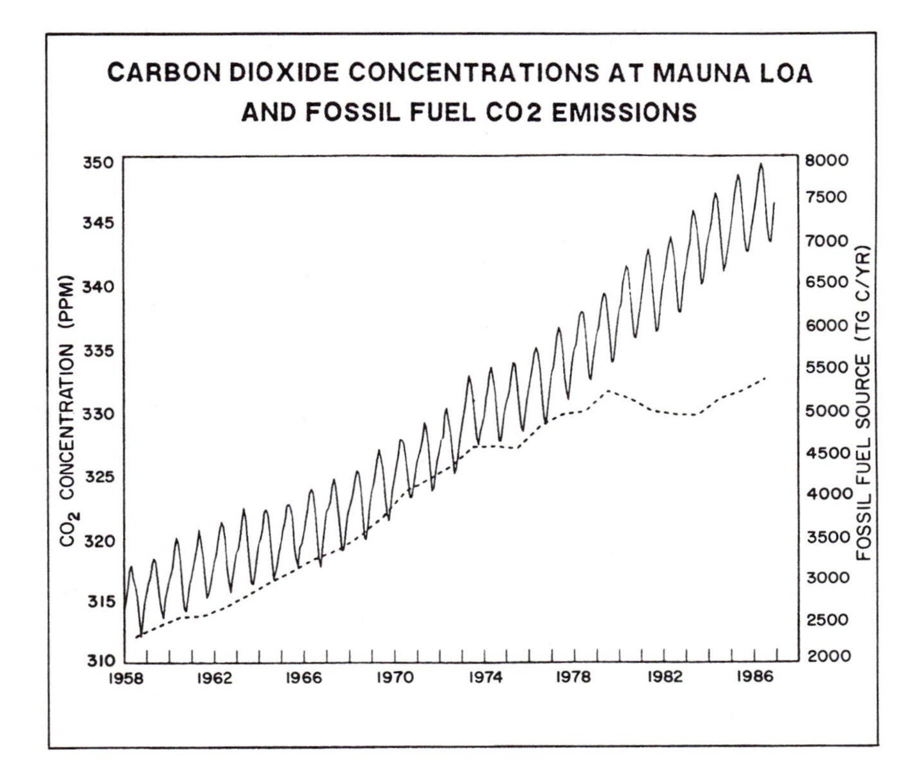

Figure 11.2: *The solid line depicts monthly concentrations of atmospheric CO_2 at Mauna Loa Observatory, Hawaii. The yearly oscillation is explained mainly by the annual cycle of photosynthesis and respiration of plants in the northern hemisphere. The steadily increasing concentration of atmospheric CO_2 at Mauna Loa since the 1950s is caused primarily by the CO_2 inputs from fossil fuel combustion (dashed line). Note that CO_2 concentrations have continued to increase since 1979, despite relatively constant emissions; this is because emissions have remained substantially larger than net removal, which is primarily by ocean uptake.* [Scheraga, Joel and Irving Mintzer, 1990: Introduction. From *Policy Options for Stabilizing Global Climate,* D.A. Lashof and D.A. Tirpak, Eds. U.S. Environmental Protection Agency, Office of Policy, Planning and Evaluation. Hemisphere Publishing Corp. New York.]

relatively steady. The further rise in CO_2 concentrations after 1980 is believed to indicate that the sources of CO_2 exceed its sinks (primarily the oceans). Note that the increase in CO_2 in the atmosphere is only 58% of the estimated amount of CO_2 released by burning of fossil fuels. The remainder is believed to have been taken up by the oceans. Other researchers (e.g., DAR, 1990) conclude that the oceans may be absorbing less of the anthropogenic CO_2 emissions than is generally assumed. Even with the loss of forests due to tropical deforestation, the biosphere appears to be a very effective sink of CO_2, particularly in northern mid-latitudes. Overall there remain many gaps in our knowledge of the total budget of CO_2 so that precise budgeting of sources and sinks of CO_2 are not currently possible. Major uncertainties lie with both the rate of exchanges with deep ocean layers due to ocean circulations and biological processes (Baes et al., 1985).

Nonetheless, the evidence is strong that the bulk of the increase in CO_2 concentrations in the last century is due to anthropogenic sources. First of all, the natural variability of CO_2 over 100 year periods as deduced from ice core bubbles is only 10 ppm while the rise in CO_2 is about an order of magnitude greater. A value as large as the current concentration does not exist in the ice core history over the last 160,000 years (Barnola et al., 1987) although variations from 180 ppm to over 300 ppm were measured over periods of thousands of years. Second, the increase in CO_2 concentrations roughly parallels the rise of industrial emissions of CO_2. Third, the observed isotopic trends of C^{13} and C^{14} agree with those expected from fossil fuel and biosphere emissions of CO_2 (Houghton et al., 1990). While the bulk of the anthropogenic sources of CO_2 are related to burning of fossil fuels which is concentrated in the industrialized countries, another important anthropogenic contribution to CO_2 is the cutting down of the earth's forests. This source is shared by underdeveloped and developed countries alike. While CO_2 makes the largest contribution to anthropogenic-related greenhouse warming, other trace gases also contribute to the problem.

11.2.2 Methane

Methane is the most abundant hydrocarbon in the atmosphere and its concentration is currently increasing at about 1% per year. Based on trapped gas analysis of ice cores, Craig and Chou (1982) and Graedel and Crutzen (1989) have estimated that methane (CH_4) has increased from about 0.7 ppm to around 1.7 ppm over the last 300–400 years, with the greatest contribution to the increase occurring in the last 100 years. Between 1986 and 1990, methane increased at an average annual rate of 0.8%. It is estimated that methane contributes about 20% to the current increase in the greenhouse effect. Principle sources of methane seem to be enteric fermentation in ruminant animals, release from organic-rich sediments beneath shallow water bodies and rice paddies, biomass burning, fossil fuel production, and landfills. Figure 11.3 illustrates estimates of the various natural and human-related contributions to tropospheric methane. The human contribution is expected to grow. In south and southeast Asia, for example, areas cultivated for rice are expected to increase from the 877,120 km^2 in 1980 to 1,245,990 km^2 in 2000 (Global Change, 1990).

11.2.3 Nitrous oxide

Nitrous oxide (N_2O) is the most important nitrogen containing greenhouse gas. Estimates of its concentration is about 300 ppb (Note: parts per billion compared to parts per million for CO_2 and CH_4) and that it has increased by 5-10% since pre-industrial times and is currently increasing at a rate of about 0.25% per year.

11.2.4 Chlorofluorocarbons

Chlorofluorocarbons or CFCs came into major use in the 1960s and increased rapidly in concentrations (10-15%/yr), but since the mid-1970s their concentrations have actually declined somewhat (Chemical Manufacturers Association, 1983). The most abundant CFC's are CF_2CL_2, $CFCL_3$ which have concentrations of the order of 392 and 226 ppt (Note: parts per trillion!). Their importance lies with the fact that they mainly absorb terrestrial radiation in the atmospheric window (7 to 13 μm), that they are about 20,000 times more effective on a per molecule basis of absorbing terrestrial radiation than is CO_2, and the concentration of CO_2 is so large that many absorption bands are nearly saturated such that greenhouse warming varies logarithmically with concentration of CO_2, whereas it varies linearly with CFC concentrations (Shine, 1991).

11.2.5 Ozone

Ozone is another greenhouse gas which has a relatively short lifetime. Its concentration is affected by other gases such as carbon monoxide and nitrogen oxides (NOx), which contribute to the chemistry of the atmosphere and thereby affect greenhouse warming.

11.2.6 Summary

Figure 11.4 illustrates an estimate of the contribution of the various trace gases to greenhouse warming from 1980 to 1990. By the 1980's CO_2 contributed roughly 55% of the trace-gas greenhouse warming, followed by methane at 15%, CFC's at 24%, and nitrous oxide at 6% (Houghton et al., 1990). Keep in mind, however, the largest contributor to greenhouse

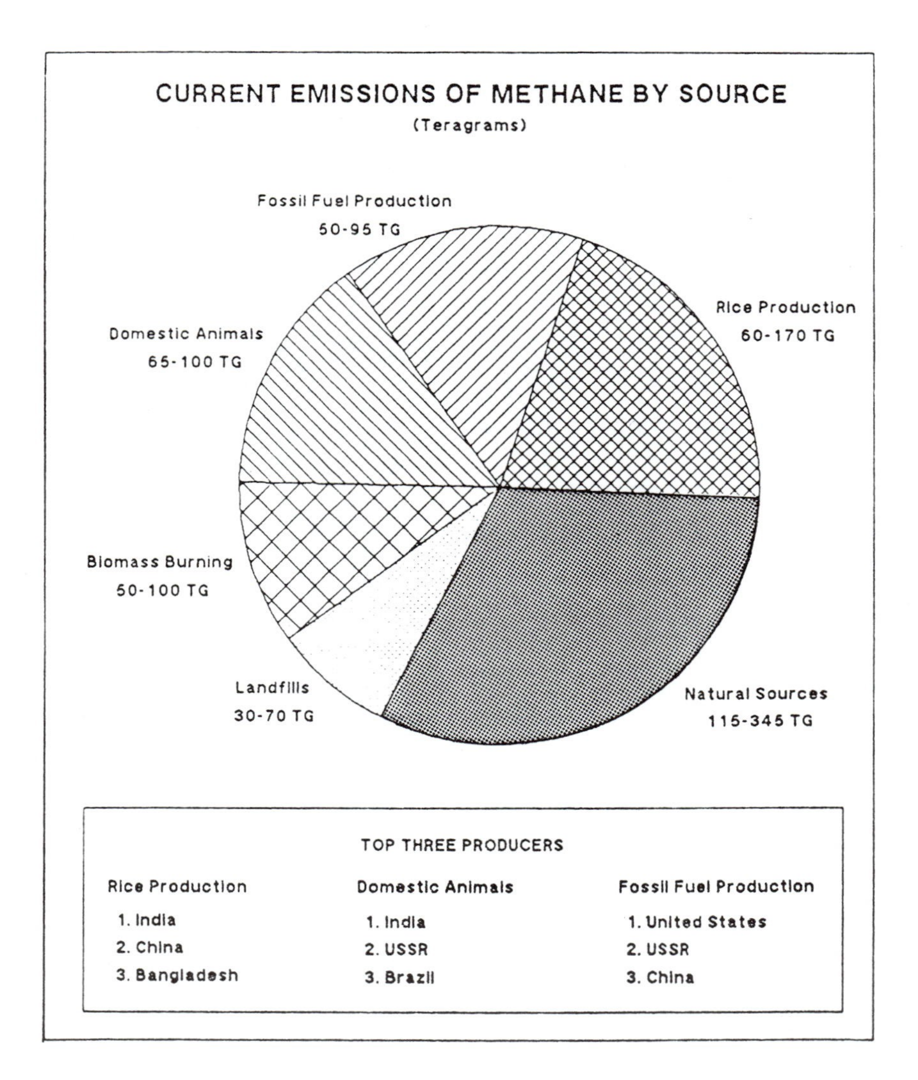

Figure 11.3: *Human activities in the agricultural sector (domestic animals, rice production and biomass burning) and the energy sector (fossil fuel production) are the major sources of atmospheric CH_4. Natural sources, from wetlands, oceans, and lakes, may contribute less than 25% of total emissions.* [Fung, I. and Michael Prather, 1990: Greenhouse Gas Trends. From *Policy Options for Stabilizing Global Climate.* D.A. Lashof and D.A. Tirpak, Eds. U.S. Environmental Protection Agency, Office of Policy, Planning and Evaluation. Hemisphere Publishing Corp. New York.]

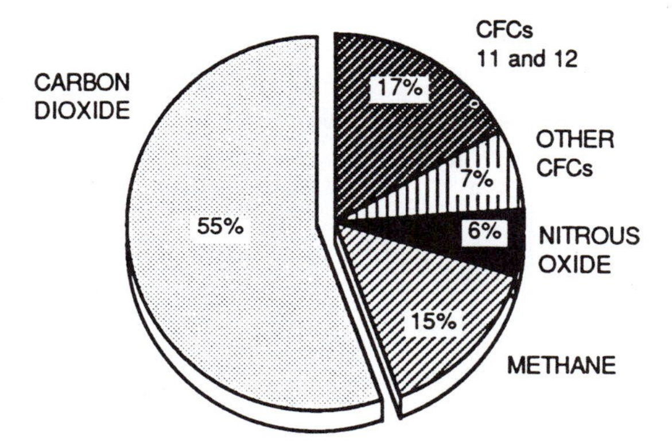

Figure 11.4: *The contribution from each of the human-made greenhouse gases to the change in radiative forcing from 1980 to 1990. The contribution from ozone may also be significant, but cannot be quantified at present.* [From *Climate Change. The IPCC Scientific Assessment*, 1990: J.T. Houghton, G.J. Jenkins, and J.J. Ephraums, Eds. Cambridge University Press, Cambridge, MA. ©Intergovernmental Panel on Climate Change.]

warming is the highly variable constituent, water vapor. Figure 11.5 illustrates the relative warming by the various greenhouse gases as a function of concentration. Note the different abscissas for the various chemical species. As noted previously, the contribution to greenhouse warming by CFCs varies linearly with increases in concentration of those chemicals. The contribution of CO_2 and CH_4 to greenhouse warming, on the other hand, vary roughly with the square root of the concentrations due to saturation of the absorption lines. Thus CFCs could readily catch up to the greenhouse warming effects of CO_2 if their concentrations continue to rise.

Recently, Hayden (1992) presented evidence that volatile biogenic hydrocarbons emitted from vegetation during the growing season may be an overlooked important greenhouse gas. He suggests that the effect of this trace gas in semi-arid and arid areas with vegetation on the reduction of cooling at night is from one to five times larger than the expected warming from a doubling of CO_2 concentrations. Minimum temperatures in these locations are as much as 25°C higher than they would otherwise be.

Projections of trace gas concentrations over the next century are quite variable. Trabalka (1985) estimates that fossil fuel consumption will result in a doubling of CO_2 in the atmosphere sometime during the 21st century. Most climate models use a doubling of CO_2 as an input to examine the climatic effects of greenhouse gases. In regard to strategies for

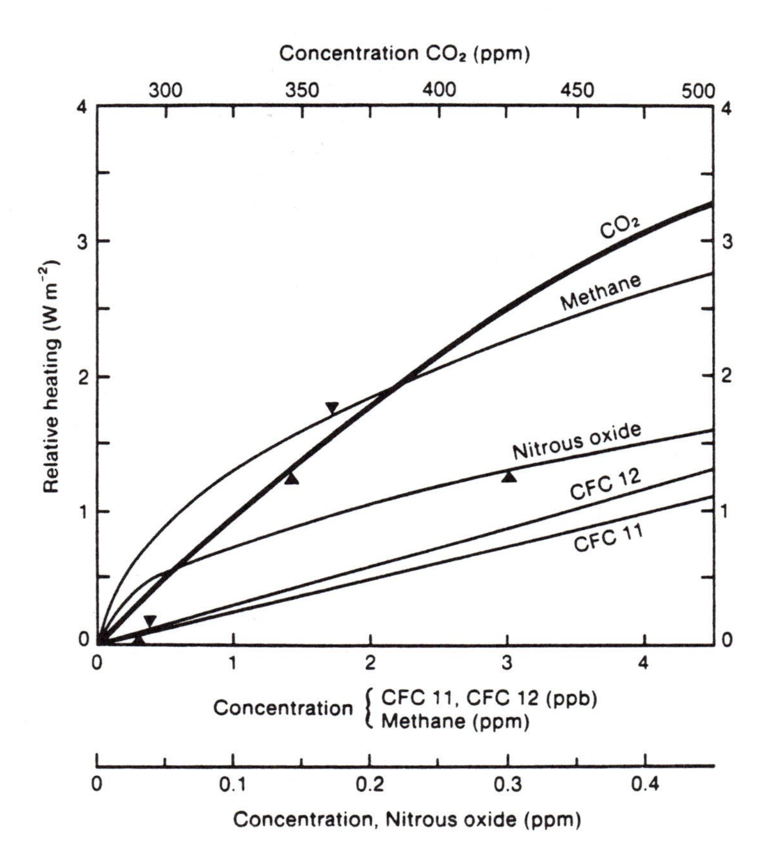

Figure 11.5: *Greenhouse heating due to trace gases (in watts per square meter), showing (top scale) concentration of* CO_2 *(in parts per million; note that the baseline value is 275 ppm and the curve shows the extra heating due to increases above that level. The current heating due to* CO_2 *is about 50 W* m^{-2}*), (middle scale) concentration of CFC11 and CFC12 (parts per billion) and of methane (parts per million), and (bottom scale) concentration of nitrous oxide (parts; per million). The triangles denote 1985 concentrations.* [From Mitchell, J.F.B., 1989: The "greenhouse" effect and climate change. *Rev. Geophys.*, **27**, 115-139. ©American Geophysical Union.]

reduction of anthropogenic emissions of greenhouse gases, it seems that a reduction of industrial sources of CO_2, CH_4, and CFCs is quite possible in the next century. *A number of sources of greenhouse gases, however, particularly methane and impacts of deforestation on the CO_2 budget, are directly related to human activities involving basic food and fiber production. Therefore as long as world population continues to rise at current rates, there will be no substantial reduction in greenhouse gas production. The most fundamental step to achieving a reduction of greenhouse gases, therefore, is getting the earth's population under control (i.e., zero population growth)!*

11.3 The Role of Feedback Processes

In order to estimate the climatic effects of increasing concentrations of greenhouse gases, a variety of climate models are used. They run the entire spectrum of models summarized in Chapter 8. Likewise the global average temperatures estimated vary quite widely as well. The simplest models used are the global energy balance models and the one-dimensional radiative convective models. Schlesinger and Mitchell (1987) summarized the predicted surface temperatures with the various radiative-convective models for a doubling of CO_2 concentrations. As shown in Table 11.1 the values of global-averaged surface temperatures are all positive and range from 0.48 to 4.2° C. The magnitudes depend on the nature of formulation of the radiative-convective models, especially the amplification and damping of the basic radiative responses to CO_2 increases by positive and negative feedback processes. Unfortunately many of the feedback processes cannot be handled adequately in a one-dimensional model. Therefore, three-dimensional GCMs are required to make quantitative estimates of the feedbacks. Some of the important feedbacks are due to: (1) water vapor, (2) surface albedo, (3) ocean circulations, (4) clouds, (5) biochemical processes, and (6) vegetation respiration and photosynthesis. In the next sections we will examine each of these feedbacks relative to the total problem of global warming.

11.3.1 Water vapor feedbacks

As noted previously, water vapor is the principle greenhouse gas. Therefore, any changes in water vapor concentration in response to warming by the other greenhouse gases would substantially alter the net greenhouse heating. As the atmosphere and the ocean warm in response to enhanced greenhouse warming, more water vapor evaporates from the ocean and land surfaces. The higher water vapor content of the atmosphere causes further greenhouse warming, which causes more evaporation. There are other possible feedbacks associated with higher moisture

Table 11.1: Surface Temperature Change Induced by a Doubled CO_2 Concentration as Calculated by Selected Radiative-Convective Models. [From Schlesinger, M.E., and J.F.B. Mitchell, 1987: Climate model simulations of the equilibrium climatic response to increased carbon dioxide. *Rev. Geophys.*, **25**, 760-798.]

Reference	$\Delta T_s^\circ C$
Manabe and Wetherald (1967)	1.33–2.92
Manabe (1971)	1.9
Augustsson and Ramanathan (1977)	1.98–3.2
Rowntree and Walker (1978)	0.78–2.76
Hunt and Wells (1979)	1.82–2.2
Wang and Stone (1980)	2.00–4.20
Charlock (1981)	1.58–2.25
Hansen et al. (1981)	1.22–3.5
Hummel and Kuhn (1981a)	0.79–1.94
Hummel and Kuhn (1981b)	0.8–1.2
Hummel and Reck (1981)	1.71–2.05
Hunt (1981)	0.69–1.82
Wang et al. (1981)	1.47–2.80
Hummel (1982)	1.29–1.83
Lindzen et al. (1982)	1.46–1.93
Lal and Ramanathan (1984)	1.8–2.4
Somerville and Remer (1984)	0.48–1.74

contents. Clouds can form, and as we will see in the following discussion, this can lead to both warming and cooling of the atmosphere through their interactions with radiation. Moreover, as water vapor condenses in clouds, the latent heat released further warms the atmosphere. This has led several one dimensional modelers to assume that the relative humidity of the atmosphere remains constant since both the temperature and moisture content of the atmosphere will rise in response to greenhouse warming (Möller, 1963; Manabe and Wetherald, 1967; Ramanathan, 1981). Simulation of the moisture feedback in GCMs requires realistic models of the surface hydrological budget (rainfall and evapotranspiration of land and ocean surfaces) and of the response of the ocean to heating of the atmosphere. Unfortunately, the responses of the various GCMs to simulated surface energy budgets are extremely diverse and the primary cause of that diversity is related to the formulations of their hydrological cycles (Randall et al., 1992). The fluxes of moisture at the ocean surface, for example, is a function of the ocean surface temperature as well as the surface wind stresses. Thus, ocean surface temperatures in tropical regions can warm appreciably in response to greenhouse warming, but in regions of weak winds, such as over the intertropical convergence zone, only weak vertical moisture fluxes will be experienced. Moreover, crucial to water vapor feedback is the vertical distribution of moisture into the middle and upper troposphere. This, in turn, is related not only to local convective transports, but also to the strength of regional circulations such as the Hadley and Walker circulations (Philander, 1990) both of which transport moisture over long distances horizontally and vertically. Thus, while the moisture feedback can amplify globally-averaged surface air warming by as much as a factor of three (Ramanathan, 1981), consideration of realistic cloud feedbacks and realistic ocean/atmospheric circulation responses requires more skill in the simulation of climate responses than is possible in current models.

11.3.2 Surface albedo feedbacks

Changes in surface snow and ice coverage is the most studied feedback to greenhouse warming, while other surface albedo feedbacks are related to changes in surface vegetation coverage. The so-called 'ice-albedo' feedback relates melting of sea ice and snow cover to greenhouse warming. Snow and ice reflect more solar radiation than open water and bare soil or soil covered by vegetation. As a result, a warming planet is expected to reduce sea ice cover appreciably, cause earlier seasonal melting of snow, and a retreat in glaciers. It is estimated that these effects will positively amplify greenhouse warming by 10 to 20% globally (Hansen et al., 1986; Lian and Cess, 1977), but it can have a 2- to 4-fold amplification in polar oceans and near the sea ice margins. Major uncertainties in modeling the ice-albedo feedback are related to distinguishing between the albedo of old snow and fresh snow, identifying albedo effects of thick versus thin sea ice, and in simulating the strong influence of ocean circulations on snow and sea ice cover. In addition, cloud cover can mask albedo changes associated with increases or decreases of snow and sea ice, and increases in precipitation at higher latitudes can increase snow cover even if warming occurs. There are indications, for example, that changes in cloud distribution associated with snow cover changes can produce a reversal of the sign of this feedback (Cess et al., 1991). Likewise, snow albedo changes over forested regions can be partially masked.

An even more complicated feedback in surface albedo is associated with changes in vegetation coverage. If the tundra-boreal forest boundary should shift poleward in response to a warming planet then this would cause a decrease in albedo and represent a positive feedback. If, on the other hand, deserts increase in semi-tropical areas due to reductions in rainfall, this could create a negative effect. Likewise, it is possible that enhanced CO_2 concentrations will serve as a fertilizer (King et al., 1985; Houghton, 1987; 1988; Idso, 1988) and result in increased biomass coverage causing a decrease in albedo and a positive feedback. The increase of biomass, however, would provide an enhanced atmospheric sink of carbon, at least until new biomass decays. Cooper (1988) has suggested that if the forest biomass is limited by leaf area index and moisture transport, then the biomass coverage may not increase with CO_2. In summary, because the surface vegetation feedback is strongly coupled to the entire hydrological response, which is the major uncertainty in geophysical feedback, it can have both a positive and negative amplification to the greenhouse signal with unknown amplitude.

11.3.3 Ocean feedbacks

Ocean feedbacks affect almost every aspect of climate change ranging from global responses to regional responses (e.g., see the sensitivity experiment reported in Randall et al. (1992), in which changes in latent heat flux from the oceans in 19 separate GCMs from a prescribed 4°C globally uniform sea surface warming resulted in drastically altered downwelling infrared radiation at the earth's surface, as well as dramatic differences between models). As mentioned previously, the amplitude of the moisture- and ice-albedo feedbacks are strongly influenced by the response of ocean circulations to a warming atmosphere. One of the most important effects of the ocean is the large amount of heat it is capable of storing. As a result, the time that it takes the atmosphere/ocean system to respond to greenhouse warming is largely controlled by the ocean response. If exchanges of heat with the deep ocean are small, then the upper mixed layer of the ocean which is 70 to 100 m deep, will respond to a warming atmosphere on the time scale of decades. If, on the other hand, the rate of exchange of heat with deep ocean layers is greater, it may require time scales on the order of a century or more for the upper levels of the ocean to respond appreciably to greenhouse warming. This will substantially delay the time-scale that greenhouse warming will be detectable in the earth/atmospheric system. Estimates of the rate of exchange of heat with deep ocean layers is made by examining the transfer of passive materials such as radioactive substances into deep ocean layers. The problem with this approach is that heat is not passive, it causes changes in ocean stratification which affects vertical heat transfer and the strength of the atmospheric winds which also drive heat exchanges with the surface and deeper ocean layers. Even the equilibrium responses of GCMs to an imposed doubling or quadrupling of CO_2 vary depending upon the way in which the GCMs are interfaced with the ocean. Thus the average surface temperature change to an imposed doubling of CO_2 varies from 0.2° C for a model with a non-interacting ocean, to 1.3 to 3.9° C for models with a swamp ocean, to 3.5 to 4.1° C with an ocean mixed layer (Schlesinger, 1983).

Not only is the ocean a large reservoir of heat, it is also a large reservoir of CO_2. The ocean contains as much as 50 times the amount of carbon as resides in the atmosphere. The simplest feedback is that as the ocean warms, the solubility of CO_2 decreases and more CO_2 is released from the ocean to the atmosphere creating a positive feedback. Kellogg (1983) has also discussed a number of complicated feedback loops involving ocean circulations and CO_2. One of these, illustrated in Figure 11.6, shows that if the atmospheric circulation weakens in tropical regions (as simulated by some GCMs) then upwelling of deep ocean water will be reduced. The reduced supply of nutrients from the deeper ocean layers will diminish photosynthesis which will reduce ocean uptake of CO_2. This would create a positive feedback to greenhouse warming. Another hypothesized chain of events, shown in Figure 11.7, is related to the rate of vertical exchange of sea water at high latitudes. If as suggested by some models (e.g., Manabe and Wetherald, 1980) the upper levels of the ocean warms at high latitudes in winter, then the stability of the ocean would increase and the vertical exchange of surface sea water with deeper ocean layers (downwelling) would diminish. Because the rate of uptake of CO_2 by the oceans is controlled by the rate of exchange of near surface water with deep ocean water, the reduced rate of downwelling of surface water in a warmed atmosphere would create a positive feedback in that CO_2 accumulation in the atmosphere would be enhanced.

Kellogg (1983) also summarized several potential ocean feedbacks that involve changes in biomass in the arctic and the tropics (Figures 11.8 and 11.9, respectively). He proposes that if the arctic pack ice disappears in the summer in response to a warmer ocean/atmosphere, this will create a more temperate and moist arctic climate near the ocean which would favor poleward spread of forests with their ability to uptake CO_2 and cause a negative feedback to greenhouse warming. He also suggests that a warmer earth will experience more rainfall and enhanced soil moisture in the subtropics, which will likewise increase biomass in the subtropics and create a negative feedback to greenhouse warming by reducing CO_2 contents.

Another possible feedback involving warming of the oceans is the DMS-albedo feedback that we discussed in Chapter 9. The hypothesis proposed by Charlson et al. (1987) is that DMS production will be enhanced over the warmer tropical and subtropical waters causing an increase in CCN and cloud albedo; a negative feedback. At this time this hypothesis is too speculative to estimate its relative importance to greenhouse warming. These are just a few examples of potential feedback loops associated with ocean responses to greenhouse warming. Some of the feedbacks are positive, some are negative and some are difficult to even assess a sign. As one can guess it is extremely difficult to place any quantitative value to any of these feedback loops associated with the CO_2 budget, which introduces considerable uncertainty in predicting

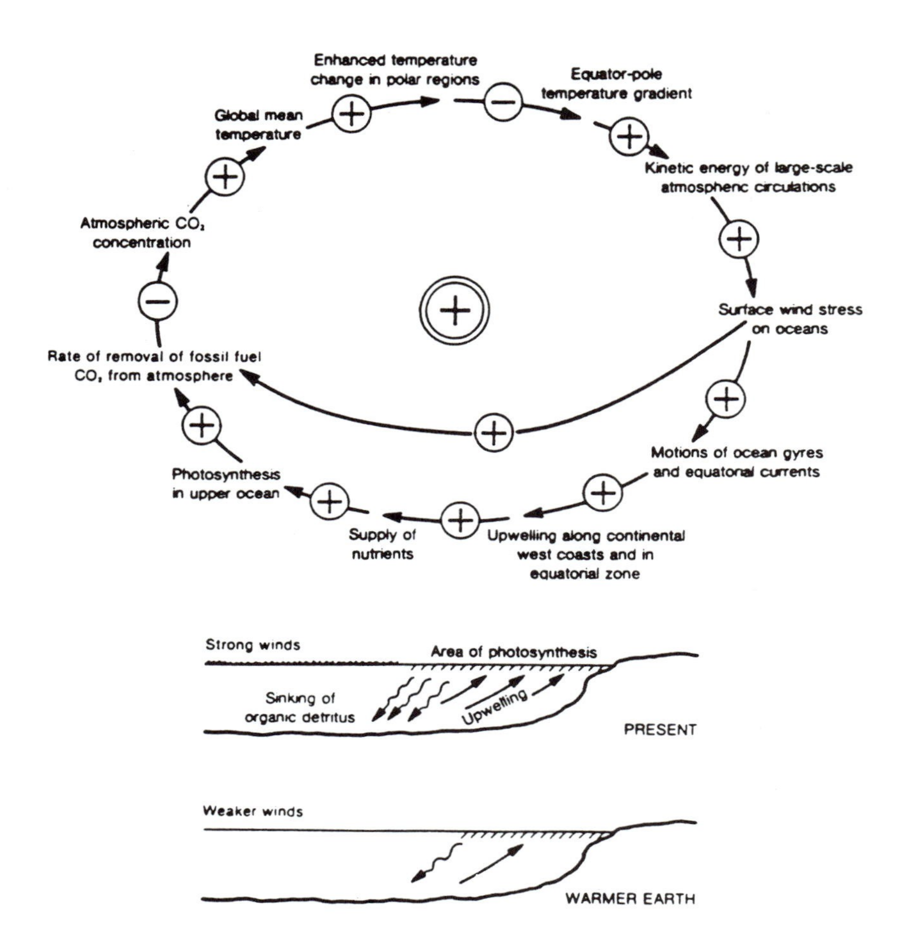

Figure 11.6: *Carbon dioxide – ocean circulation – upwelling feedback loop. The sketch at bottom (as in subsequent figures) illustrates the physical processes involved.* [From Kellogg, W.W., 1983: Feedback mechanisms in the climate system affecting future levels of carbon dioxide. *J. Geophys. Res.,* **88,** 1263-1269. ©American Geophysical Union.]

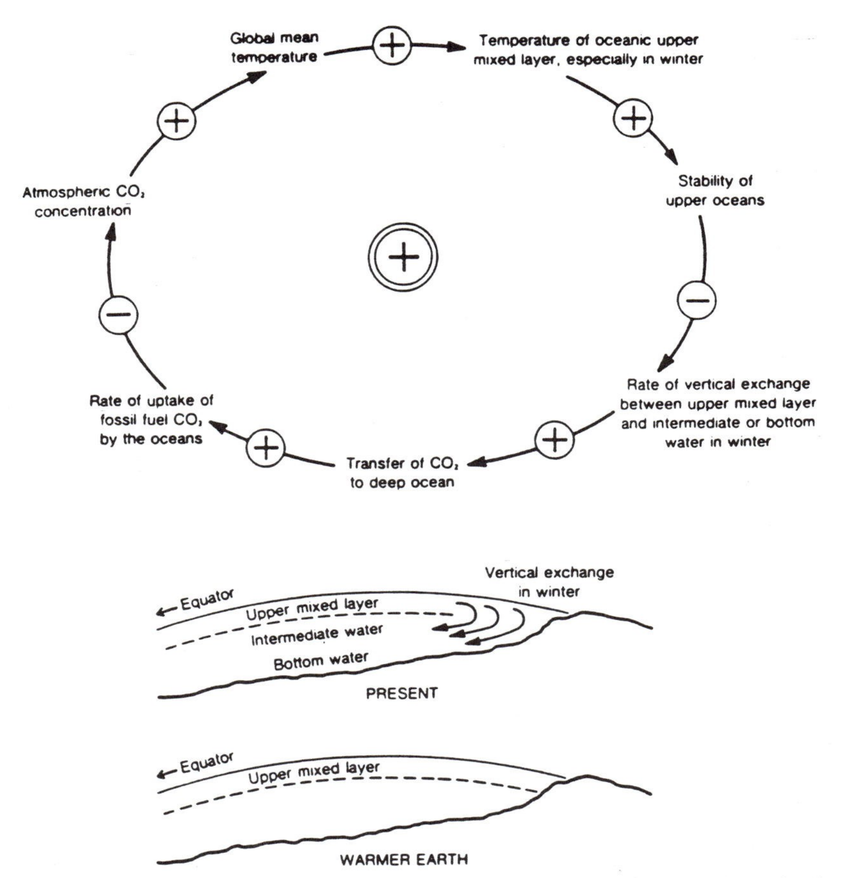

Figure 11.7: *Carbon dioxide – ocean stability – winter downwelling feedback loop.* [From Kellogg, W.W., 1983: Feedback mechanisms in the climate system affecting future levels of carbon dioxide. *J. Geophys. Res.,* **88**, 1263-1269. ©American Geophysical Union.]

the longer-term trends in CO_2 levels in the atmosphere. The thermodynamic and hydrodynamic feedbacks from the ocean to increases in CO_2 are potentially more predictable, but require complicated, coupled atmospheric/ocean GCMs which are still in an early stage of use and require enormous amounts of computer resources (Randall et al., 1992).

11.3.4 Cloud feedbacks

We have seen that one of the major positive feedbacks to CO_2 or other greenhouse gas warming is that as the ocean surface and ground temperatures rise, an increase flux of water vapor into the atmosphere is expected. In a cloud-free atmosphere, enhanced water vapor content provides a strong enhancement of greenhouse warming. In response to

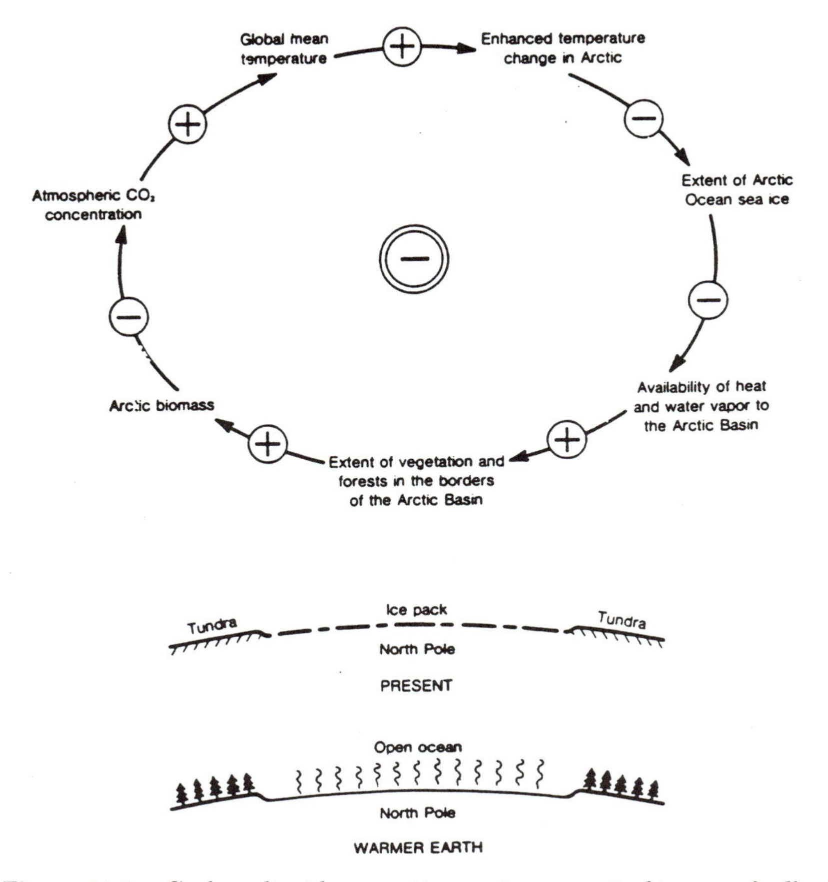

Figure 11.8: *Carbon dioxide – arctic sea ice – arctic biomass feedback loop.* [From Kellogg, W.W., 1983: Feedback mechanisms in the climate system affecting future levels of carbon dioxide. *J. Geophys. Res.,* **88**, 1263-1269. ©American Geophysical Union.]

the enhanced water vapor content of the atmosphere, however, one would also expect that cloud coverage would increase and some clouds would become wetter or be optically thicker. Moreover, clouds alter the stability of the atmosphere in response to their release of latent heat and their associated rising and sinking motions vertically and horizontally transfer heat, moisture, momentum, and various particles and trace gases in the atmosphere. In this section we will examine the radiative feedback of clouds as well as the feedback associated with changes of atmospheric stability and transport.

We also found that the global-average radiative effects of high clouds are to warm the atmosphere while low clouds cool the atmosphere and middle-level clouds are in near balance between cooling by reflection of solar radiation and warming by absorption of longwave radiation. Av-

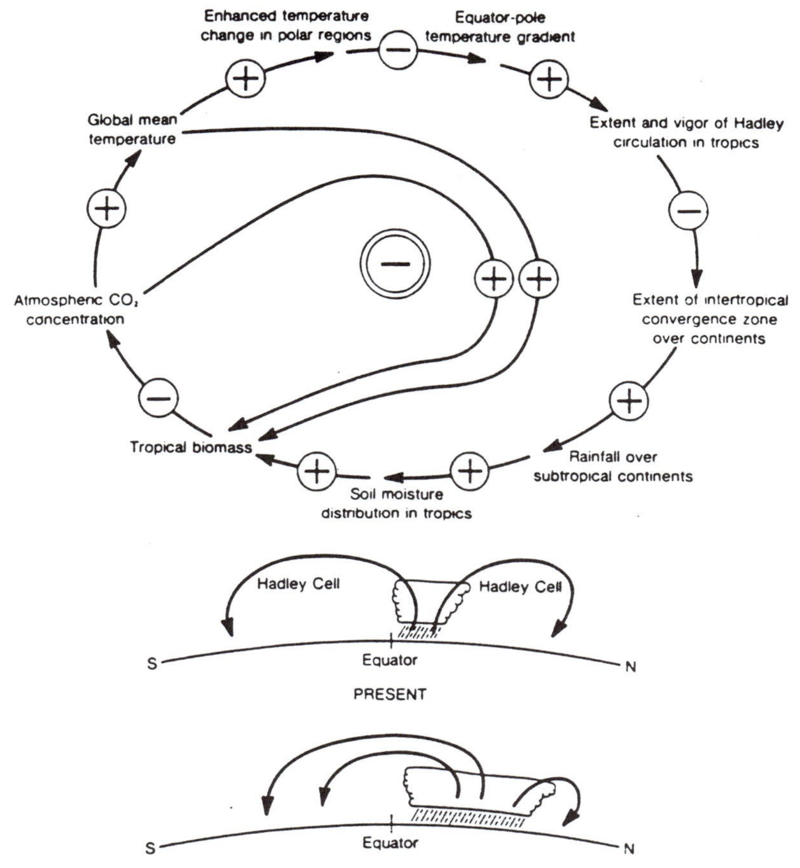

Figure 11.9: *Carbon dioxide – rainfall distribution – tropical biomass feedback loop.* [From Kellogg, W.W., 1983: Feedback mechanisms in the climate system affecting future levels of carbon dioxide. *J. Geophys. Res.,* **88,** 1263-1269. ©American Geophysical Union.]

eraged over the entire earth, the enhanced albedo of clouds is slightly greater than their greenhouse warming, so clouds radiatively cool the atmosphere compared to a cloud-free earth (Ramanathan et al., 1989a,b; Randall et al., 1989). In the case of an atmosphere/ocean system warmed by greenhouse gases, what is the feedback of clouds to that response? Unfortunately, clouds are forced by relatively small-scale atmospheric motions, thus climate models must parameterize clouds in rather crude ways, with the cloud cover being specified based on climatology or parameterized as a function of relative humidity and as a by-product of convective parameterization schemes. Even in the more sophisticated GCMs which have a prognostic equation for cloud liquid water or ice-water amounts, the horizontal and vertical resolution of those models is not sufficient to simulate the major ascending motions leading to cloud formation. As a result, the simulated feedbacks of clouds on the greenhouse theory vary from model to model, depending on the nature of the cloud/radiative scheme and resolutions of the models. For example, Hansen et al. (1984) found that clouds enhance simulated CO_2-induced greenhouse warming while Manabe and Wetherald (1980) found that the cloud changes were self-compensating in their GCM such that they had little effect on the global radiation budget. Subsequently, Wetherald and Manabe (1986; 1988) found that in their enhanced CO_2 simulations, clouds produced a positive feedback. This feedback was caused by: (1) a reduction of cloud amount in the middle to upper troposphere, (2) an increase of cloud amount around the tropopause, and (3) an increase in cloud amount near the earth's surface in high latitudes. Their results are strongly dependent upon the skill of their cumulus parameterization scheme since the positive feedback is largely a result of predicted increases in high cloud amounts, in response to enhanced convective heating in tropical and middle latitudes. The convective heating, in turn, causes an upward shift of the relatively stable layer of air where the convective towers detrain moist air. Unfortunately, cumulus clouds remain as one of the most uncertain parameterizations in GCMs and other large-scale numerical prediction models.

The above studies, however, did not consider possible changes in cloud liquid water paths in response to enhanced greenhouse gas concentrations. Petukhov et al. (1975), Paltridge (1980), and Charlock (1981; 1982) each considered possible changes in liquid content of clouds in response to enhanced concentrations of greenhouse gases and found clouds caused negative feedbacks. In a rather clever use of empirical studies that suggest that cloud liquid water contents increase with air temperature, Somerville and Remer (1984) concluded from calculations with a globally-averaged radiative-convective model, that clouds serve as a *thermostat* to stabilize the climate against radiative forcing such as that caused by greenhouse gases. Without including changes in cloud optical thicknesses, they calculated that doubling of CO_2 would warm surface temperatures by 1.74° C. When a cloud optical depth feedback was included, however, the amplitude of warming was reduced to less than 0.5° C. They did not, however, consider high, thin cirrus clouds whose response to warming would either weaken or strengthen the amplitude of negative feedback of clouds depending on their optical thicknesses. As noted by Randall et al. (1989), cumulus anvil clouds exert a very powerful influence on tropical convection by radiatively destabilizing the upper troposphere, and by trapping terrestrial radiation. They argue that the skill in predicting the effects of those clouds in GCMs, while being qualitatively correct, are not sufficiently accurate to be relied upon quantitatively in considering their strong influence on simulated climates. Their conclusion is particularly valid when one considers the important role that organized mesoscale systems play in producing long-lasting upper tropospheric anvil clouds in both the tropics and middle latitudes. These systems produce large areas of optically-thick, stratiform-anvil clouds in response to detrainment of rising moist air from deep convective towers as well as cloud formation in slowly ascending mesoscale motion. Current cloud parameterization schemes in GCMs do not consider the organized mesoscale motions in their estimates of anvil cloud cover or optical thicknesses.

It must be recognized, however, that convective clouds are not uniformly distributed throughout the diurnal cycle. For example, some cloud systems such as mesoscale convective complexes and tropical cloud clusters exhibit a well-defined nocturnal maximum in their frequency of occurrence (see Cotton and Anthes, 1989). Thus if a warmed ocean creates more tropical cloud clusters, then their net impact would be to cause a positive feedback rather than function as a thermostat, since they would have little effect on solar radiation. It is important, therefore to consider the diurnal variations of cloudiness when examining the impacts of clouds on climate.

In another study, Roeker et al. (1987) perturbed a GCM with an explicit hydrological budget including cloud liquid water prediction by increasing the solar constant. Like Somerville and Remer, they found that changes in cloud liquid water paths in a warmer atmosphere created a strong negative feedback on surface temperatures.

One of the most outspoken critics of the greenhouse theory is Lindzen (1990). We share with Lindzen his concern that the apparent unanimity of the scientific community with respect to the influence of greenhouse gases on climate is not healthy for meteorology nor society as a whole. Lindzen argues that the greenhouse effect is not discernable in the temperature records from natural variability and that the theory is flawed with regard to the treatment of clouds. We consider his theoretical hypothesis here and discuss the temperature record more fully in the following text. The basis of his theoretical argument is that the modeling of upper level cloud cover responses to GCMs is in error. He argues that deep convective clouds should produce a net drying of the upper troposphere due to the drying influence of compensating subsidence. He bases his argument on the behavior of simple cumulus parameterization schemes that he has developed. While drying effects of compensating subsidence does indeed take place in the middle troposphere, it appears that the moistening effects of detrainment of water substance from cumulus towers and stratiform-anvil clouds of mesoscale convective systems overrides the drying effects. As a result, a general moistening in the upper troposphere associated with deep convection is observed (Rind et al., 1991).

Another interesting feedback from clouds has been hypothesized by Mitchell et al. (1989). They performed two simulations of the response of their GCM to a doubling of CO_2: (1) using procedures common to GCMs in which the presence of clouds is determined by the relative humidity in the model but the liquid water paths are prescribed from climatological data, and (2) with the liquid water content of clouds explicitly predicted including a simple parameterization of ice-phase precipitation. They found that the global-averaged surface warming for a doubled CO_2 case was reduced from 5.2° C to 2.7° C when cloud water was predicted. This strong negative feedback was related to a reduction in the depletion of cloud water content in a CO_2 warmed atmosphere due to a reduction in ice-phase precipitation formation in middle latitudes. As a result, cloud cover increased and the stronger albedo caused a reduced rate of greenhouse-induced warming than in the version of the model using relative humidity as the primary parameter determining cloud amounts.

Several researchers have attempted to evaluate cloud feedbacks to a warming atmospheric/ocean system by examining the observed relationship between sea surface temperature (SST) anomalies and changes in cloudiness and net radiative budgets. Ramanathan and Collins (1991), for example, diagnosed changes in cloudiness and net radiation associated with warm SST anomalies during the 1987 El Niño. They inferred that in warm ocean regions where SSTs were less than 300 K, the net effect of enhanced water vapor and cloudiness resulted in a positive greenhouse effect. However, when SSTs exceeded ~300 K, they found that cirrus clouds formed by the flux of water substance into the upper troposphere by cumulonimbus clouds, became optically thick and, as a result, they reflect more solar radiation than is absorbed and re-radiated downward by terrestrial radiation. They argue that the highly reflective, optically-thick cirrus clouds acts as a *thermostat*, which prevents further warming of the oceans. They suggest that the implications to greenhouse warming is that "it would take more than an order-of-magnitude increase in atmospheric CO_2 to increase the maximum SSTs by a few degrees in spite of a significant warming outside the equatorial regions."

In another study, Peterson (1991) examined the relationship between sea surface temperature anomalies and anomalies of high, middle, and low-level cloudiness using satellite data. He found that over much of the tropics and the south Pacific convergence zone, high clouds increased with warm SST anomalies, while in subtropical stratocumulus regions, low cloud coverage decreased with positive SST anomalies. Averaged over the entire region he sampled, total cloudiness increased with positive SST anomalies. Because the coverage of optically thick low-clouds decreased while optically-thin high clouds increased over regions of warm SSTs, he calculated that the average, net radiative flux to space decreased in response to warm SST anomalies. He concluded that his results provide observational evidence that clouds provide a positive feedback loop to global warming scenarios.

In each of these studies, characteristic regional/general circulation patterns develop in response to the SST anomalies. Circulations, such as the Hadley and Walker circulations (Philander, 1990), supply moisture to the deep convective clouds and are, therefore, primarily responsible for the alterations in cloudiness. Under the slow greenhouse warming scenario, one would expect different responses in both ocean and atmospheric circulations than in their studies. One cannot be assured that the atmospheric circulations that coincide with either Ramanathan and Collins' or Peterson's inferred SST anomalies have any resemblance to the atmospheric circulations associated with a greenhouse warming scenario. Thus, the cloudiness changes associated with a greenhouse warming scenario could be quite different from the cloudiness distribution in today's climate that is associated with SST anomalies in an El Niño event or in Peterson's data set.

There are other effects of clouds that can create positive or negative feedbacks to greenhouse gas-induced warming. Some of these are associated with changes in the stability of the troposphere including the height of the tropopause. In response to a warming climate, for example, deep convective overturning will raise the height of the tropopause in the tropics, yielding a colder tropopause. The temperature of the tropical tropopause has a strong influence on the water vapor content of the stratosphere. The colder the tropopause, the lower the water vapor content of the stratosphere, since the tropopause acts as an effective trap of moisture as more water will be condensed and precipitated out of the air. As a result, with a colder tropopause, there will be little absorption of terrestrial radiation in the stratosphere causing a cooling effect on surface temperatures. Overall, simulation of the feedback of clouds on climate change can be considered to be the 'Achilles heel' of climate models since those feedbacks can be quite large and cause a major moderating influence on greenhouse gas-induced warming. Unfortunately, most of the cloud processes occur on scales considerably below the resolution of current GCMs. Arking (1991), in a recent review article, concludes that "1) clouds *may* have a strong influence on climate change, but 2) we are far from knowing the magnitude, and even the sign, of this influence."

11.3.5 Vegetation feedbacks

If vegetation growth is enhanced because of an enrichment in carbon dioxide, the importance of this greenhouse gas to global warming would be reduced since vegetation has a significant role in determining the CO_2 content of the atmosphere (Leith, 1963). Soil organic carbon resulting from this active exchange with the atmosphere contains about two-thirds of the carbon in the terrestrial ecosystem and has long residence times (over 1000 years) according to Post et al. (1982). King et al. (1985), documents that soybean transpiration efficiency, and therefore soybean yield, would increase if CO_2 concentrations were elevated. B.A. Kimball, Sherwood B. Idso, and associates at the U.S. Water Conservation Laboratory and Western Cotton Research Laboratory in Phoenix, Arizona used laboratory and field experiments to investigate the impact of enriched carbon dioxide atmosphere on plant growth and productivity. Among their conclusions, they found that orange tree growth was greatly stimulated. During the first year the mean cross-sectional areas of the trunks of the CO_2-enriched trees were 102% greater than the control trees (using the current atmosphere) at 60 cm above the soil surface. After two years of growth, the total trunk plus branch volume of the CO_2-enriched trees was 2.79 times that of the control orange trees. Dr. Kimball concluded that yields, in general, will increase on the order of 33% with a doubling of atmospheric CO_2 concentrations. Ed Glenn of the Environmental Research Laboratory at the University of Arizona has proposed planting salt-tolerant vegetation (referred to as halophytes) in coastal areas and inland salty areas as a mechanism to remove CO_2 from the atmosphere through the growth of this vegetation. Ahmad (1990) has suggested the planting of 4,650,000 km^2 of trees in desert areas to capture 2.9 billion tons of CO_2 per year where the water would come primarily from vast available underground aquifers in the Sahara, India, and elsewhere. If these responses are characteristic of natural and other agricultural crops, a substantial portion of the anthropogenic carbon dioxide emissions are absorbed in biomass.

11.3.6 Other feedbacks

There are other feedbacks associated with tropospheric and stratospheric chemistry and terrestrial biology. For example, on a warmer earth the absolute humidity of the air is likely to be greater and as a result, temperature-dependent or moisture-dependent chemical reactions will be affected. Reactions involving the formation of methane and ozone will be slowed down thus decreasing their concentrations (Lashof, 1989). This will produce a weak negative feedback. In another example, climate model calculations suggest that tropospheric warming will be associated with stratospheric cooling which will affect chemical reactions important to stratospheric ozone concentrations. Unfortunately, the sign and magnitude of this effect depends on the altitudes where temperatures are decreased, as well as emissions of halocarbons and nitrous oxide. Neither the sign or magnitude of these feedbacks can be estimated accurately (Ramanathan et al., 1987). As noted previously, terrestrial biota feedbacks are associated with changes in vegetation albedo and with the carbon budget. Increases in grassland and desert area, as a result of changes in precipitation and temperatures or other human activity in subtropical regions could partially compensate for decreased albedo at higher latitudes (Lashof, 1989). Garratt (1992), for example, reports on GCM results which yield an average decrease in continental precipitation of 1 mm day^{-1} in response to an average albedo increase of 0.13. The most important feedbacks of vegetation to the atmosphere are the evapotranspiration and the water balance over long time periods (Graetz, 1991), as well as the landscape albedo.

Overall there are many potential feedbacks to a greenhouse gas-induced climatic change, many of which are small in magnitude and of uncertain sign. Those feedback processes associated with water vapor, clouds, oceans, the biosphere and changes in ice/snow cover are not small, however, and introduce considerable uncertainty in estimates of the overall climatic consequences of increases in greenhouse gas concentrations.

11.4 Equilibrium Climatic Change Estimates

11.4.1 Surface temperature responses

Estimates of the equilibrium climatic response to enhanced concentrations of greenhouse gases are typically made by instantaneously increasing the concentration of greenhouse gases such as doubling CO_2 concentrations and then integrating a model until a steady-state solution is obtained. This has been done with simple one-dimensional radiative-convective models, two-dimensional zonally-averaged models, and three-dimensional general circulation models of varying complexity. As noted previously, estimated global-averaged surface temperature increases for a doubled CO_2 atmosphere with one-dimensional radiative-convective models range from 0.48 to 4.20° C. Estimates of equilibrium responses to a doubled CO_2 concentration varies from 1.3 to 3.9° C for GCMs coupled to swamp ocean models, to 3.5 to 5.2° C for GCMs coupled to mixed layer ocean models (Schlesinger and Mitchell, 1987; Wilson and Mitchell, 1987; Cess et al., 1990). The GCMs generally yield stronger estimated responses to a doubled CO_2 change than one-dimensional models because of the strong feedbacks from ice-albedo and moisture increases. As mentioned previously, inclusion of complicated (more realistic?) cloud feedbacks has nearly halved the estimated equilibrium surface temperature response to a doubled CO_2 concentration (Mitchell et al., 1989). How further refinements in model resolution and physics are likely to affect those numbers is anyone's guess, but it is unlikely that equilibrium estimates of global average surface temperature changes will exceed +2° C.

11.4.2 Other temperature changes

General circulation models are capable of simulating other responses to enhanced greenhouse gases (see reviews by Schlesinger and Mitchell, 1987; Mitchell, 1989). For example, because of enhanced emission of longwave radiation to space, they predict that the stratosphere will cool.

Also surface warming at high latitudes is greater than the global average in winter but smaller in summer. This is largely due to the ice-albedo feedback, since ice and snow cover are less extensive in the warmed atmosphere. Furthermore, the warming at high latitudes in winter is confined to the shallow layer beneath the low-level temperature inversion whereas in summer (and lower latitudes) the heating is mixed through the depth of the troposphere yielding less surface warming. They furthermore predict that the surface warming is least in the tropics and exhibits little seasonal variation. This is because of the greater depth the heating must be mixed in the tropics throughout the year and because moist convection is very efficient is transporting moisture to midlatitudes where it participates in releasing the latent heat of condensation (Manabe and Wetherald, 1975). These results are fairly consistent among the various GCMs, with the most consistent being an overall warming of the troposphere and cooling of the lower stratosphere.

11.4.3 Changes in surface hydrology and precipitation

There is more variability among the various GCMs in their simulation of changes in surface hydrology and precipitation than of temperature. This is because soil moisture is dependent upon rainfall and rainfall, in turn, depends on small-scale atmospheric phenomena driven by physiographic features such as orography, land-sea contrasts, landuse patterns, and transient weather systems (Mitchell, 1989). In general the GCMs agree that soil moisture increases in high latitudes in winter. They disagree, however, on the continental-scale changes in soil moisture in the northern hemisphere summer. Some models predict drying in middle latitudes while others predict moistening of the soil. Washington and Meehl (1984) argue the differences are due to the amount of soil moisture in the control simulation. This, in turn, affects the timing of snow melt and the amount of moisture in the soil in summer. As far as precipitation is concerned, the GCMs agree that the global-average precipitation increases in proportion to the warming. However, in the tropics some models predict regions of enhanced precipitation while others predict decreases in those regions. The differences are due to differing simulated behavior of the main tropical rainbands (Schlesinger and Mitchell, 1987). There is general concurrence among the GCMs that precipitation increases at high latitudes throughout the year, that it increases in midlatitudes in winter, and there are little changes in rainfall in the subtropical arid regions.

11.5 Transient Climatic Responses

Estimates of climatic changes to increased concentrations of greenhouse gases has been largely accomplished with models in which the concentration of CO_2 and other greenhouse gases are instantaneously increased (i.e., doubled or quadrupled) and the equilibrium climate response of the model is then examined. The concentrations of greenhouse gases have been slowly building up over a time scale of a century and more, and, furthermore, because of the large heat capacity of the oceans, the atmospheric ocean system has a large thermal lag in its rate of heating. We must ask, therefore, if the equilibrium calculations are really representative of the real transient response to increasing greenhouse gas concentrations? In order to examine the transient responses to greenhouse gases we need to consider a coupled atmospheric-ocean system since the oceans play a very important role in controlling the rate of warming of the earth system and furthermore, has a major influence on the zonal and regional patterns of climate change. The simplest approach to interfacing an ocean model to an atmospheric climate model is to interface an energy balance atmospheric model to a simple box diffusion model of the ocean (Hoffert et al., 1980; Harvey and Schneider, 1985; Wigley and Raper, 1987). In these models heat transports within the ocean are parameterized by vertical eddy mixing through an eddy diffusion model and vertical heat transport by specified vertical velocities. Latitudinal heat transports are assumed to occur instantaneously such that a single temperature rise characterizes the whole globe. The temperatures of the downwelling sea surfaces in polar regions and upwelling deep water are specified. The transient predictions of such models are highly dependent upon these specified parameters as well as those built into the atmospheric model. The parameters are calibrated for today's observed climate and estimates of observed rates of vertical exchange of passive materials in the ocean. Because heat does not move in the atmosphere or the ocean similar to passive materials, the transient predictions of such models are of dubious value.

Another approach to a transient estimate of climate change is to couple a full atmospheric GCM to an ocean model in which all horizontal heat transport is fixed at values determined from observations from the present climate (Hansen et al., 1988). This approach would be realistic as long as the ocean does not respond with surprise changes in ocean circulations in response to enhanced greenhouse warming. Unfortunately, the limited numerical experiments with fully-coupled atmospheric-ocean models suggests that the ocean circulation does significantly change in response to greenhouse gas caused circulation changes.

The most realistic simulation of the transient response to increased greenhouse gases is to use fully coupled general circulation models of the atmosphere and the ocean. These are very computationally expensive models since the ocean must be set up on a numerical grid of comparable dimensions to the atmospheric model, and furthermore, ocean models must be run for a time scale of 100's of years because of their very slow adjustment times. Only very limited experience has been obtained with such coupled models (e.g., Stouffer et al., 1989 using the GFDL GCM; Washington and Meehl, 1989, using the NCAR CCM; Houghton et al., 1990), therefore the state of the science is very much in its infancy. Thus far the ocean models have very crude horizontal grid spacing (~500 km) and few vertical levels. Moreover, the added nonlinearity of the coupled atmospheric/ocean system makes it much more difficult to simulate today's climate. In fact it is currently necessary to introduce arbitrary adjustments to the models such as a flux correction of one or more variables at the sea-air interface in order to reproduce the current climate reasonably well (Meehl, 1990). These adjustments are much larger in magnitude than expected changes in fluxes due to greenhouse warming. Even so, the models still produce tropical sea-surface temperatures that are too cold and high-latitude temperatures that are too warm (Washington and Meehl, 1989). It is thus premature to interpret those model calculations too seriously quantitatively or even qualitatively. Nonetheless it is instructive to examine some of the transient responses of those models to increasing greenhouse gas concentrations to help us understand just how complicated the atmospheric-ocean system really is.

Both Washington and Meehl (1989) and Stouffer et al. (1989) performed transient coupled ocean/GCM numerical experiments with a specified 1% per year increase in CO_2, with the former introducing the CO_2 over a 30-year period and the latter over a 70-year period. This increase in CO_2 roughly corresponds to the estimated radiative forcing of all greenhouse trace gases. As described in Houghton et al. (1990), the GFDL model exhibited a standard deviation of ten-year-averaged air surface temperatures of ±0.08° C which is less than the observed temperature variability over a 100-year period. The response to CO_2-induced greenhouse warming must be greater than that amount to be meaningful. At the end of 30 years the NCAR model predicted about 0.7° C warming while the GFDL model predicted slightly less than 1° C global-averaged surface warming relative to their control simulations. This is less than half the value predicted for the instantaneous doubling of CO_2 but the amount of CO_2 at that time is only a 30% increase in CO_2 relative to the control. At 70 years the GFDL model predicted less than 2.5° C warming compared to about 4° C warming for the equivalent instantaneous doubling experiment. Thus both models agree that there is a greater lag in the amount of warming in the coupled ocean system relative to an uncoupled ocean GCM.

Both models also exhibited considerable hemispheric and regional variation in the response to greenhouse warming in comparison to equilibrium model responses. These differences are due to simulated changes in the ocean circulations. The ocean circulations are driven by wind stress which, in turn, is driven by changes in wind speed and direction and in thermodynamic stability of the atmosphere immediately above the air-water interface. Moreover, the ocean circulations are also driven by thermohaline circulations or circulations induced by spatial gradients in the density of the water. The density of water is a function of both temperature and of salinity. Thus, warm ocean water is less dense. Likewise ocean water low in salt content is less dense. For example, water flowing out of the warm, salty Mediterranean sea is very dense and actually undercut warmer, less salty water in the Atlantic. In our current climate, the warm Gulf Stream flows northward towards the British Isles, where it moderates the climate over northern Europe. This current flows northward to replace cold, dense deep water that flows southward from Arctic regions while spreading out across the ocean bottom. Both the NCAR and GFDL models responded to the enhanced greenhouse warming by producing increased rainfall in the summer and fall months at high latitudes, and thereby forming a layer of fresher, less dense water on the surface of the North Atlantic. The strength of the thermohaline circulation thus weakens and the westerly winds also weaken. This reduces the flow of warmer water across the Atlantic, causing cooler surface temperatures than in the equilibrium model simulations or in the current climate. This response, while physically plausible is dependent upon many details of the models (i.e., precipitation parameterization schemes, vertical and horizontal resolution of both the atmospheric and ocean models) and thus should be viewed as one of many complicated feedback scenarios that could occur in a coupled atmospheric-ocean system.

In the southern hemisphere, Stouffer et al. (1989) noted that the CO_2-induced surface warming decreases with the absolute value of the latitude whereas the reverse occurred in the northern hemisphere. Moreover, the equilibrium experiments did not exhibit such a difference. They found the rate of warming was especially slow in the Antarctic circumpolar ocean. This is partly due to the larger surface area of the southern ocean yielding a much larger surface inertia. In addition, a deep wind-driven cell or ocean gyre is present in the southern ocean that contain deep downwelling and upwelling components that rapidly mix the surface heating through a deeper layer of the ocean than in the northern hemisphere. Stouffer et al. also note that a strong thermohaline cell exists in the vicinity of Antarctica which also contributes to stronger vertical mixing of heat in the southern oceans. As a result the thermal inertia of the southern oceans is much greater than the northern oceans both because of their greater area and greater depth of vertical mixing of heat. Therefore, a smaller amplitude warming is anticipated in the southern than in the northern hemisphere. Overall there remains considerable uncertainty with regard to the response of the atmospheric/ocean system to a gradual increase in greenhouse gas concentrations. Clearly, the amount of greenhouse gas induced warming expected over the next 30 or 70 years will be considerably less than estimates with equilibrium response models and we can anticipate that there will be both qualitative and quantitative differences in the regional and hemispheric patterns of those responses.

In Section 11.4 we speculated that given the potential strength of negative cloud feedbacks, the most likely equilibrium response to a doubling of CO_2 would be about 2° C surface warming. Considering the transient response, we speculate that at the time that CO_2 doubles in concentration, the expected surface temperature rise will be probably no more than 1° C. It must also be recognized that if greenhouse gas concentrations immediately leveled off, the thermal inertia of the oceans would result in continued warming for periods of several decades.

11.6 Regional Responses

Thus far the focus of most of the research on the impacts of enhanced concentrations of anthropogenically-produced greenhouse gases on climate has been of impacts on global average temperatures and precipitation, with some emphasis on impacts on the scale of hemispheres. Few of us are really concerned about whether the global-average surface temperature rises by 0.5° C or 3.0° C. What we are really concerned about is how will that affect me in the region I live in or how will it affect the economy, food, and energy production in my country? In this section we will examine the attempts and prospects for assessing regional impacts of greenhouse warming. Here we refer to regional scale as being subcontinental in scale, such as western Europe or the eastern United States, or northern Australia. This section differs from Part II, where we examined the human-induced regional changes in climate and weather which are caused by human activities on the regional scale rather than globally. Some examples of the particular regional responses that we might be concerned with include the following.

- Impact on surface precipitation and its variability.
- Impact on agriculture, which, in turn, is dependent on precipitation, soil moisture, temperature and its extremes (frosts, maximum temperatures, strong winds), sunshine amounts, and hail damage.
- Impacts on air quality. Will there be more stagnant highs capable of producing high concentrations of pollutants and photochemical production of pollutants?
- Impact on energy production and consumption. Will precipitation changes affect hydroelectric power production or the amount of sunshine affect solar radiation? Will summer temperatures be warmer causing greater energy consumption by air conditioning?

- Impact on water levels such as the Great Lakes and coastal areas. Lake water variability is also a sensitive indicator of climate variations because it represents the residual of a water balance between precipitation, evapotranspiration, and runoff (Stockston, 1990).

Let us examine the approaches to investigate the potential for regional impacts of greenhouse gas-induced climate change, and the reliability of those methods. Kellogg (1982) summarized the various methods as:

1. reconstructions of regional climate of warmer periods in the distant past;
2. examination of observed regional changes of weather and climate associated with warm episodes from the climatic record of the last 100 years or so; and
3. analysis of GCM grid point data on regional scales and, in particular, coupled GCM-ocean models and regional-scale models nested to GCMs.

11.6.1 Reconstructions from paleo-climates

Kellogg (1982) gave an example of estimating regional responses to greenhouse warming by examining the Altithermal Period that occurred 4500 to 8000 years ago during which the earth was generally several degrees warmer. Using the distribution of pollen types and spores found in ancient lakes and bog sediments, and inferred lake levels and stream flows, Kellogg (1977) and Butzer (1980) inferred regions of above and below-normal precipitation. While their separate analyses did not agree in large regions of earth, they did agree that central North America was drier during this period, whereas Mexico, north and east Africa, much of India, and western Australia were wetter. Aside from the uncertainties associated with extracting climatic data from such indirect techniques, the major concern about the use of this technique is that the major driving force for the inferred global warming during this period must have been quite different than that caused by current enhanced greenhouse gas concentrations. Therefore, changes in the regional distribution of climate could be quite different if the warming was caused by changes in the earth's orbital parameters or reduced volcanic activity, etc.

11.6.2 Inferences from the recent climatic record

As summarized by Kellogg (1982), the premise of this approach is that periods when it was unusually warmer in the northern polar regions, might have characteristic anomalies in the general circulation, temperature, and precipitation patterns that resemble a greenhouse-warmed earth. This result is based on GCM simulations (e.g., Manabe and Wetherald, 1975) which indicate that the warming would be concentrated at high latitudes. Kellogg (1982) summarized studies by Williams (1979) and Wigley et al. (1979). Maps of the inferred anomalies in precipitation are shown in Figure 11.10. Both Williams and Wigley et al. found that these periods

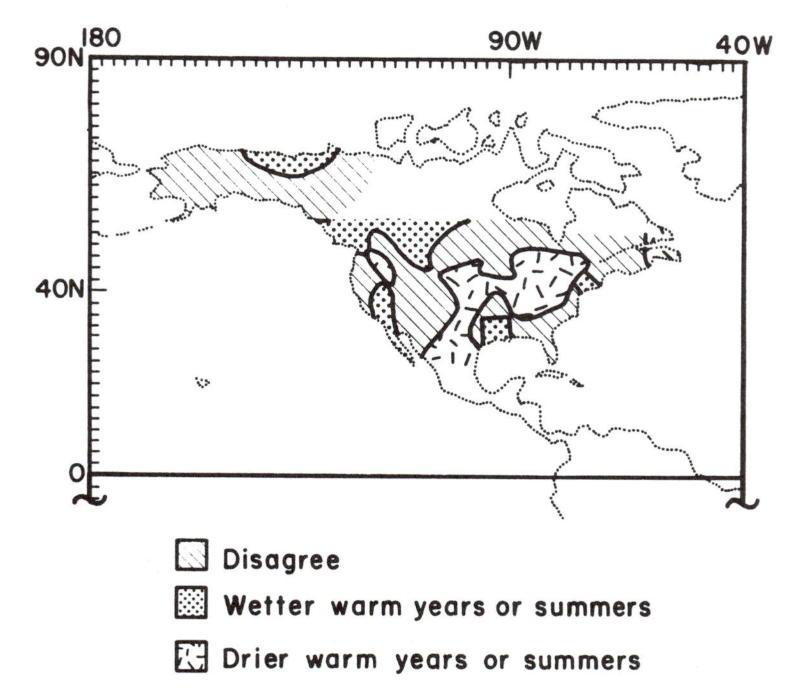

Figure 11.10: *Areas where the maps of Williams (1979) and Wigley et al. (1979) agree as to deviations from the long-term mean precipitation over North America. Williams' results apply to the summertime, whereas Wigley et al. apply to annual means.* [From Kellogg, W. W., 1982: Precipitation trends on a warmer earth. In *Interpretation of Climate and Photochemical Models, Ozone, and Temperature Measurements*, R.A. Reck and J.R. Hummel, Eds., Institute of Physics, New York, 35-46.]

corresponded to below average precipitation in the midwestern United States. Elsewhere in North America the two analyses differ, but Kellogg suggested that it may be due to the fact that Williams examined only summer data while Wigley et al. examined annual averages. Namais (1980) found that over the United States, warm years, as inferred from atmospheric thickness data, are as likely to be associated with above normal precipitation as below. He also suggested that hemispheric temperatures are not dependable indicators of temperatures over the United States. As a result he concluded that zonal and hemispheric averages are by no means adequate indicators of regional temperature and precipitation fields.

A problem with using recent climatic data is that the time scale of these anomalies is an order of magnitude less than those anticipated with a slow increase in greenhouse gas concentrations. Like paleo-climate estimates, the causes of recent past climate anomalies are different than greenhouse gas forcing, so that regional-scale responses are likely to vary depending on the causes of those forcing phenomena.

11.6.3 Regional estimates from GCMs

Another approach to estimating regional impacts of greenhouse warming is to examine the regional temperature, precipitation, and soil moisture fields that are simulated by GCMs (e.g., see Dickinson, 1991; Giorgi and Mearns, 1991) along with regional changes in the carbon storage in the soil and vegetation (Burke et al., 1991). Typically GCMs have grid-truncation scales or grid spacing on the order of 300 to 400 km. This means that the only features reliably modeled are on scales of 1200 to 1600 km and greater (Pielke, 1984). Thus they are capable of resolving features on the scale of continents and possibly some subcontinental-scale features, although the reliability of those simulated data become more questionable as the scales of interest approach the truncation scale of the models. Furthermore, while different GCMs produce reasonably consistent global-averaged, and hemispheric-averaged simulated temperature and precipitation changes in response to greenhouse warming, they produce significant differences on subcontinental scales (Grotch, 1988). That is, the regional-scale predictions of the models are much more sensitive to details of formulations of cloud parameterizations, soil moisture formulations and initial conditions, computational procedures, etc. Most attempts at examining the regional impacts of greenhouse gas-induced warming have been performed for equilibrium calculations with GCMs for a doubled CO_2 scenario. However, as indicated in Section 11.5, coupled atmospheric/ocean models are creating considerably different regional responses, depending on alterations of ocean currents and variations of the depth of vertical mixing of heat into the oceans. It is thus quite unrealistic to examine regional scale impacts of greenhouse gases in anything but a transient, coupled atmospheric/ocean model. Even well into the interiors of continents, the precipitation fields are strongly influenced by the transport of moisture from distant oceans by circulations fields that respond to the sea surface temperature fields affected by transient ocean currents.

A number of scientists have speculated that greenhouse warming will increase sea-surface temperatures (SSTs), which will in turn, increase the frequency and intensity of tropical cyclones. Emanuel (1987), for example, has hypothesized that tropical cyclones behave much like a

Carnot heat engine in which the strength of the storm is largely controlled by SSTs and the temperature of the top of the storm. Because greenhouse warming is expected to both increase SSTs and, by virtue of raising the tropical tropopause, decrease the temperature of the tops of storms in the tropics, he has speculated that greenhouse warming would increase the maximum sustainable central pressure drop, or increase the intensity of tropical cyclones. Observationally, we know that a certain threshold SST is required before tropical cyclones form (Merrill, 1988). The relationship between SSTs and storm intensity is not unique, however, because there are many other environmental factors which influence storm intensity and the frequency of storms (see reviews by Anthes, 1982; Cotton, 1990; Pielke, 1990).

Recently, Broccoli and Manabe (1990) examined the impact of doubling CO_2 on the frequency, duration, and number of days of tropical cyclones, or *cyclone-like disturbances* in their GCM. The model was run with two different resolutions, one having a 4.5° by 7.5° latitude-longitude spacing and the other having 2.25° by 3.75° spacing. Two sets of experiments were run, one with a fixed climatologically derived cloud distribution, and a second in which cloud amount was allowed to vary in accordance with the relative humidity of the model, but the cloud optical properties including liquid water paths were specified. They showed that the model could reproduce many of the observed larger-scale features of tropical cyclones. Moreover, many features of the global distribution of tropical cyclones was well represented, although considerably fewer storms were simulated in the North Pacific than observed, and storms were simulated in the South Atlantic where no storms are observed. The lack of storms in the South Atlantic is believed to be due to the fact that SSTs are below the critical values needed to support tropical cyclones. They found that the simulated frequency, duration, and number of storm days, was quite different for the fixed cloud and variable cloud experiments. In the fixed cloud case the model predicted an increase in storm frequency, duration, and number of storm days, whereas in the variable cloud experiment, the opposite occurred. This set of GCM numerical experiments shows just how difficult estimating regional responses to greenhouse warming really is and how important clouds are to determining those responses. One would expect that GCM simulations with cloud parameterizations that allow feedbacks due to changes in cloud optical properties and feedbacks due to changes in ocean circulations would produce considerably different responses in tropical cyclone frequency, duration, and number of storm days. At the moment it is pure speculation as to how tropical storms will respond to greenhouse warming.

An extension of the GCM approach is to nest a regional-scale model with a GCM (e.g., Dickinson et al., 1989; Giorgi, 1990). In this approach, GCM grid point data are used to specify boundary conditions for the regional-scale model. For example, Dickinson et al. nested a 60 km mesh mesoscale model to a 500 km mesh GCM. With its higher resolution, the mesoscale model was better able to resolve terrain features and thus better simulate orographically-forced wintertime precipitation in the mountainous western United States. The authors noted, however, that the simulated soil moisture and subsurface drainage for 3-5 day integration periods were strongly dependent upon the initial GCM soil moisture. They suggested that the mesoscale model may have to be run for periods of several months or longer to develop its own GCM-independent soil moisture fields. Moreover, the precipitation and clouds simulated by the mesoscale model are strongly dependent on the moisture, and vertical motion fields supplied by the GCM particularly during warm seasons and regions less strongly influenced by orography. Thus, if the global model makes errors in its gross depictions of the thermal, moisture, and circulations on regional scales, the mesoscale model will follow suit and provide higher resolution and more dramatic errors in its simulated cloud and precipitation fields! Furthermore, because of the higher variability on regional scales one cannot rely on a single simulation of a regional-scale model response to a single GCM simulation. A more meaningful (and costly approach) would be to first run an ensemble of GCM realizations in which each GCM initial state is perturbed by a small amount in regard to its initial soil moisture fields, sea-surface temperatures, ice and snow cover, and initial circulations. Each simulation from the perturbed initial field would be a realization from which ensemble-averaged, large-scale data could be computed. Moreover, a regional-scale model could be one-way nested to the GCM grid point data for selected periods of interest (i.e., winter season after 30 years of slow increase in CO_2) and rerun for each independent GCM realization. In this way an ensemble of regional-scale responses to the GCM realizations could be performed and both average and variances of regional-scale precipitation, temperatures, runoff, etc. could be computed. This would assist in determining the uncertainty in estimating the regional-scale responses to greenhouse warming.

In summary, estimating the regional-scale impacts of global climate change is an extremely difficult task and state-of-the-art numerical prediction is not ready for realistic, quantitative assessments of those impacts.

11.7 Assessing Cause and Effect

Are we now experiencing anthropogenically-produced greenhouse warming? This is a difficult question to answer just as it has been difficult to assess whether cloud seeding had any significant effect on surface rainfall. One must analyze the surface temperature record and then determine whether the observed variations are real (or due to sampling and instrument errors), and if they cannot be explained by *natural variability*. Only if the observed temperature trends exceed the *natural variability* of temperature can we be confident that human causes are involved in these changes. In this section we will first examine the sources of data and their quality, then the analyzed surface temperature records, followed by the analysis of upper air data and other data sources that could be used for a multivariate analysis of temperature trends.

11.7.1 Adequacy of surface temperature data

Ellsaesser et al. (1986) performed a very thorough review of surface data sources and problems with their representativeness and homogeneity. Land surface temperatures have been observed over a sufficiently large area to calculate global means for only a century. During that time, sensors have undergone significant changes, sites for measuring temperature have been moved, and the areas surrounding the stations have undergone substantial urbanization causing heat island influences on the sampled temperatures (Karl and Jones, 1989). Even in rural areas, the weather stations are often affected by the encroachment of asphalt airport runways, parking lots, and roads all of which yield a warm bias relative to the desired natural vegetation coverage of surrounding underlying surfaces. Ratner (1962) suggested that the dozens of extreme temperature records that had been recorded around the time of the publication of his paper were not the result of a sudden change in climate, but of a change in exposure for a large part of the network. These effects could easily yield uncertainties in temperature records on the order of $\pm 1°$ C, with a large potential of a warm bias. The problem of obtaining meaningful long-term temperature trends over the oceans is much greater than over land. Ship coverage was very sparse until this century; also sensors and methods of sampling have changed. Overall, we must be cautious in interpreting the measured surface temperature trends because the amplitude of measurement errors is of the same magnitude as the expected changes in global-averaged temperature produced by greenhouse warming and the magnitude of corrections and adjustments that have been made to those data are of similar magnitude (Lindzen, 1990).

11.7.2 Inferred changes in surface temperatures

Figure 11.11a shows the combined land and sea surface temperatures for the period from the late nineteenth century to the present. Most of the global warming that has occurred over the last century, however, did so before the recent rise in greenhouse gas concentrations. Between 1919 and 1986, global mean surface temperatures rose only 0.15° C, yet greenhouse gas concentrations increased 30% (Karl et al., 1989). From 1921 to 1979,

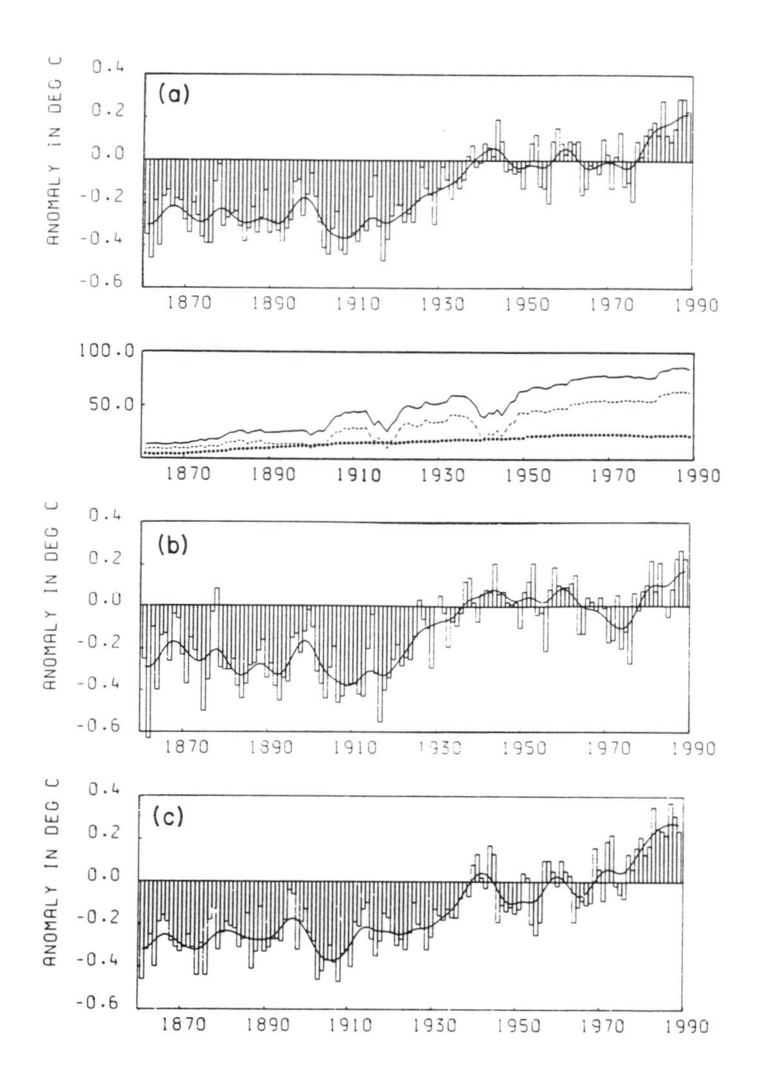

Figure 11.11: *Combined land air and sea surface temperatures, 1861-1989, relative to 1951-1980. Land air temperatures from P.D. Jones and sea surface temperatures from the UK Meteorological Office Farmer et al. (1989). Sea surface temperature component is the average of the two data sets. (a) Globe, (b) Northern Hemisphere, (c) Southern Hemisphere. Percentage coverage of the data is shown for Figure 11.11a, expressed as a percentage of total global surface area for land (dotted line) and ocean (dashed line) separately, and for combined data (solid line) plotted annually. 100% coverage would imply that all 5° × 5° boxes had data in two or three months in each season of the year.* [In *Climate Change. The IPCC Scientific Assessment.* J.T. Houghton, G.J. Jenkins, and J.J. Ephraums, Eds. Cambridge University Press, Cambridge, MA. ©Intergovernmental Panel on Climate Change.]

the global average land temperatures actually cooled. The data suggest that the earth has experienced a general increase of temperature from the period of the late nineteenth century to the present. The amount of warming, for the decades of 1880-1900 and the decade 1980-1990 is 0.45° C. The corresponding combined land and sea surface temperatures over the northern hemisphere exhibits a rapid rise in temperatures in the 1920s to 1930s, a leveling off and even cooling trend from the 1940s to 1976, followed by a sharp rise to the present. The overall average warming for the northern hemisphere is 0.42° C for the entire period. In contrast, the southern hemisphere experienced a more gradual rise in surface temperatures in the 1920s to the 1930s, followed by a period of little temperature change from the 1940s to the middle 1960s and a gradual increase in temperature to the present. The overall rise in surface temperature in the southern hemisphere is 0.48° C, which is slightly greater than the northern hemisphere.

Putting aside our concern about the representativeness of the sensors and sampling errors, do these data support the hypothesis that the temperature changes are due to anthropogenically-caused greenhouse warming? If we use the same level of objective criteria for evaluating cause and effect that has been used with the analysis of cloud seeding experiments, the answer is clearly no! First of all we must ask, are the analyzed temperature trends for the period greater than *natural variability* of temperature? Unfortunately the period of reliable record almost exactly coincides with the period of enhanced industrialization. Thus to assess *natural variability* we must have a global temperature record that goes back in time for periods of many centuries and perhaps even a thousand years or more. To do so requires going to other, less direct data sets in which temperature trends are inferred from paleoclimatological analyses. These records, which have a noise level far greater than our direct sensor measurements, nonetheless suggest that mean surface temperatures warmer than the present have occurred repeatedly without evidence of strong CO_2 forcing (Ellsaesser et al., 1986). Moreover, there are numerous other ways that human activity could have affected climate as discussed in other chapters in Part III.

Are the hemispheric and regional variations in surface temperatures consistent with simulated responses of GCMs to enhanced greenhouse gases? As shown in Figure 11.11b and c the Southern Hemisphere experienced a more consistent, gradual rise in mean surface temperatures than did the Northern Hemisphere. This is consistent with the expected greater thermal inertial over the southern hemisphere due to the greater surface coverage of oceans there and to the deeper vertical mixing in the southern oceans. Inconsistent with the GCM simulations, however, is the fact that the southern hemisphere has experienced more warming during the period than the northern hemisphere. The coupled ocean/atmospheric models suggest (e.g., Washington and Meehl, 1989) that the southern hemisphere should lag behind the northern hemisphere in its rate of warming due to enhanced greenhouse gases. Figure 11.12 illustrates calculated zonally-averaged, land surface temperatures and sea surface temperatures from the 1870s to the present. The Figure shows

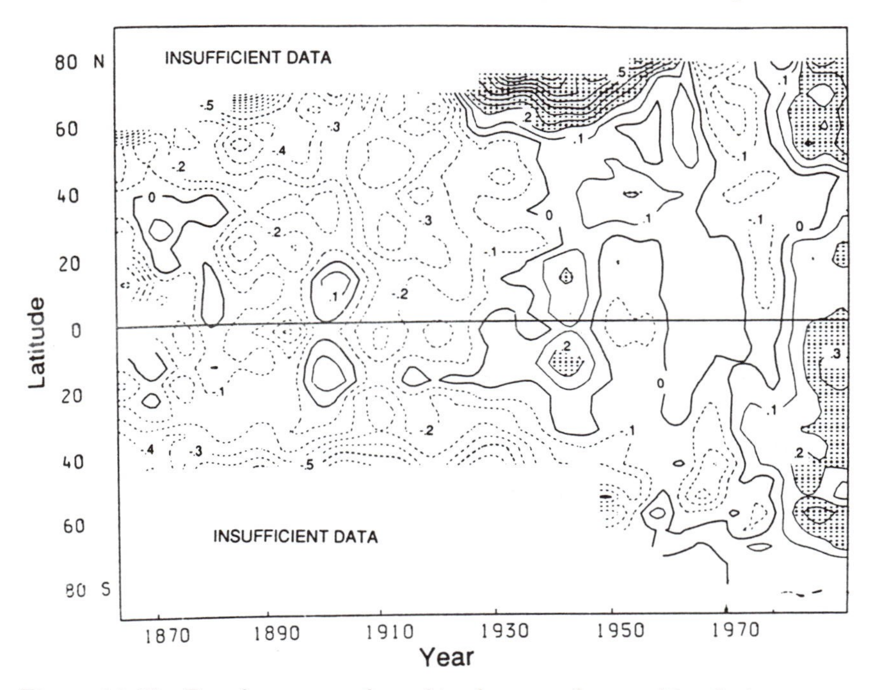

Figure 11.12: *Zonal average of combined sea surface and land air temperature data, 1861-1989. Land air temperatures from P.D. Jones and sea surface temperatures from the UK Meteorological Office.* [From Folland, C.K., T.R. Karl, K.Ya. Vinnikov, 1990: Observed climate variations and change. In *Climate Change. The IPCC Scientific Assessment.* J.T. Houghton, G.J. Jenkins, and J.J. Ephraums, Eds. Cambridge University Press, Cambridge, MA. ©Intergovernmental Panel on Climate Change.]

how variable surface temperatures are over the period. During the period 1920 to 1940 warming occurred in most zones, except the northern part of the southern oceans. Weak cooling is evident from 1950 to 1970 first in the southern hemisphere midlatitudes, followed by the tropical regions. Stronger cooling is evident in the northern hemisphere throughout the 1950 to 1970 period. Warming first appeared in the southern hemisphere in the early 1970s followed by warming in the northern hemisphere at all latitudes. While there is some evidence of stronger warming at high latitudes in the northern hemisphere relative to the southern hemisphere consistent with equilibrium responses of GCMs to enhanced greenhouse gases, this response is by no means a consistent, gradual rise as one might expect from the model results.

11.7.3 Other "fingerprints" of global warming

As discussed at the beginning of Section 11.7, the attempt to determine if variations in the observed global-averaged surface temperature record can be explained by human-induced greenhouse warming is similar to trying to see if cloud seeding resulted in increased rainfall on the ground. In both cases we are examining the potential response of a single variable to the hypothesized forcing function. As we have seen in cloud seeding, this approach requires large data records in order to determine if the signal (i.e., rainfall or temperature) exceeds the *natural variability* of the system. Often in cloud seeding the experiments were not run long enough to be immune from the 'chance' occurrence of a few greatly above or below normal events which then dominated the average statistics. The situation is not much different in attempting to determine if anthropogenically-induced greenhouse warming is the cause of observed changes in global mean temperatures. For short records (in this case a century) a few anomalous years can strongly affect the average behavior and therefore prevent unambiguous detection of cause and effect. This is clearly the situation we are in now with respect to global warming.

An alternate approach that is being used in recent years to detect seeding-induced changes in precipitation is also being applied to detecting greenhouse-induced warming. The approach is called a multivariate analysis, in which several different variables are hypothesized to respond to enhanced concentrations of greenhouse gases. These could be changes in temperature at a number of levels in the atmosphere, or temperature changes with latitude, or other alterations such as in rainfall amount as predicted by GCMs. This is analogous to identifying people by their fingerprints since if a set of variables all respond in a way that is consistent with the predictions of a GCM, then the greenhouse signal can be uniquely identified without having to analyze centuries of data. In order to apply the fingerprint method of detecting anthropogenically-induced greenhouse warming, we use GCM simulations as guidance in determining which variables are likely to have a consistent response to greenhouse warming and are likely to yield a high signal to noise ratio. One problem of this approach is that since we must be dependent upon upper air data to obtain additional response variables, the length of the time record where good global coverage of upper air data is available is quite limited being on the order of 30 years or so.

Angell (1988), for example, examined upper air sounding data to identify such features as warming throughout the troposphere, cooling in the lower stratosphere, and stronger warming at polar latitudes in winter; all features that have been simulated by GCMs for the doubled CO_2 scenario. He found that during the period there is a marginally significant increase in global mean temperature throughout the troposphere and a significant decrease in global mean temperature in the lower stratosphere during the latter part of the period (1973-1987). He also found that there was greater surface warming at both north and south polar latitudes in winter during the 30 year period, but on an annual average there was a slight decrease in surface and upper air temperatures in the north polar and middle latitudes. The latter decreases are inconsistent with model predictions. While the trend for warming in the troposphere to cooling in the lower stratosphere is generally consistent with simulated greenhouse warming, this trend could have been caused by other factors and the level at which the reversal from warming to cooling is observed to occur is lower than simulated by the GCMs. This could be due to the poor vertical resolution of the GCMs, however.

More recently, Angell (1990, personal communication) reported unusually warm surface air temperatures in the Arctic in the Spring of 1990 of 4°C above average and 2.5°C higher than previously observed during 30 years of records. The Arctic stratosphere, in contrast, was abnormally cold with values 6°C below normal in the 9 to 21 km range. Using 63 measurement sites around the earth, Angell found that the March-May 1990 values were 0.3°C warmer than reported previously during this 30 year period of record. Angell also found that the global increase in air temperature recorded between the 1960s and 1980s is reduced by about a third when effects of El Niños are removed from the data and if El Niños are assumed independent of global temperature change. He finds that when adjusting the temperature record to account for the El Niño effect, 1988 was the warmest year in the period 1958 to 1989. The World Meteorological Organization reported in July 1990 that the global average annual surface temperature in 1989 was 0.23°C greater than the 1951-1980 average.

There are other response variables that one could examine, such as global mean atmospheric moisture content and sea level. The GCMs also predict the greatest greenhouse warming to occur at high latitudes. However, as shown in Figures 11.13 and 11.14 reproduced from the NOAA Environmental Digest (1990), neither sea ice extent in the Arctic or Antarctic areas, nor monthly snow cover data for North America and Eurasia show any systematic trends, although permafrost temperature records show a warming of 2° C to 4° C in northern Alaska (La Brecque, 1989b). Michaels et al. (1990), for instance used 1000-500 mb thickness data over North America to show that the lower half of the troposphere has not warmed appreciably. Remember, however, that North American represents only a small percentage of the Earth's surface.

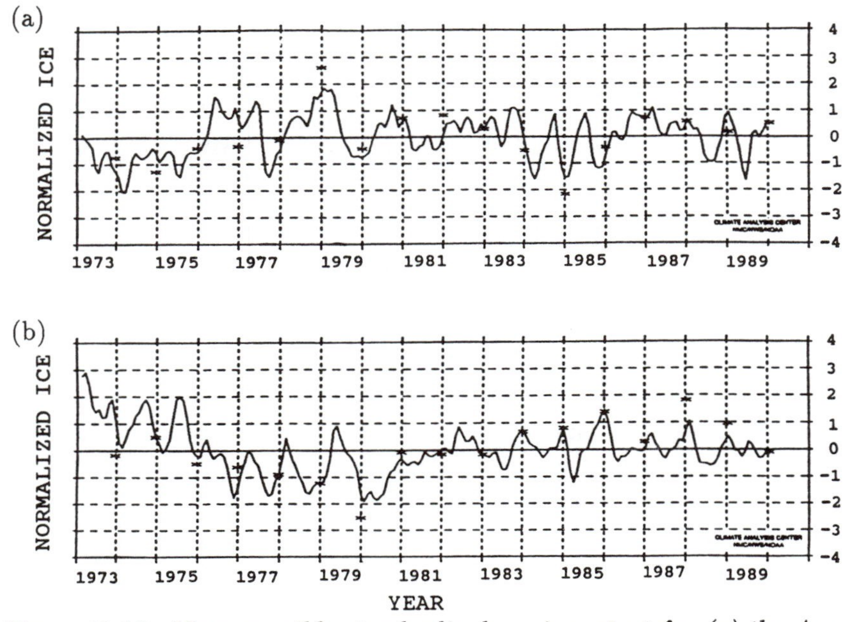

Figure 11.13: *Mean monthly standardized sea-ice extent for (a) the Arctic, and (b) the Antarctic, for 1973 to 1989. The December standardized values for each year are marked by an *. Solid lines represent a 3-month running mean. Courtesy of the Climate Analysis Center, NOAA National Weather Service.* [From NOAA Environmental Digest, 1990: Selected environmental indicators of the United States and the Global Environment. September 1990, Office of the Chief Scientist, 66 pp. plus appendices.]

11.8 Status of the Greenhouse Hypothesis

The greenhouse hypothesis is clearly a physically consistent hypothesis that has strong theoretical support for its relevance to global climate change. The major uncertainties with the theory are related to the various feedback processes, a number of which either delay the onset of global warming or weaken it to such an extent that it may never be detectable from the noise of the *natural variability* of climate. While the analysis of a number of potential response variables to greenhouse-induced climatic changes suggest we could be experiencing the early symptoms of greenhouse warming, it is premature to state categorically that a greenhouse effect is already being observed (Angell, 1988). Many greenhouse theory supporters point out that the 1980s were the warmest period on record, and that world-wide high temperature records are falling every year. One must keep in perspective, however, that we have a very short period of global temperature records compared to the time-scales of *natural cli-*

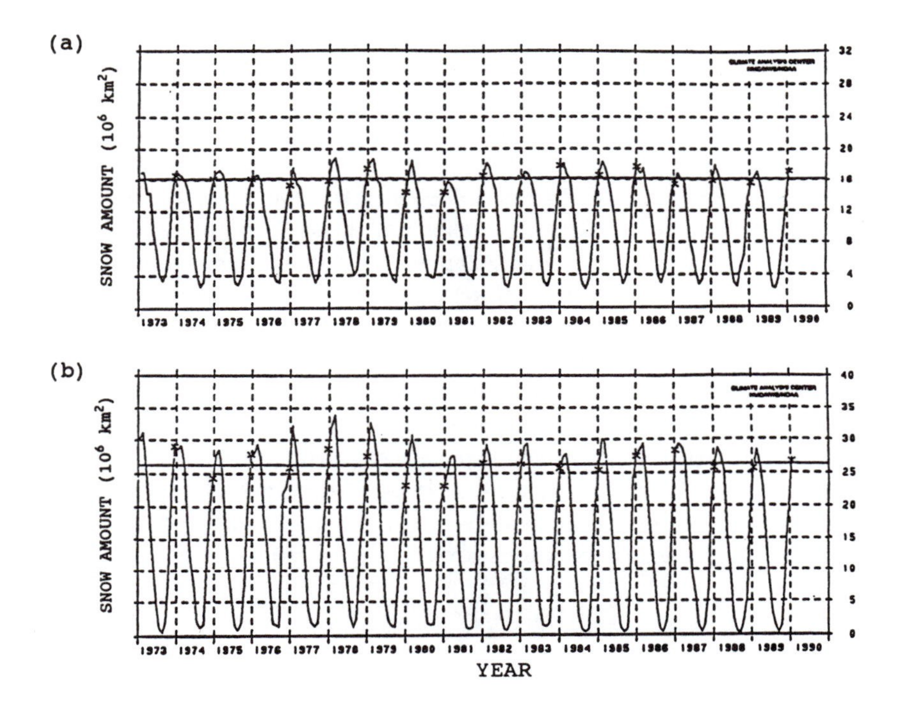

Figure 11.14: *Monthly extent of snow cover (in 10^6 km^2 for (a) North America, and (b) Eurasia. The December snow cover extent is designated by an * for each year. Solid horizontal lines represent 1973-1989 mean snow cover extent. Courtesy of the Climate Analysis Center, NOAA National Weather Service.* [From NOAA Environmental Digest, 1990: Selected environmental indicators of the United States and the Global Environment. September 1990, Office of the Chief Scientist, 66 pp. plus appendices.]

mate variability and even those records contain a great deal of 'noise' due to instrument errors, biases caused by siting, etc.

Some proponents also claim that a higher frequency of severe weather and flash floods or tropical cyclones in a given season is evidence that we are experiencing the consequences of human-induced greenhouse warming. These claims are no more credible than the cloud seeder's claim that a given seeded cloud produced more rain than a nearby similar cloud or that seeding once a week in New Mexico produced heavy rainfall with a once a week peak in the central United States! Houghton et al. (1990) predicted that if the current rate of increase of greenhouse gas emissions continues through the next century, then global mean temperatures should rise 0.3° C per decade with an uncertainty of 0.2° C to 0.5° C per decade. This would result in the global mean temperature being 1° C above the present value by 2025 and 3° C before the end of the twenty-first century. Considering the uncertainties in model simulations (really not predictions) due to feedback processes associated with clouds and the oceans, and due to *natural variability* of climate to a variety of internal and external (to the earth) phenomena, these predictions should be viewed as pure speculation.

In summary, we should not underestimate the difficulty in understanding the effect of greenhouse gas emissions on climate change. It will probably take decades of modeling and data analysis before the many scientific issues are satisfactorily resolved. The question is, with our capricious approach to management of science in the United States and in other developed countries, will funding in climate change be maintained at sufficiently high levels for a long enough period of time to resolve these issues or will the climate programs experience the crash that the science of weather modification experienced? We are confident, however, in stating that the claims that variations in weather and climate occurring now and in the next several decades are a direct result of the human production of greenhouse gases is extremely anthropocentric!

Chapter 12

Human Impacts on Biosphere Forcing of Climatic Variability

In Chapter 6, we discussed changes in land surface processes as a result of human activities which have apparently altered local and regional weather and climate. This chapter will investigate the possible cumulative effect of these landscape changes as well as other biospheric forcing on global climate.

12.1 Landuse Changes

Estimates of the earth's landscape which have been disturbed from their natural state vary according to how the disturbance is defined. In terms of global cultivated land, Dudal (1987) indicates that 14,610,000 km^2 of a potential cultivated coverage of 30,310,000 km^2 are presently being utilized. Since the earth's land surface covers 133,920,000 km^2, this indicates that 10.9% of the landscape is cultivated, with the potential level reaching 22.6% coverage.

This value of land disturbance due to human activities is an underestimate, however. Human activities also include domestic grazing of semi-arid regions, urbanization, drainage of wetlands, and alterations in species composition due to the introduction of exotic trees and grasses. In the United States, for example, 426,000 km^2 (4.2% of the total land area) have been artificially drained (Richards, 1986).

In China, of the 2,000,000 km^2 in the temperature arid and semi-arid grassland regions, hundreds of thousands of km^2 have been degraded due to overgrazing and the overextension of agriculture, often to the extent that desertification has occurred (Committee on Scholarly Communication with the People's Republic of China, 1992).

The influence of vegetation on climate includes it's influence on albedo, water-holding capacity of the soil, stomatal resistance to water vapor transfer, aerodynamic roughness of the surface and effect on snow cover. These effects are seasonally varying and are essential in GCM simulations of the effect of vegetation removal on climate change (Rind, 1984).

The sensitivity of global climate to even small changes in surface properties has been discussed in Pielke (1991). As described in that paper, an increase of the average albedo of the land surface of the earth as viewed from space of 4% would result in a lowering of the earth's equilibrium temperature by 2.4°C. A decrease of this average albedo by 4% would increase the equilibrium temperature by 2.4°C. This is on the same order as the estimates of a global *surface* equilibrium temperature change in GCM doubled CO_2 radiative-convective model calculations of 0.5°C to around 4°C (see Table 11.1).

The simple analysis presented in Pielke (1991) ignores how such a temperature change would be distributed with height or geographically. Nonetheless even if the change of equilibrium temperature is reduced by one-half as a result of these effects, the importance of the landscape energy budget to global climate is obvious.

There have been almost no global quantitative evaluations of changes in the heat and moisture fluxes at the surface due to human activity. One pioneering work that has been completed in this area is Otterman (1977) who estimated the change in the earth's surface albedo over the last few thousand years due to overgrazing. He has concluded that overgrazing results in a higher albedo of the trampled, crumbled soil than in the original steppe where there was dark plant debris accumulating on a crusted soil surface (also see Chapter 6 for a discussion of overgrazing). Otterman estimated that the current earth surface albedo is 0.154 whereas before anthropogenically-caused grazing (6000 years before the present), this albedo value was 0.141. This albedo change could have resulted in an average cooling effect of 0.77° in the earth's equilibrium temperature. Hartmann (1984) suggested that a shift of precipitation between land and ocean areas at low latitude can also lead to planetary albedo changes as a result of alterations in vegetation and through changes in cloud cover distribution resulting from these changes in land surface albedo. He concluded that this albedo change could result in an approximately 5°C global mean temperature change without any other feedbacks.

So far we have discussed a change in a mean parameter (i.e., the average land surface albedo). As shown in Chapter 6, using satellite observations and numerical model simulations, there is considerable spatial structure in landscape on scales of tens of kilometers and smaller. This heterogeneity of landscape will result in fluxes of moisture and heat to the atmosphere over a range of spatial and temporal scales as illustrated, for example, in Figures 12.1 and 12.2. In those Figures, surface heat

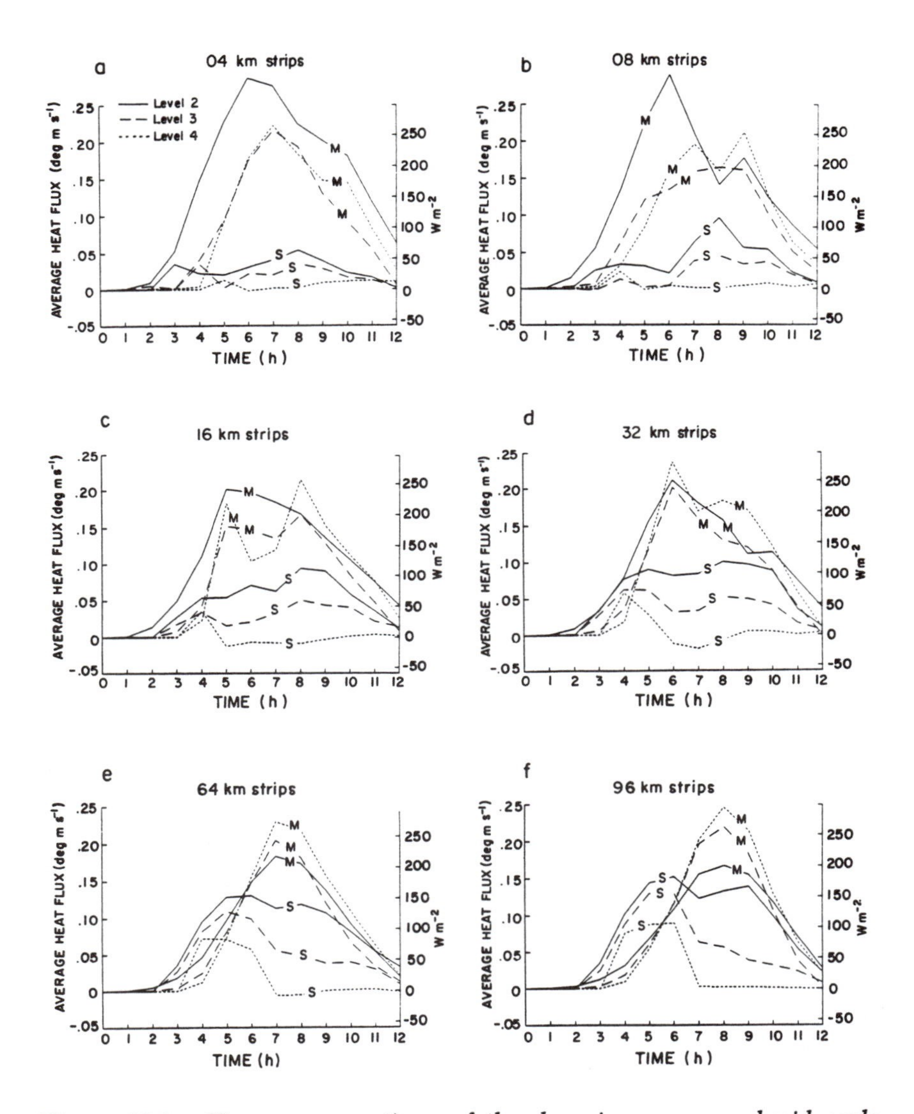

Figure 12.1: *Time cross sections of the domain average subgrid scale vertical heat flux* (S) *and the domain average explicit vertical heat flux due to mesoscale circulations* (M) *for model levels 2–4 (150 m, 315 m, 497 m). The surface boundary is (a) 4 km alternating land and water strips, (b) 8 km strips, (c) 16 km strips, (d) 32 km strips, (e) 64 km strips, and (f) 96 km strips. The initial time is 0600 LST.* [From Pielke, R.A., G. Dalu, J.S. Snook, T.J. Lee, and T.G.F. Kittel, 1991b: Nonlinear influence of mesoscale landuse on weather and climate. *J. Climate*, **4**, 1053-1069.]

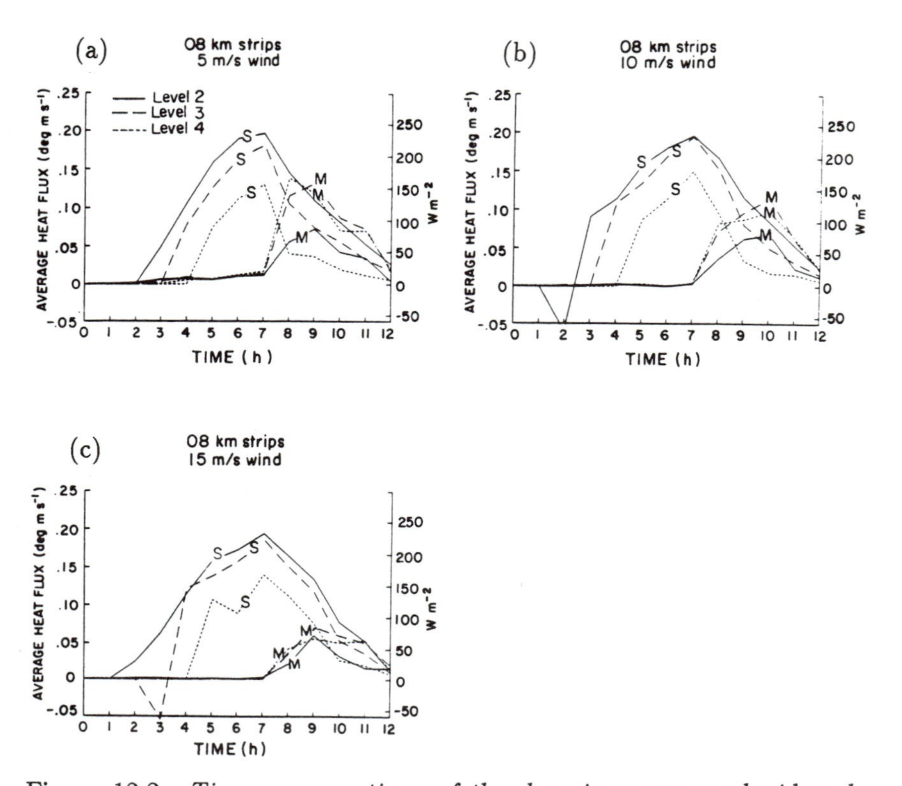

Figure 12.2: *Time cross sections of the domain average subgrid scale vertical heat flux* (S) *and the domain average explicit vertical heat flux due to mesoscale circulations* (M) *for model levels 2–4 (150 m, 315 m, 497 m) with an 8 km alternating land and water surface boundary. Model simulations were initialized with a uniform horizontal wind of (a) 5 m* s^{-1}*, (b) 10 m* s^{-1}*, and (c) 15 m* s^{-1}. [From Pielke, R.A., G. Dalu, J.S. Snook, T.J. Lee, and T.G.F. Kittel, 1991b: Nonlinear influence of mesoscale landuse on weather and climate. *J. Climate*, **4**, 1053-1069.]

patches of various horizontal sizes are inserted into a numerical model simulation of the diurnal heating of the boundary layer. The spatial gradients in sensible heat flux to the atmosphere that result from the spatially varying surface heating create significant mesoscale heat fluxes to the atmosphere, particularly when the large-scale prevailing wind is weak (Figure 12.2). With a horizontally homogeneously heated surface, these mesoscale heat fluxes, of course, would not appear.

Cloud cover changes as a result of landscape variability and alterations in heat and moisture fluxes could magnify the impact of these mesoscale fluxes on global climate change. Two types of landscape variability must be evaluated – those due to human activity (e.g., landuse), and those which occur naturally (e.g., due to spatially variable precipitation, geological formations, terrain, etc.). Figures 6.2, 6.6, and 6.7 present examples of this variability. Figures 12.3 and 12.4 from Rodriguez (1992) show examples of albedo variations resulting from spatial variations in surface geology, vegetation, and snow cover (in the winter) for horizontal scales of 1 km and larger. In the winter scene (Figure 12.3), as reported in Pielke et al. (1990d), the average albedo at the time of the satellite image (1100 LST) was 38.8% with a standard deviation of 11.2%. The range of values were from 18% to 84%. In the summer image, the average albedo was 20.3% with a standard deviation of 3.9% and a range of values of 12.6% to 49.8%. These large horizontal variations in albedo would be expected to produce major variations in sensible heat flux to the atmosphere during the day, thus resulting in large mesoscale heat fluxes such as simulated in Figure 12.2.

Similarly, a spectral analysis of topography shows large horizontal variations at scales smaller than GCM grid resolution. Figures 12.5 and 12.6 present terrain spectra evaluated for the Blue Ridge Mountains of Virginia (Pielke and Kennedy, 1980) and for the Colorado Front Range (Young and Pielke, 1983; Young et al., 1984). In Virginia, most of the spatial structure is for horizontal scales larger than 2 km, while in Colorado there is substantial variability even at the highest resolution of the terrain data. This small-scale variability needs to be represented in GCMs as it influences terrain wave drag (e.g., such as described by Baines and Palmer, 1990 and mountain-valley thermally forced circulations).

The GCMs discussed in Chapter 7 have thus far ignored most types of landscape variability on scales smaller than their grid resolution. Vegetation effects, for example, are treated using a simple biome[1] representation at each grid point, although the need to include spatial variability is recognized. As reported in Henderson-Sellers (1990), a component of the Global Energy and Water Cycle Experiment (GEWEX) will conduct field experiments to test the accuracy of one-dimensional parameterizations of atmosphere-vegetation interactions and to assess mesoscale variability in the atmosphere due to land surface variations.

The current parameterizations used in GCMs include the Simple Biosphere (SiB) model (Sellers, 1991) and the Biosphere-Atmosphere Transfer Scheme (BATS) summarized in Dickinson and Henderson-Sellers (1988). In mesoscale models, the Land Ecosystem-Atmosphere Feedback (LEAF) model has been used (Lee, 1992). All three of these parameterizations are summarized in Lee et al. (1991). While each of this parameterizations consider the moisture and heat fluxes between the soil, vegetation, and the lower atmosphere, a "representative" plant characteristic must be used. Whether such an idealization is reasonable despite the substantial heterogeneity of landscape on GCM (and mesoscale) grid scales (Avissar and Verstraete, 1990) is not known.

[1]Biomes represent broad ecological classifications of vegetation.

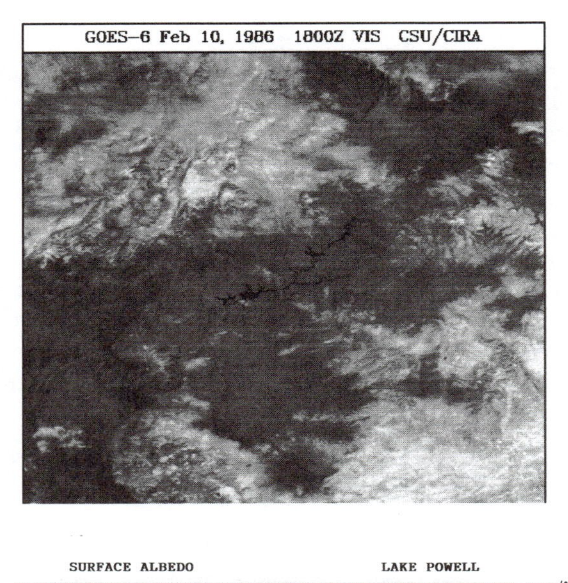

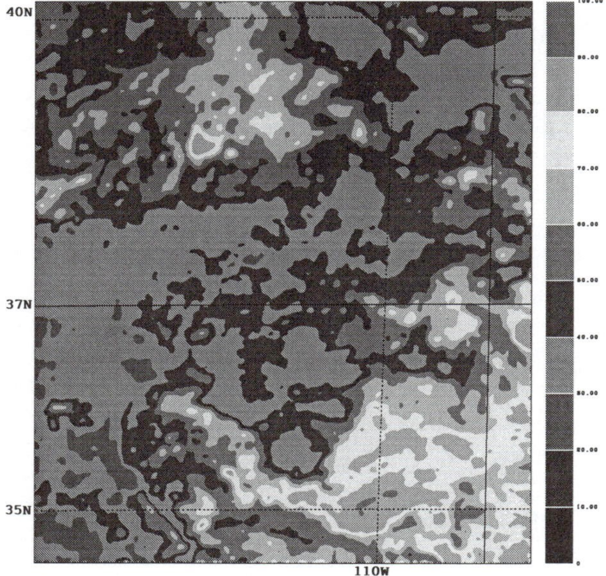

Figure 12.3: *(a) GOES-6 visible satellite image for February 10, 1986 at 1100 LST. (b) Smoothed surface albedo distribution for February 10, 1986 at 1100 LST (contouring every 10%).* [From Rodriguez, H., 1992: Influence of mesoscale variations in land surface albedo on large-scale averaged heat fluxes. M.S. Thesis, Department of Atmospheric Science, Colorado State University.]

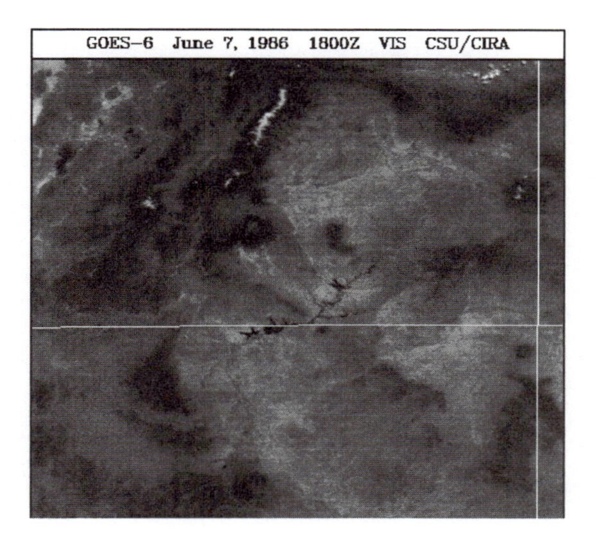

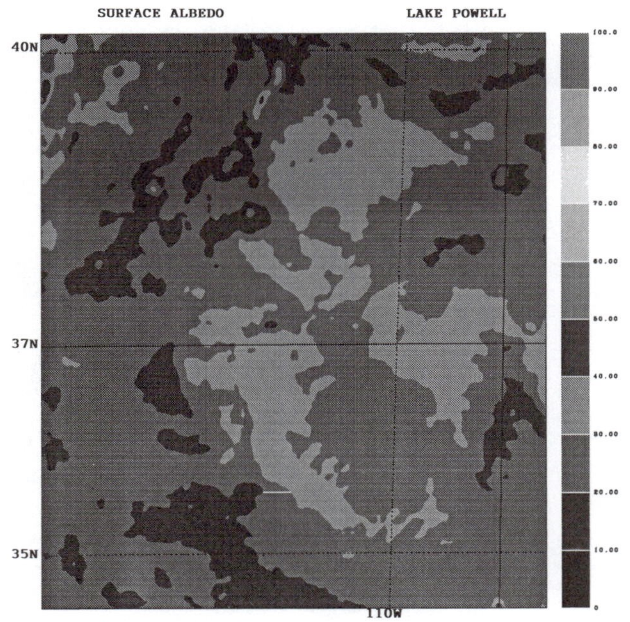

Figure 12.4: *(a) GOES-6 visible satellite image for June 7, 1986 at 1100 LST. (b) Smoothed surface albedo distribution for June 7, 1986 at 1100 LST (contouring every 10%).* [From Rodriguez, H., 1992: Influence of mesoscale variations in land surface albedo on large-scale averaged heat fluxes. M.S. Thesis, Department of Atmospheric Science, Colorado State University.]

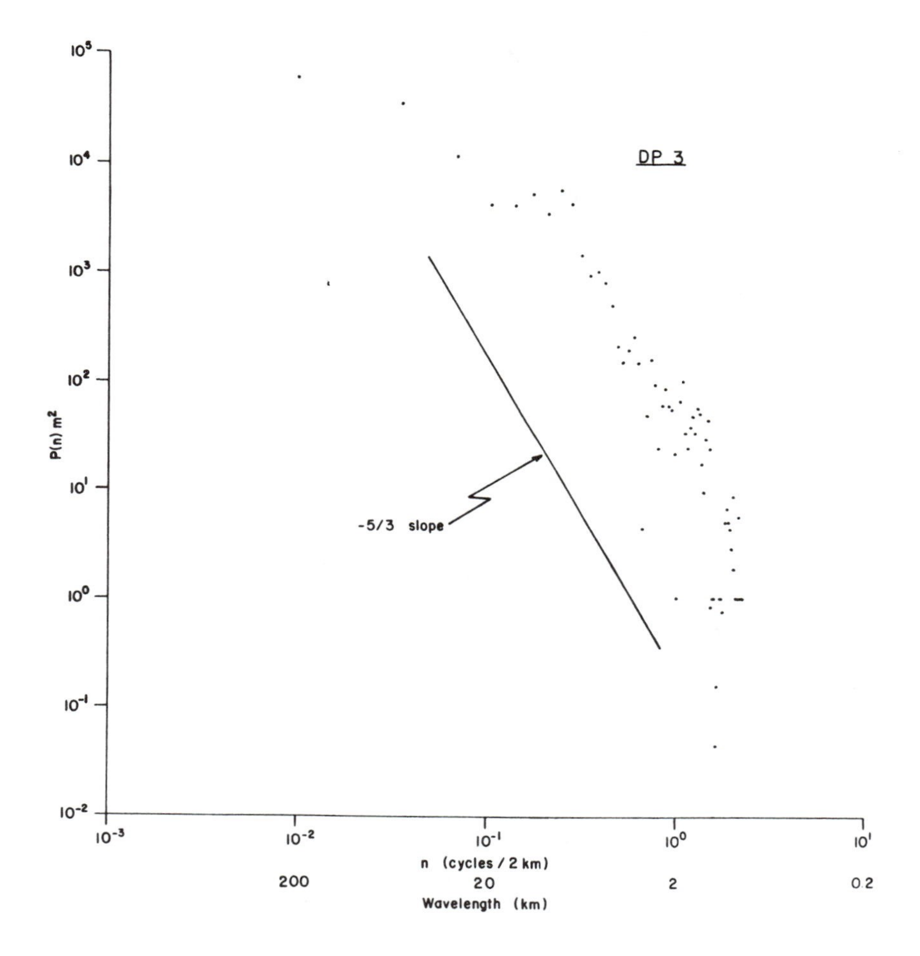

Figure 12.5: *Blue Ridge terrain spectra for a cross section through central Virginia on a northwest and southeast line near Waynesboro, Virginia.*[From Pielke, R.A. and E. Kennedy, 1980: Mesoscale terrain features, January 1980. Report # UVA-ENV SCI-MESO-1980-1, 29 pp.]

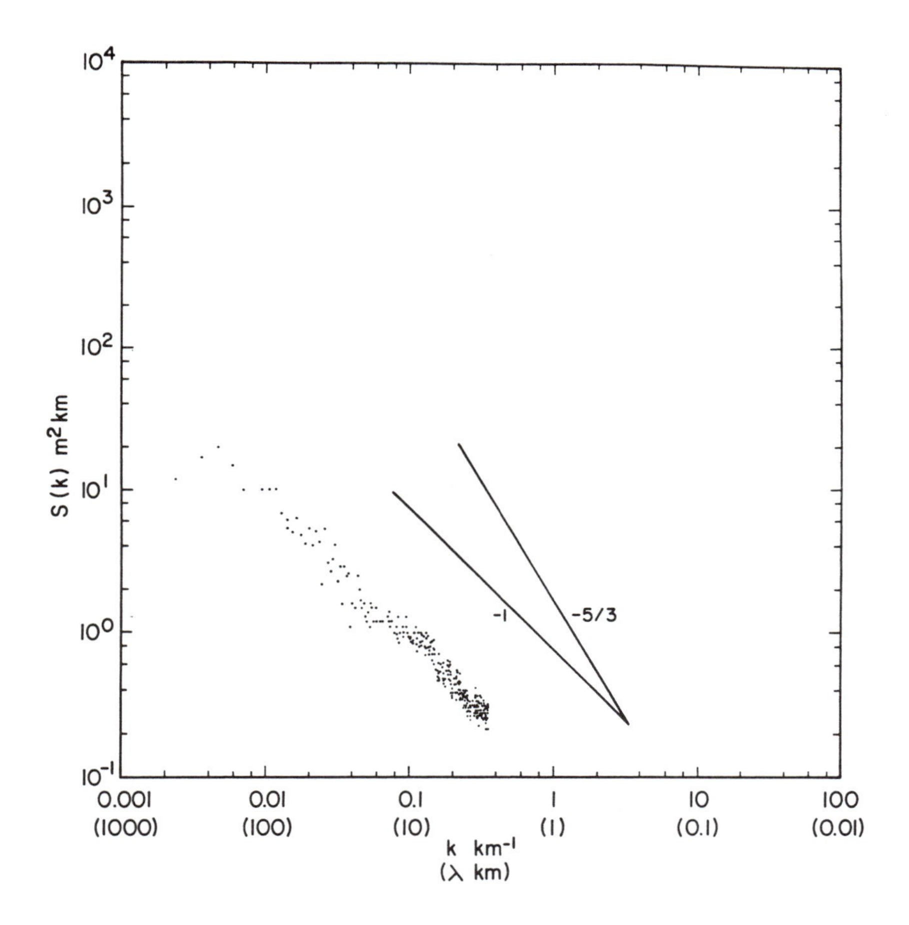

Figure 12.6: *Average terrain height variance spectra for the multiple cross section. Steamboat Springs, Colorado data set. Data were taken from 20, 253 km long, vertical sections centered along 40° 30′ N at Steamboat Springs in northwestern Colorado.* [From Young, G.S. and R.A. Pielke, 1983: Application of terrain height variance spectra to mesoscale modeling. *J. Atmos. Sci.*, **40**, 2555-2560.]

The response of the biosphere to changes in climate and to trace gas composition also needs to be explored. Current ecosystem models (e.g., see Pielke et al., 1991b) use climatological data or atmospheric model information as specified input. In reality, of course, the biospheric response to climate can be such as to alter the subsequent climate as has apparently occurred with the extreme case of desertification.

12.2 Atmospheric-Vegetation Interactions

The atmosphere and biosphere are coupled, dynamic systems. On the long time scales, the major vegetation biomes of the earth propagate in response to climate change. During the Pleistocene, for example, spruce and fir forests covered what is now northern Florida, as a vast continental ice sheet covered much of northeastern North America. In the eastern plains of Colorado, large sand dunes developed about 3000 to 1500 years before the present (B.P.) at a time of greater aridity than exists today (Muhs, 1985). Human agricultural systems are also affected. Neumann and Sigrist (1978) conclude that in Babylonia about 3800 to 3650 years B.P. the barley harvest was 10 to 20 days earlier than it is now due to a warmer climate, while cooler weather from the years 2600 to 2400 years B.P. resulted in a harvest 10-20 days later than under the current climate in this region.

On shorter time scales, Schwartz and Karl (1990) report that the seasonal upward progression of temperature in the eastern United States is temporarily halted as vegetation greens in the spring. At least a 3.5°C reduction in surface daily maximum temperature occurred over the subsequent two week period after the beginning of leafing in agricultural inland areas of the central and eastern United States, as contrasted with the two weeks prior to first leaf. This occurred even when the overlying atmosphere had the same lower tropospheric average temperature. On a hemispheric scale, the seasonal pulsations of carbon dioxide concentrations are clearly evident in Figure 11.2 as vegetation extracts CO_2 in spring and summer in the vast land areas of the northern hemisphere. The importance of plants and organic soils as a longer-term sink for CO_2 is becoming better recognized since it has been found that oceans are absorbing less of this gas than previously thought (*DAR Bulletin*, 1990). Since only about one-half of the anthropogenic increase of CO_2 gas can be accounted for in the atmosphere, much of the remainder must be entering the biosphere.

On daily time scales, vegetation interacts with the atmosphere through its direct influence on the partitioning between latent and sensible heat fluxes, as discussed in Chapter 6. During the day, over regions of transpiring vegetation, huge fluxes of moisture to the atmosphere and a significant reduced surface heating effect occurs. At night the stoma on the leaves close and respiration of oxygen dominates over photosynthesis and the uptake of carbon dioxide by plants.

The interaction of the biosphere and the atmosphere as a mechanism to minimize the net change in either environment in response to externally (or anthropogenically imposed) forcing is a form of the Gaia hypothesis introduced by Lovelock (1988). Lovelock has introduced the term "geophysiology" to describe this integrative self-regulation of a planet. He states in his 1990 *Nature* article that

> "The geophysical and geochemical evolution of the terrestrial planets is progressive and towards states like those of Mars and Venus now. During this evolution there will be a period when conditions are favorable for life and, to survive, the organisms must become abundant enough to affect and become coupled to the geochemical evolution: mere adaptation is not enough. If they fail, planetary conditions will continue to change inorganically until the point is reached when life is impossible."

Thus, according to the Gaia hypothesis, it is the development of life forms and their interaction with the physical world that sustains a world which continues to be hospitable for life.

Lovelock (1990) offers the following predictions of the Gaia hypothesis and their current status of validation.

- Mars will be lifeless as determined by atmospheric evidence (1968): strongly confirmed in the 1977 Viking mission (Hitchcock and Lovelock, 1967).

- Organisms make compounds that can transfer essential elements from ocean to land surfaces (1971): dimethylsulphide (DMS) and methyl iodide found (Lovelock et al., 1972) (see Chapter 9 where the influence of DMS on cloud formation is discussed).

- Climate regulation by control of CO_2 through biologically enhanced rock weathering (1981): micro-organisms greatly increase rock weathering (Lovelock and Whitfield, 1982; Schwartzman and Volk, 1989).

- Climate regulation through cloud density control linked to algal sulphur gas emission (1987): still under test (Charlson et al., 1987).

- Oxygen has remained at 21$\pm$5% of the earth's atmosphere for the past 200 million years (1973): oxygen levels are regulated by fires and by phosphorus cycling (Lovelock, 1988; Kump, 1988).

- Ancient earth atmospheric chemistry dominated by methane (1988): still under test (Watson and Lovelock, 1983; Lovelock, 1988).

A quarterly journal entitled *Gaian Science* is being published in the United States to promote geophysiological research. The May-July 1991 issue contains an interview of James Lovelock.

The equilibrium in ocean salinity has also been proposed as support for the Gaia hypothesis. As summarized in Mann (1991), geological weathering and runoff into the oceans should have made the oceans more saline over time. Geological evidence shows, however, that salinity has remained at less than 10% of saturation for hundreds of millions of years. Mann suggests that excess salts are removed into salt flats in the ocean floor by salt-tolerant bacteria.

Lovelock cautions, however, that while perturbations to the equilibrium can be masked (e.g., through increased low-level cloudiness, thereby increasing the earth's albedo as anthropogenic greenhouse gases increase), if the perturbation is sufficiently large, catastrophic climate change can occur when the interactive system is overwhelmed.

There are suggestions, however, that even small perturbations to the interactive geophysiological earth may result in significant changes in the resultant planetary biospheric and atmospheric characterization. Watson and Lovelock (1983) presented the Daisyworld model as an idealized example of the Gaia hypothesis. In the Daisyworld model, white and black daisies evolve in land coverage in response to changes in solar output, with the white daisies (e.g., greater albedo) preferentially developing when the sun's output increases. Their model assumes a continuous birth and death of daisies. In Zeng et al. (1990) the Daisyworld model was expressed as a discrete model since the birth and death of daisies occurs over a finite generation time associated with a season. This discrete model was then used to demonstrate that if the relation between the successful germination of daisies and temperature were sufficiently nonlinear, a chaotic response as a function of generation would result. Figure 12.7 presents an example of the equilibrium temperature and area coverage of white and black daisies as a function of generations along with the frequency spectra of the model. The latter result, as well as an evaluation of the model, mathematically demonstrated that a chaotic response results. The implication for the actual atmosphere-ecosystem interaction is that man-caused landscape changes could result in very substantial irregular changes in the global climate even for small perturbations if the interactions are sufficiently nonlinear. While this result does not refute the Gaia hypothesis, it does demonstrate that a more realistic form of the Daisyworld model than used by Watson and Lovelock (1983) is sensitive to the choice of initial conditions and *is not resistant* to small perturbations.

The importance of biological interactions with the atmosphere is only now becoming recognized. The ability of vegetation to effectively utilize increased carbon dioxide to increase global biomass may result in reduced carbon dioxide warming from what otherwise would occur. We need to determine, however, the impact of deforestation and conversion of

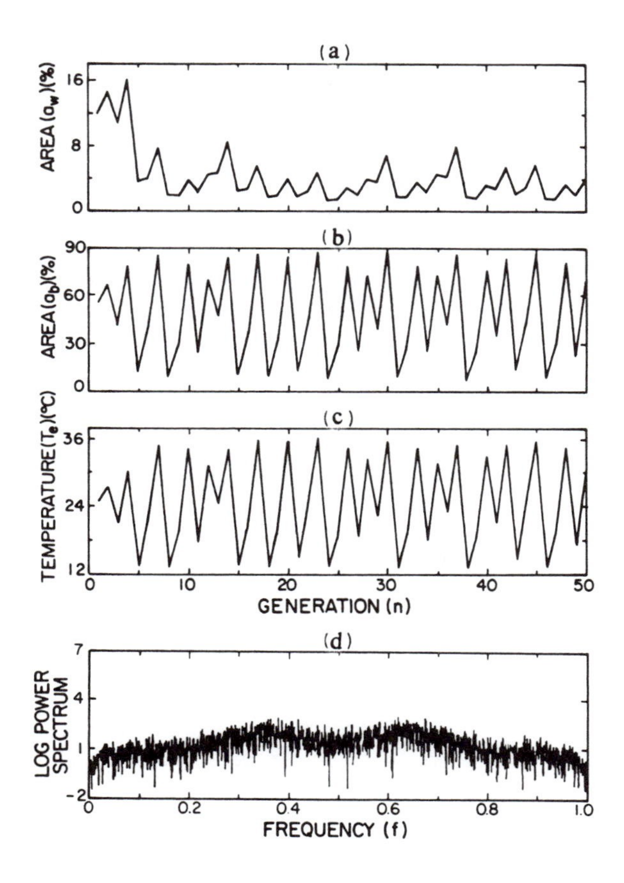

Figure 12.7: *Variation with generation of (a) percent of land area of white daisies; (b) percent of land area of black daisies; (c) surface temperature; and (d) spectrum of temperature variation as a function of frequency from the Daisyworld model.* [From Zeng, X., R.A. Pielke, and R. Eykholt, 1990: Chaos in Daisyworld. *Tellus*, **42B**, 309-318.]

vegetated land to asphalt, roads, industrial centers, etc. on this potential storage mechanism for carbon. Also, how are methane and nitrogen oxide inputs to the atmosphere from industrial and agricultural practices influenced by the presence of plants? Does enhanced nitrogen permit enriched fertilization for plant growth?

Considerable research is needed to satisfactorily answer these important questions. At present we do not even have an inventory of current landuse and land cover type at sufficient spatial resolution (e.g., 1 km intervals), much less what the land cover was prior to human disturbance. The IGBP-DIS project, initiated an effort in 1991 to obtain the *current* inventory, but it will be several years before this information will be available. To our knowledge, however, there have been no complete global studies as to the extent anthropogenic activities have altered the landscape of our planet. Without that knowledge, we cannot evaluate the importance of landuse on climate change as contrasted with climate change due to anthropogenic emissions of gases and aerosols.

Epilogue

Throughout this book a number of common themes reappear. Some of these common themes were apparent to us before we began preparing the book, and in fact, they were a major motivation for us to write the book. Others, became more and more apparent as we performed the literature survey and analysis, and prepared the written text. Some of these common themes are:

- the importance of natural variability,
- the dangers of overselling,
- the capricious administration of science, and
- scientific credibility and advocacy.

Let us examine each of these common themes more fully.

E.1 The Importance of Natural Variability

We have seen that our ability of determining if human activity *causes* some *observed or hypothesized effect*, such as changes in local rainfall or global climate, is strongly dependent upon the *natural variability* of the system. This has certainly been true of determining if cloud seeding actually is capable of causing increases of rainfall in specified target areas. However, it is also true in assessing if anthropogenic greenhouse gas emissions, or deforestation, or release of CCN have any significant impact on global climate. While the time and space scales are very different, nonetheless the bottom line in examining potential human-caused effects is: *are these effects large enough in magnitude to be extricated from the 'noise' of the natural variability of the system?* As we have seen there are few, if any, cases in which we can answer this question affirmatively. Ice cores have shown, for example, that a switch from an ice age climate to a non-ice age environment can occur over only a few decades (La Brecque, 1989a) without human intervention.

E.2 The Dangers of Overselling

We have seen that funding of the science of weather modification underwent a period of rapid rise, followed by an abrupt crash. One of the leading causes of that crash, we believe, is that the program was oversold. The claims that only a few more years of research and development will lead to a scientifically-proven technology that will contribute substantially to water management and severe weather abatement, were either great exaggerations, or just false. This is largely because we greatly underestimated the complexity of the scientific and technological problems we were (and still are) faced with. This tendency to oversell can even be taken to the absurd such as the "pork-barrel" funding of the University of Alaska in Fairbanks of more than 57 million dollars over the last several years through the efforts of Alaskan Senator Ted Stevens (Cohen, 1991). This funding has as one of its objectives the harnessing of the Aurora Borealis for electric power generation – a concept which has no scientific basis.

The same can be said about human impacts on global climate. There are many scientists who are claiming that the short-term (periods of year-to-year, or decades) variations in weather and climate are clear evidence that we are experiencing the effects of anthropogenic greenhouse emissions. Moreover, many claim that the 'forecasts' being made by global climate models, represent realistic expectations of global-averaged changes in temperature and rainfall in the next decade or century. In our opinion, both of these claims represent overselling of the climate program. These claims appear and are discussed in both the professional literature (e.g., Schneider, 1990; Titus, 1990a,b; 1991; IPCC, 1991; Kellogg, 1991) and in the lay press (e.g., Brooks, 1989; Schneider, 1989; Thatcher, 1990; Bello, 1991; Luoma, 1991; UCAR/NOAA, 1991). Titus (1990a), for example, proposes the rerouting of the Mississippi River to save coastal Louisiana! As an example of such extreme claims to mitigate anthropogenically caused global warming, a 1991 National Academy Press report (NAS, 1991) has considered the insertion of 50,000 100 km^2 mirrors in space to reflect incoming sunlight. Such gross global climate engineering represents a close analog to the exaggerated claims in weather modification which were made in the 1960s and 1970s. Short-term variations of weather and climate are clearly within the *natural variability* of climate to the extent that we can realistically assess it. Moreover, the models are not really 'forecast' models. They are simply research models designed to simulate the responses of hypothesized anthropogenic changes to weather and climate, *other things being the same.* Besides having many limitations in their physical/chemical parameterizations, they are not designed to simulate (or predict) the consequences of many other natural factors affecting climatic change. That is because we simply do not know enough about all the processes of importance to climatic change to include them in any quantitative forecast system. What it amounts to is that many scientists are grossly underestimating the complexity of interactions among the earth's atmosphere, ocean, geosphere, and biosphere. These problems are so complex that it may take many decades, or even centuries, before we have matured enough as a scientific community to make *credible predictions* of long-term climate trends and their corresponding regional impacts. Even then, we may find that the uncertainty level of those predictions due to outside (the earth) influences may be so large that those predictions are not useful for social planning.

E.3 The Capricious Administration of Science and Technology

In the United States, as well as many developed countries, science and technology, is often poorly administered. As we have seen in weather modification, administration of many programs is fragmented among a number of basic and mission-oriented agencies, all of which compete for funding at national and state levels. This competition amongst the agencies often leads to the greatly exaggerated claims that many of the scientific and technological issues will be solved in the next five to ten years.

In addition, because many of these agencies are mission-oriented, their job is to examine the impacts of human-induced changes on weather and climate on energy, air quality, water resources, or agriculture. Their job is not to advance the fundamental scientific issues regarding the behavior of the earth system, but to get on with the business of evaluating the impacts of anthropogenic activity on their programs. As a result, they are often looking for shortcuts to bottom line answers that can probably only be obtained through meticulous, often time consuming scientific research.

Moreover, national governing bodies (legislatures, presidents, etc.) all work on time scales of two, four, or six years, and want to be able to identify impacts of their programs on time scales of their tenure. If significant progress is not made on those time scales, then often funding in those programs is reduced, if not curtailed, and new, competing programs are brought to the forefront. This results in shortsighted funding in science and technology in which programs are begun and before they reach maturity they are curtailed, then the rush is on to get on the bandwagon with the latest fad. As we have discussed in this book, however, the scientific and technological problems associated with furthering our understanding of human impacts and natural variability and feedbacks of weather and climate are so complicated and multifaceted that many of the issues will not be resolved on time scales of decades or possibly centuries. *Thus, programs associated with the investigation of human impacts on weather and climate require sustained, stable national funding at a high level.* A view supporting this idea has been recommended in the Policy Statement of the American Meteorological Society on Global Climate Change (1991) and in the report, Global Climate Change: A New Vision for the 1990s (Michaels, 1990), which was produced by a group of climate scientists in the fall of 1990 who questioned the overselling and shortsighted perspective of current climate change government policy.

E.4 Scientific Credibility and Advocacy

In this book we have examined human impacts on weather and climate ranging from purposeful attempts at changing weather through cloud seeding, to inadvertent modification of weather and climate on regional or mesoscales, to human impacts on global climate. We have seen that the evidence is very suggestive that human activity can have potentially significant impacts on weather and climate over a broad range of space and time scales. Furthermore, as long as the population of the human species continues to rise, the prospects for causing major changes in weather and climate become increasingly likely. We must emphasize that with few exceptions, the scientific evidence is not conclusive that human activity is causing observed changes in weather and climate nor that it will do so in the next century. With that in mind we ask, *should scientists be actively involved in advocating that we apply cloud seeding techniques to enhancing rainfall, or reducing emissions of greenhouse gases to alleviate greenhouse warming?* Certainly the scientists are the best informed with regard to the consequences of human activity and, one could say, that if the informed scientist does not take an advocacy role in recommending that action be taken, then no one else will.

Such a position is not without its dangers, however. For example, in the 1970's we worked for the former Experimental Meteorology Laboratory of NOAA in Miami, Florida, which was involved in developing and testing dynamic cloud seeding techniques to enhance rainfall. During that period, the state of Florida experienced a severe drought and asked the laboratory director to assist the state by directing a rainfall enhancement project. The director asked her three assistant administrators, one of which was Cotton, whether we should respond to the state's request by providing them operational cloud seeding support. I (Cotton) was the lone dissenter and argued that we were scientists who were involved in the objective assessment of the impacts of dynamic cloud seeding. As a result our participation in an operational program and obvious role as advocates of applying dynamic seeding, would jeopardize our credibility as truly objective scientists and therefore adversely affect both the program and the individual scientists. The same can be said with regard to advocates of major disruptive societal changes with regard to greenhouse emissions.

Some might argue that the risk of losing one's scientific credibility is purely a personal one and must be weighed against the potential societal gains by taking immediate action to relieve drought or reduce greenhouse warming. In fact, the adverse impacts extend far beyond those affecting the individual scientist. Loss of scientific credibility is infectious and can, therefore, propagate through an entire scientific discipline and even to the scientific community as a whole. The fall of the science of weather modification by cloud seeding was almost certainly due, in part, to a loss of scientific credibility. The global climate change community must likewise be careful that a loss of scientific credibility does not propagate through their discipline, or the discipline of atmospheric science as well. Thus premature advocacy that action be taken now, could, in the long run, destroy the prospects for obtaining solid scientific evidence that human activity is affecting weather and climate.

E.5 Should Society Wait for Hard Scientific Evidence?

We have seen that there is little hard scientific evidence that anthropogenic activity, either advertently or inadvertently, is causing significant changes in weather and climate, particularly on the global scale. This is certainly true with respect to cloud seeding where there are only a few limited examples of where cloud seeding has been scientifically shown to be effective in enhancing rainfall. Nonetheless, there are many nations which are currently running operational cloud seeding projects. Apparently, the decision has been made in those nations and states that the benefits outweigh the risks of applying the scientifically unproven technology of weather modification by cloud seeding. The major risks, however, are limited to the possibility of creating severe weather or floods, and to increasing rainfall in one local region at the expense of rainfall in a neighboring local region. Often the decision to apply cloud seeding technology in a particular country or state is a prescription of a *political placebo* or a decision that it is better to do something than to sit idly by and do nothing as reservoirs dry up and crops wither and die due to the absence of water.

Again, the situation is not much different with respect to human impacts on global climate. We lack hard scientific evidence that anthropogenic activity is causing, or will cause, changes in global climate. Nonetheless, there is convincing evidence that CO_2 concentrations are increasing at an alarming rate (Figure 11.2). Figure E.1 shows that there is a strong rise in global CO_2 emissions since the 1950's. Figures E.2–E.8 illustrate that the United States is the largest contributor to CO_2 emissions, while the Soviet Union and the People's Republic of China are rapidly catching up. Clearly, reductions in CO_2 emissions in these countries will have a significant impact on global CO_2 emissions and reduce the chance that human activity will have a significant impact on weather and climate. Certainly there is evidence that the more developed nations are at least causing a leveling off of CO_2 emissions (see Figures E.2, E.5, E.6, and E.8).

But what are the costs? Some of the costs are in terms of reduced industrial productivity which impacts employment and the general standard of living. Another cost is associated with the impacts of using alternate energy resources. As shown in Figure E.9, the only country exhibiting a significant decline in CO_2 emissions is France. This decline in CO_2 emissions is apparently a result of France's decision to convert to heavy use of nuclear power as an alternate energy resource. In this case one is trading off the potential impacts of CO_2 emissions on global warming against the long-term problem of disposing of nuclear waste as well as the dangers of inadvertent releases of nuclear materials. Is this a wise tradeoff? Without solid scientific evidence that CO_2 emissions are causing significant changes in climate, one cannot make an objective evaluation of the relative cost of each alternative.

One of the most alarming figures is Figure E.7 which shows a sharp rise in CO_2 emissions in India, while per capita emissions remain steady. This shows a clear impact of increased population on CO_2 emissions and indicates that one of the most important steps in reducing those emissions is getting the world population stabilized. *One does not need strong scientific evidence that human activity is causing global warming to recognize a reduction in the population on earth will have long-term benefits; common sense is all that is needed!*

Indeed, population growth is a problem which is much more severe than any of the scenarios proposed to occur as a result of greenhouse gas warming. Catastrophic social upheavals are likely to result as the human density continues to increase. Bryson (1989) presents evidence that even today India is at their absolute water limit, such that below average rainfall can cause massive deaths. Fraser-Smith (1989) demonstrate that globally we are now in explosive population growth (see Figure E.10) which will be limited in some, probably undesirable fashion prior to 2020, which is well before the hypothesized major greenhouse warming would occur towards the end of the next century.

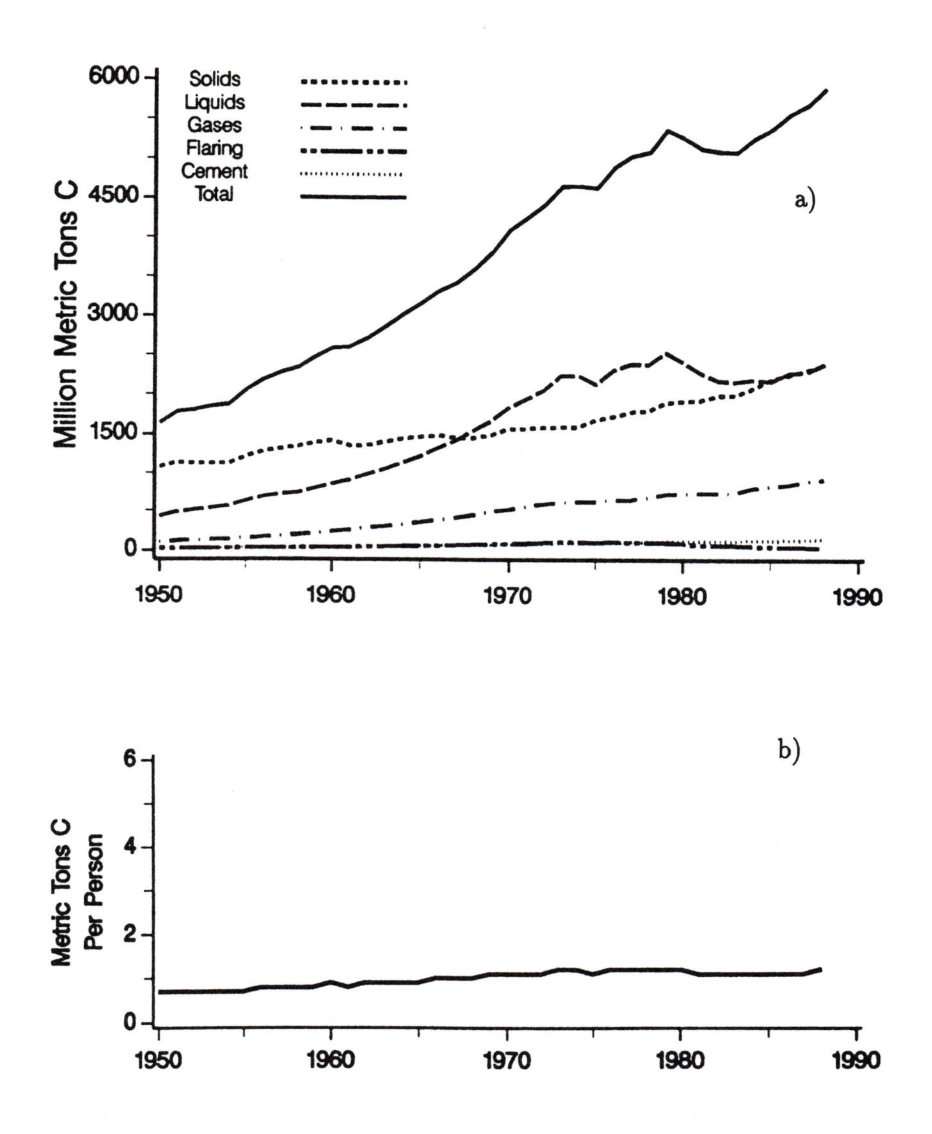

Figure E.1: *a) Global CO_2 emissions and b) Global per capita CO_2 emission estimates.* [From Marland, G., 1991. CO_2 Emissions. pp 92-133. In Boden, T.A., P. Kanciruk, and M.P. Farrell, *Trends '90: A Compendium of Data on Global Change,* Carbon Dioxide Information Analysis Center, Oak Ridge National Laboratory, Oak Ridge, Tennessee.]

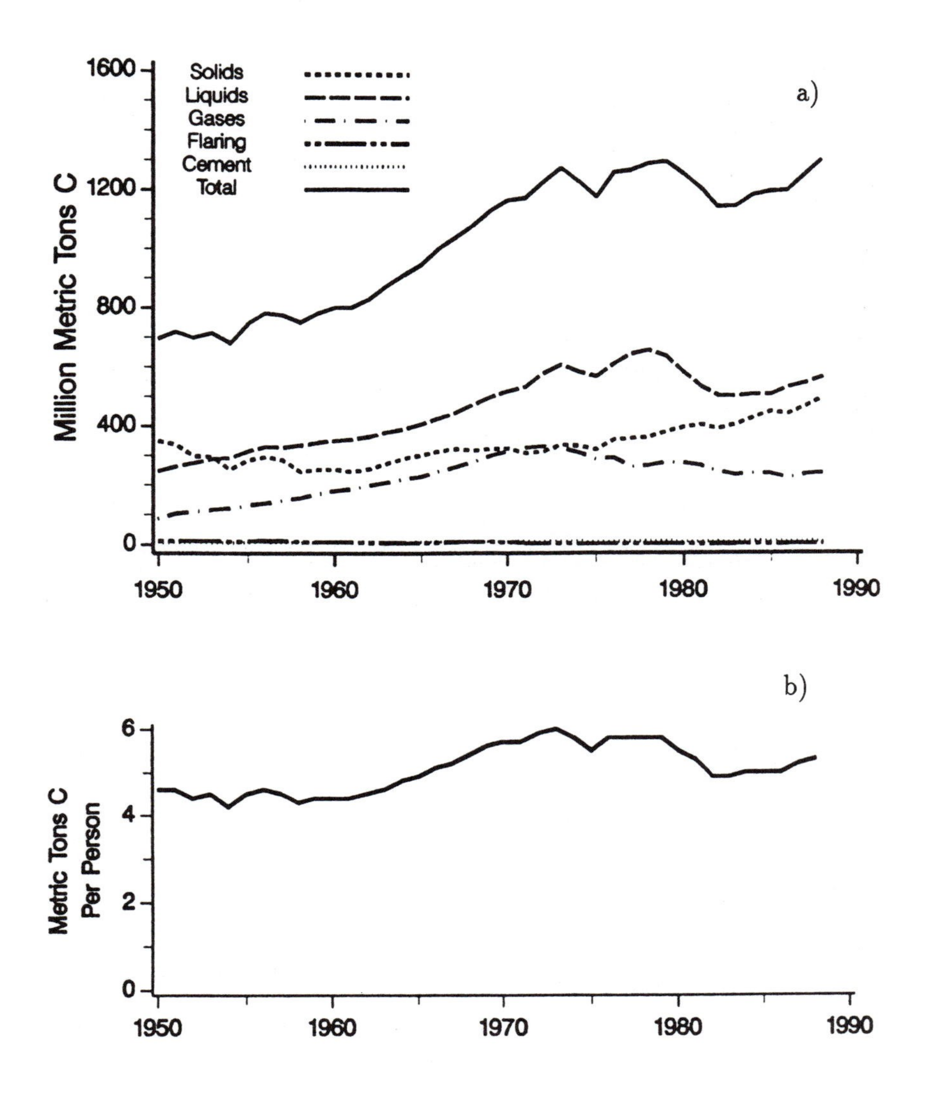

Figure E.2: *a) CO_2 emissions from the United States and b) Per capita CO_2 emission estimates for the United States.* [From Marland, G., 1991. CO_2 Emissions. pp 92-133. In Boden, T.A., P. Kanciruk, and M.P. Farrell, *Trends '90: A Compendium of Data on Global Change,* Carbon Dioxide Information Analysis Center, Oak Ridge National Laboratory, Oak Ridge, Tennessee.]

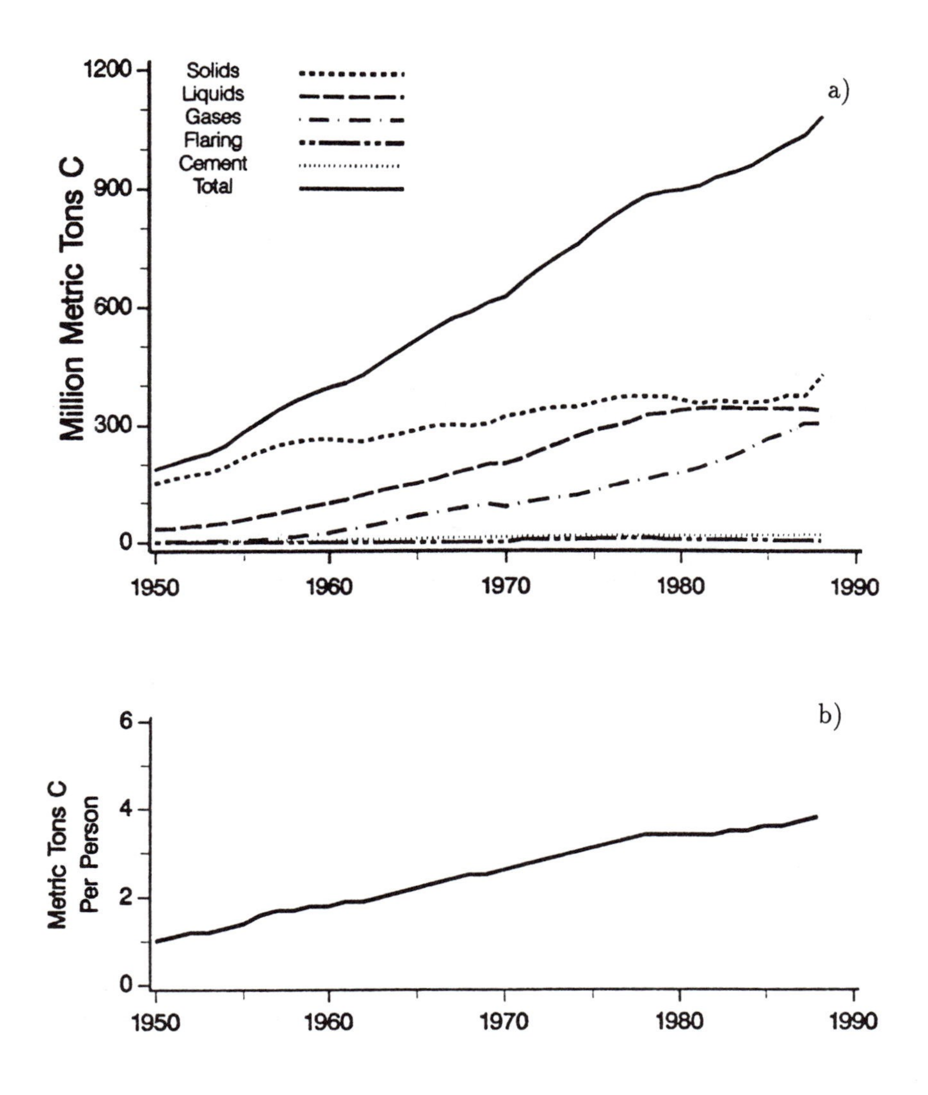

Figure E.3: *a) CO_2 emissions from the Union of Soviet Socialist Republics and b) Per capita CO_2 emission estimates for the Union of Soviet Socialist Republics.* [From Marland, G., 1991. CO_2 Emissions. pp 92-133. In Boden, T.A., P. Kanciruk, and M.P. Farrell, *Trends '90: A Compendium of Data on Global Change,* Carbon Dioxide Information Analysis Center, Oak Ridge National Laboratory, Oak Ridge, Tennessee.]

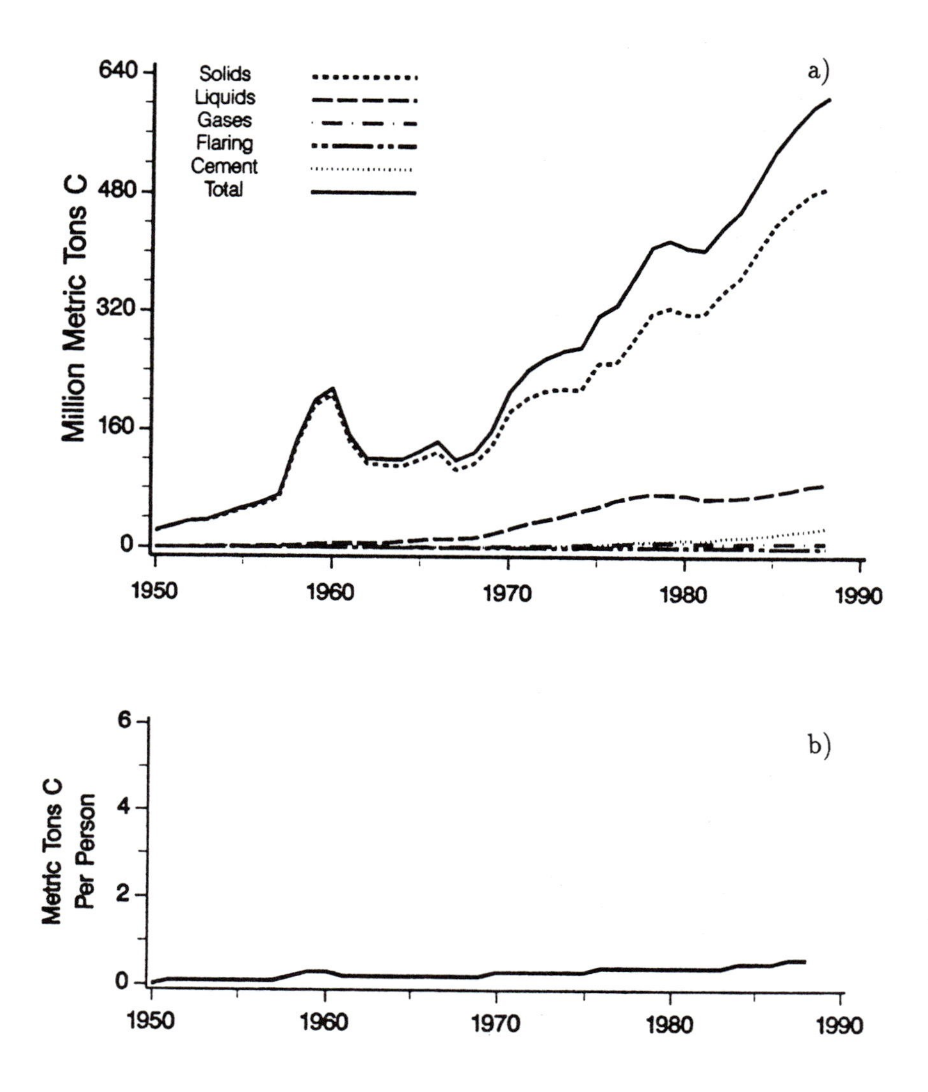

Figure E.4: *a) CO_2 emissions from the People's Republic of China and b) Per capita CO_2 emission estimates for the People's Republic of China.* [From Marland, G., 1991. CO_2 Emissions. pp 92-133. In Boden, T.A., P. Kanciruk, and M.P. Farrell, *Trends '90: A Compendium of Data on Global Change,* Carbon Dioxide Information Analysis Center, Oak Ridge National Laboratory, Oak Ridge, Tennessee.]

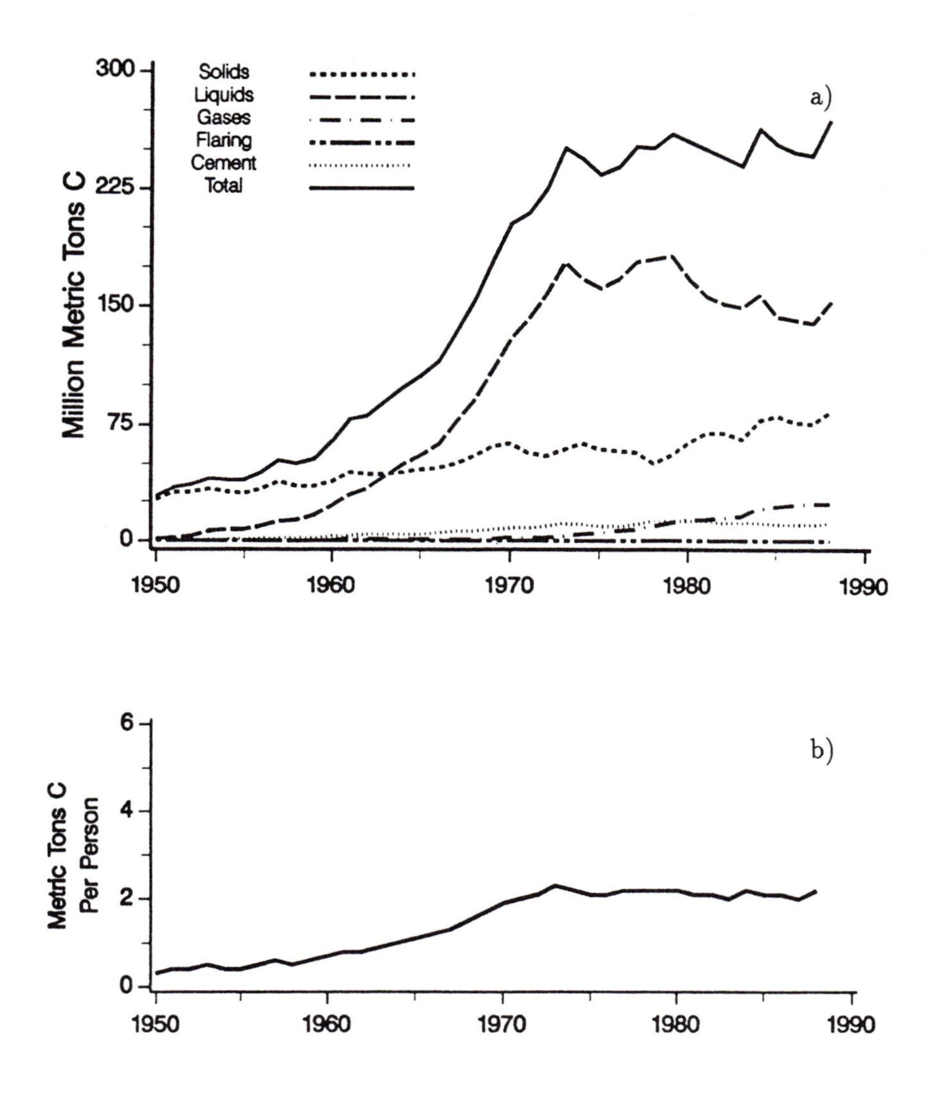

Figure E.5: *a) CO_2 emissions from Japan and b) Per capita CO_2 emission estimates for Japan.* [From Marland, G., 1991. CO_2 Emissions. pp 92-133. In Boden, T.A., P. Kanciruk, and M.P. Farrell, *Trends '90: A Compendium of Data on Global Change,* Carbon Dioxide Information Analysis Center, Oak Ridge National Laboratory, Oak Ridge, Tennessee.]

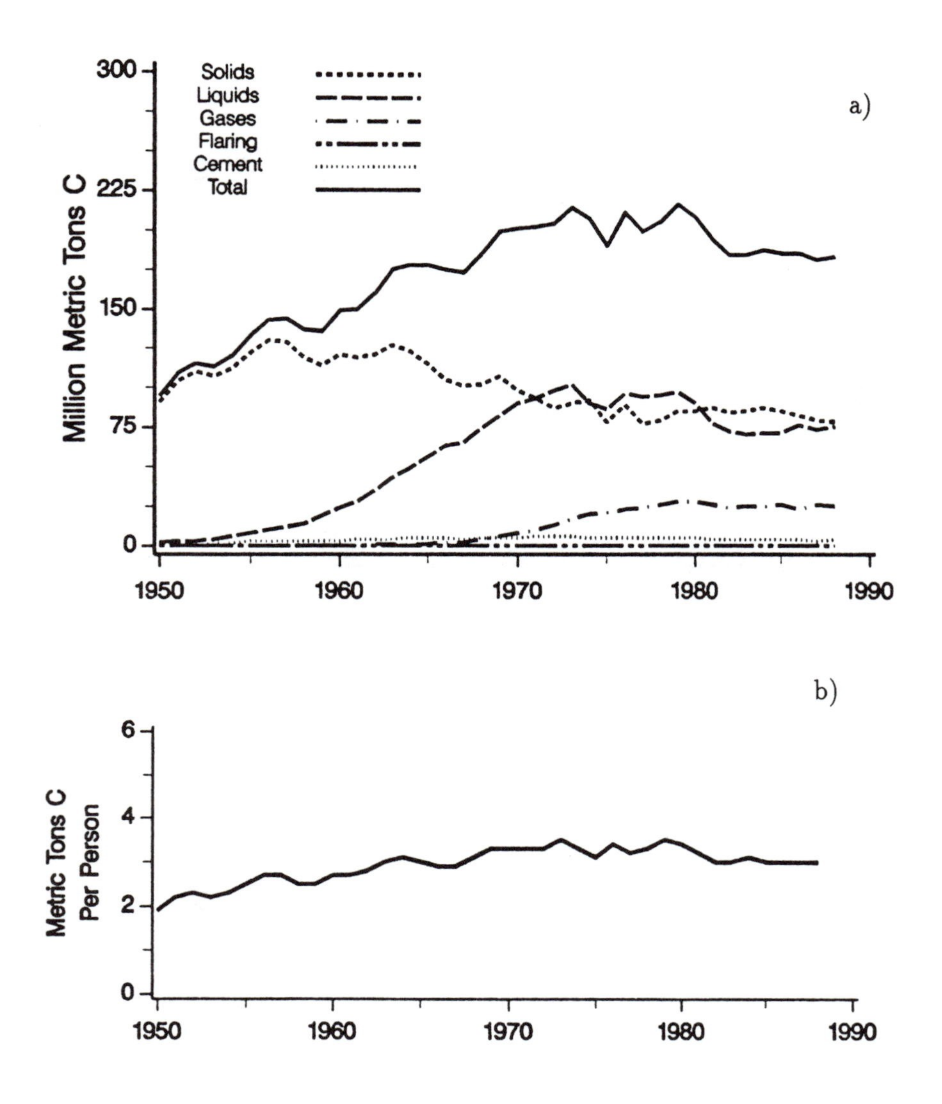

Figure E.6: *a) CO_2 emissions from the Federal Republic of Germany and b) Per capita CO_2 emission estimates for the Federal Republic of Germany.* [From Marland, G., 1991. CO_2 Emissions. pp 92-133. In Boden, T.A., P. Kanciruk, and M.P. Farrell, *Trends '90: A Compendium of Data on Global Change,* Carbon Dioxide Information Analysis Center, Oak Ridge National Laboratory, Oak Ridge, Tennessee.]

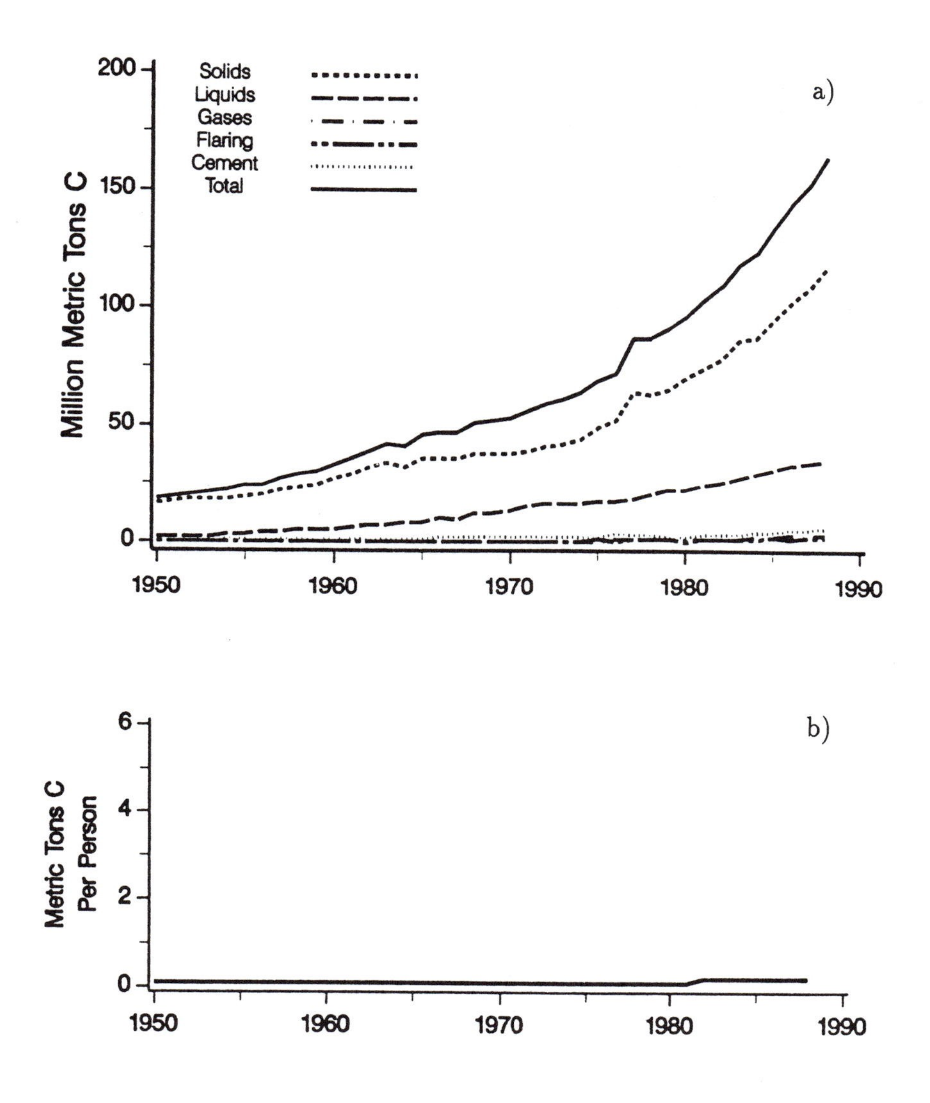

Figure E.7: *a) CO_2 emissions from India and b) Per capita CO_2 emission estimates for India.* [From Marland, G., 1991. CO_2 Emissions. pp 92-133. In Boden, T.A., P. Kanciruk, and M.P. Farrell, *Trends '90: A Compendium of Data on Global Change,* Carbon Dioxide Information Analysis Center, Oak Ridge National Laboratory, Oak Ridge, Tennessee.]

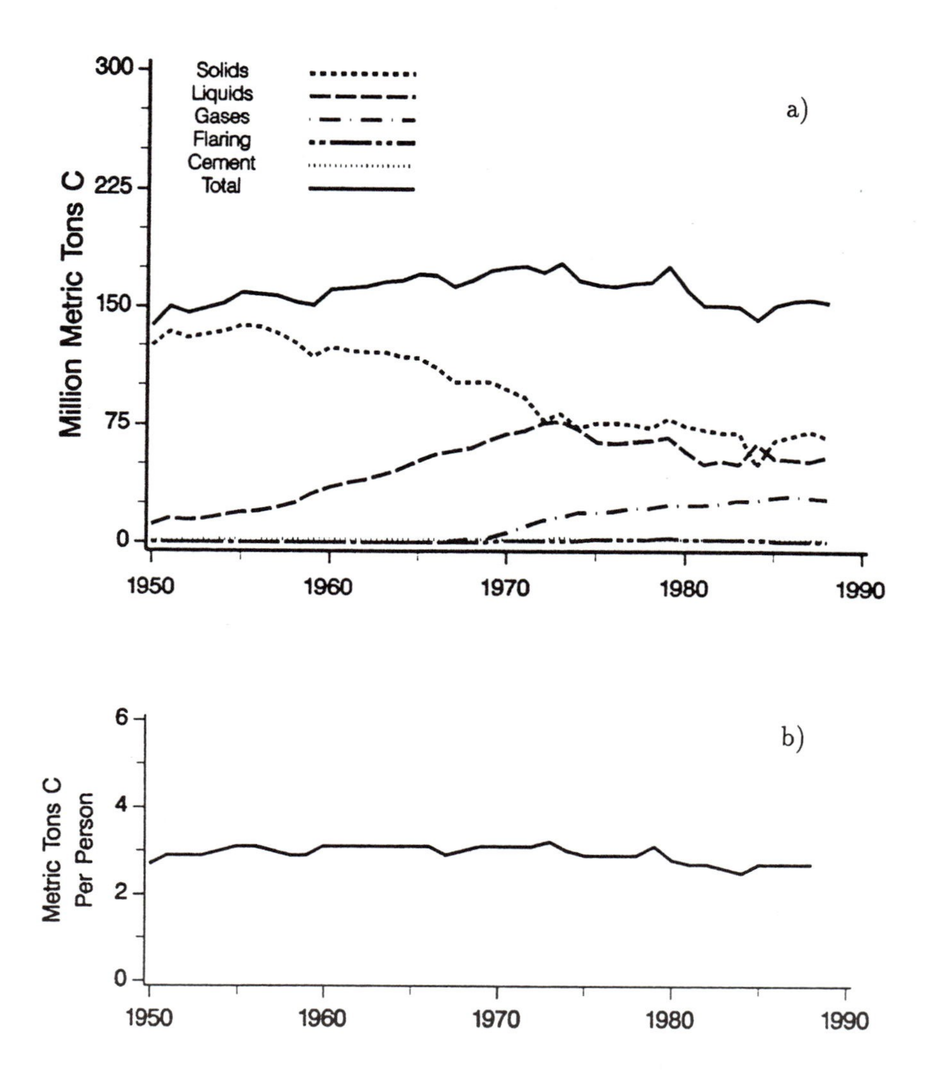

Figure E.8: *a) CO_2 emissions from the United Kingdom and b) Per capita CO_2 emission estimates for the United Kingdom.* [From Marland, G., 1991. CO_2 Emissions. pp 92-133. In Boden, T.A., P. Kanciruk, and M.P. Farrell, *Trends '90: A Compendium of Data on Global Change,* Carbon Dioxide Information Analysis Center, Oak Ridge National Laboratory, Oak Ridge, Tennessee.]

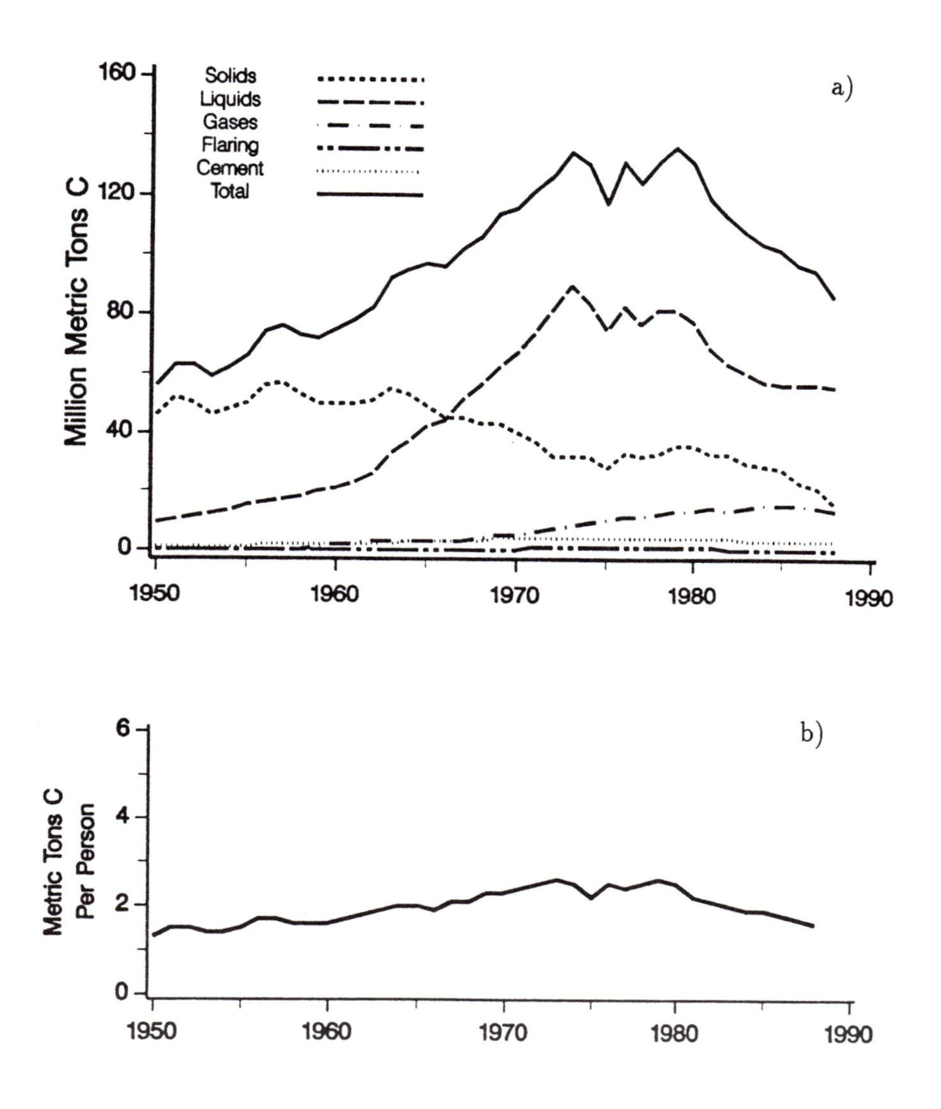

Figure E.9: *a) CO_2 emissions from France and b) Per capita CO_2 emission estimates for France.* [From Marland, G., 1991. CO_2 Emissions. pp 92-133. In Boden, T.A., P. Kanciruk, and M.P. Farrell, *Trends '90: A Compendium of Data on Global Change,* Carbon Dioxide Information Analysis Center, Oak Ridge National Laboratory, Oak Ridge, Tennessee.]

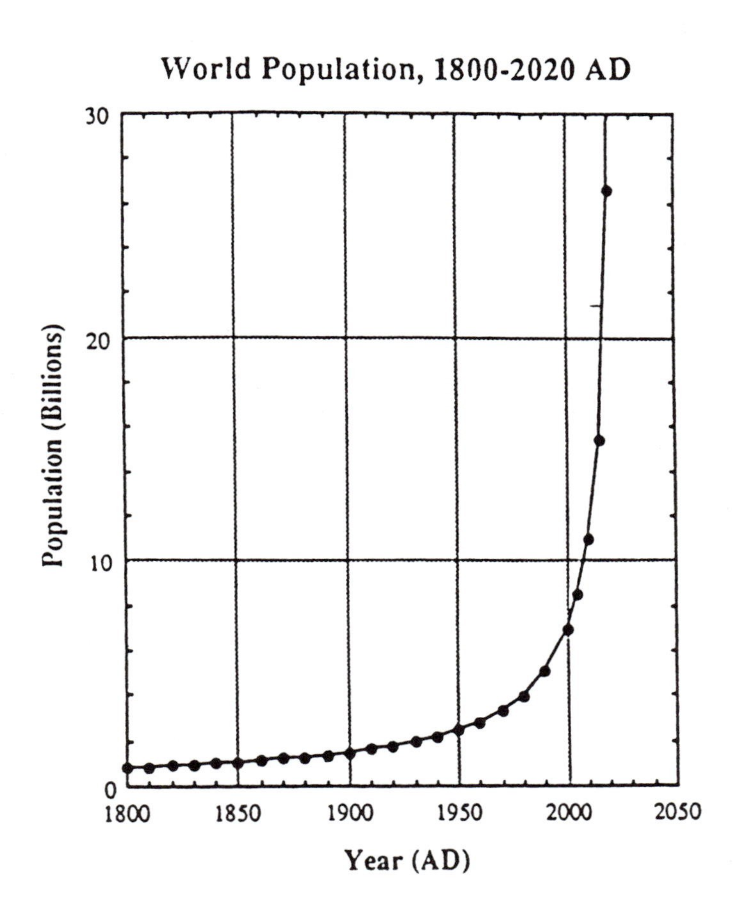

Figure E.10: *Graph of world population plotted against year for the time interval 1800-2020 A.D. according to the equation of von Foerster et al. (1960). Population numbers are shown every 10 years up to 2000 A.D. and every 5 years thereafter. Current estimates of world population are slightly higher for the 1989 population implied by the data used for the graph.* [From Fraser-Smith, A.C., 1989: Space science strategy 1990-2026 A.D. *Eos,* December 26, 1989, 1569. ©American Geophysical Union.]

References

Ackerman, T.P., K.-N. Liou, and C.B. Leovy, 1976: Infrared radiative transfer in polluted atmospheres. *J. Appl. Meteor.,* **15,** 28-35.

Ahmad, M.U., 1990: A hydrologist's plan to combat the greenhouse effect. *Water Int.*, **14**, 101-115.

Akbari, H., A. Rosenfeld, and H. Taha, 1989: Recent development in heat island studies: Technical and policy. Berkeley, California, Lawrence Berkeley Laboratory Technical Report.

Albrecht, B.A., 1989: Aerosols, cloud microphysics, and fractional cloudiness. *Science,* **245,** 1227-1230.

Aleksandrov, V.V., and G.L. Stenchikov, 1983: On the modeling of the climatic consequences of the nuclear war. Computing Center of the Academy of Sciences of the USSR, Moscow, USSR.

André, J.-C., P. Bougeault, J.-F. Mahfouf, P. Mascart, J. Noilhan, and J.-P. Pinty, 1989: Impact of forests on mesoscale meteorology. *Phil. Trans. Roy. Soc. London* **B324**, 407-422.

André, J.-C., P. Bougeault, and J.-P. Goutorbe, 1990: Regional estimates of heat and evaporation fluxes over non-homogeneous terrain. Examples from the HAPEX-MOBILHY Programme. *Bound.-Layer Meteor.*, **50**, 77-108.

Angell, J.K., 1988: Variations and trends in tropospheric and stratospheric global temperatures, 1958-87. *J. Climate,* **1,** 1296-1313.

Anthes, R.A., 1982: Tropical cyclones—their evolution, structure, and effects. *Meteor. Monogr.,* **41,** American Meteorological Society, Boston, 208 pp.

Anthes, R.A., 1984: Enhancement of convective precipitation by mesoscale variations in vegetative cover in semiarid regions. *J. Climate Appl. Meteor.*, **23**, 541-554.

Arking, A., 1991: The radiative effects of clouds and their input on climate. *Bull. Amer. Meteor. Soc.*, **72**, 795-813.

Arrhenius, S., 1896: On the influence of carbonic acid in the air on the temperature of the ground. *Phil. Mag.,* **41,** 237-276.

Ashworth, J.R., 1929: The influence of smoke and hot gases from factory chimneys on rainfall. *Quart. J. Roy. Meteor. Soc.,* **25,** 34-35.

Atlas, D., 1977: The paradox of hail suppression. *Science,* **195,** 139-145.

Aubreville, A., 1949: Climats, forêts et désertification de l'Afrique tropicale. Société d'Editions Géographiques et coloniales, Paris.

Augustsson, T., and V. Ramanathan, 1977: A radiative-convective model study of the CO_2 climate problem. *J. Atmos. Sci.,* **34,** 448-451.

Avissar, R., and R.A. Pielke, 1989: A parameterization of heterogeneous land surfaces for atmospheric numerical models and its impact on regional meteorology. *Mon. Wea. Rev.*, **117**, 2113-2136.

Avissar, R., and M.M. Verstraete, 1990: The representation of continental surface processes in atmospheric models. *Rev. Geophys.*, **28**, 35-52.

Baes, C.F., Jr., A. Björkström, and P.J. Mulholland, 1985: Uptake of carbon dioxide by the oceans. In *Atmospheric Carbon Dioxide and the Global Carbon Cycle*, J.R. Trabalka, Ed., Department of Energy, Carbon Dioxide Research Div., Washington, D.C., 81-111.

Baines, P.G., and T.N. Palmer, 1990: Rationale for a new physically-based parameterization of subgrid scale orographic effects. Technical Memorandum #169, Research Department, CSIRO Division of Atmospheric Research, Aspendale, Australia.

Balling, R.C., and S.W. Brazel, 1987: Time and space characteristics of the Phoenix urban heat island. *J. Arizona-Nevada Acad. Sci.*, **21**, 75-81.

Banta, R.M., 1985: Nuclear firestorm results (Preliminary) case: Standard atmosphere with moisture. AFGL/LYC report.

Barkström, B.R., E.F. Harrison, and R.B. Lee, III, 1990: Earth Radiation Budget Experiment. *EOS*, **71**, 297, 304-305.

Barnola, J.M., D. Raymond, Y.S. Korotkevich, and C. Lorius, 1987: Vostok ice core provides 160,000-year record of atmospheric CO_2. *Nature*, **329**, 408-414.

Barnston, A.G., W.L. Woodley, J.A. Flueck, and M.H. Brown, 1983: The Florida Area Cumulus Experiment's second phase (FACE-2). Part I: The experimental design, implementation, and basic data. *J. Appl. Meteor.*, **22**, 1504-1528.

Barrie, L.A., D. Fisher, and R.M. Koerner, 1985: Twentieth Century trends in Arctic air pollution revealed by conductivity and acidity observations in snow and ice in the Canadian high Arctic. *Atmos. Environ.*, **19**, 2055-2063.

Bastable, H.G., W.J. Shuttleworth, R.L.G. Dallarosa, G. Fisch, and C.A. Nobre, 1992: Observations of climate, albedo and surface radiation over cleared and undisturbed Amazonian forest. *Int. J. Climatol.*, (submitted).

Battan, L.J., 1969: Weather modification in the U.S.S.R.—1969. *Bull. Amer. Meteor. Soc.*, **50**, 924-945.

Beebe, R.C., 1974: Large scale irrigation and severe storm enhancement. *Proc., Symposium on Atmospheric Diffusion and Air Pollution of the American Meteorological Society.* Co-sponsored by the World Meteorological Organization. September 9-13, Santa Barbara, California, 392-395.

Bello, M., 1991: Greenhouse warming threat justifies immediate action. *National Research Council News Report*, **XLI**, No. 4, 2-5.

Berger, W.H., 1982: Climate steps in ocean history-lessons from the Pleistocene. In *Climate in Earth History.* Berger, W.H., and J.C. Crowell, Panel Co-Chairmen. National Academy Press, Washington, D.C. 43-54.

Bergeron, T., 1965: On the low-level redistribution of atmospheric water caused by orography. *Suppl., Proc. Int. Conf. on Cloud Physics,* Tokyo, 96-100.

Bibilashvili, N. Sh., I.I. Gaivoronski, G.G. Godorage, A.I. Kartsivadze, and R.N. Stankov, 1974; *Proc. WMO/IAMAP Scientific Conf. on Weather Modification,* Tashkent, 1973, p. 333.

Bigg, E.K., 1988: Secondary ice nucleus generation by silver iodide applied to the ground. *J. Appl. Meteor.,* **27,** 453,457.

Bigg, E.K., 1990a: Long term trends in ice nucleus concentrations. *Atmos. Res.*, **25,** 409-415.

Bigg, E.K., 1990b: Aerosol over the southern ocean. *Atmos. Res.*, **25,** 583-600.

Bigg, E.K., and E. Turton, 1988: Persistent effects of cloud seeding with silver iodide. *J. Appl. Meteor.,* **27,** 505-514.

Bigg, E.K., J.L. Gras, and C. Evans, 1984: Origin of Aitken particles in remote regions of the southern hemisphere. *J. Atmos. Chem.,* **1,** 203-214.

Biswas, K.R., and A.S. Dennis, 1971: Formation of rain shower by salt seeding. *J. Appl. Meteor.,* **10,** 780-784.

Biswas, K.R., R.K. Kapoor, and K.K. Kanuga, 1967: Cloud seeding experiment using common salt. *J. Appl. Meteor.,* **10,** 780-784.

Black, J.F., and B.H. Tarmy, 1963: The use of asphalt coating to increase rainfall. *J. Appl. Meteor.*, **2**, 557-564.

Bohren, C., 1989: The greenhouse effect revisited. *Weatherwise*, **42,** 50-54.

Braham, R.R., 1981: Designing cloud seeding experiments for physical understanding. *Bull. Amer. Meteor. Soc.,* **62,** 55-62.

Braham, R.R., and P.A. Spyers-Duran, 1967: Survival of cirrus crystals in clear air. *J. Appl. Meteor.,* **6,** 1053-1061.

Braham, R.R., Jr., L.J. Battan, and H.R. Byers, 1957: Artificial nucleation of cumulus clouds, clouds and weather modification: A group of field experiments. *Meteor. Monogr.* **2,** 47-85.

Braslau, N., and J.V. Dave, 1975: Atmospheric heating rates due to solar radiation for several aerosol-laden cloudy and cloud-free models. *J. Appl. Meteor.,* **14,** 396-399.

Bret, B., and P. Bougeault, 1988: Simulation de quatre journées MOBILHY à l'aide du modèle PERIDOT. *Note travail EERM,* **211,** 1-90.

Brier, G, 1955: Seven-day periodicities in certain meteorological parameters during the period 1899–1951. *Bull. Amer. Meteor. Soc.,* **36,** 265-277.

Broccoli, A.J., and S. Manabe, 1987: The influence of continental ice, atmospheric CO_2 and land albedo on the climate of the last glacial maximum. *Climate Dyn.,* **1,** 87-99.

Broccoli, A.J., and S. Manabe, 1990: Can existing climate models be used to study anthropogenic changes in tropical cyclone climate? *Geophys. Res. Lett.,* **17,** 1917-1920.

Brooks, W.T., 1989: The global warming panic. *Forbes,* **December,** 97-102.

Browning, K.A., and G.B. Foote, 1976: Airflow and hail growth in supercell storms and some implications for hail suppression. *Quart. J. Roy. Meteor. Soc.,* **102,** 499-533.

Browning, K.A., C.W. Pardoe, and F.F. Hill, 1975: The nature of orographic rain at wintertime cold fronts. *Quart. J. Roy. Meteor. Soc.,* **101,** 333-352.

Bryson, R.A., 1989: Environmental opportunities and limits for development. *Environ. Conserv.,* **16,** 299-305.

Burke, I.C., T.G.F. Kittel, W.K. Lauenroth, P. Snook, C.M. Yonker, and W.J. Parton, 1991: Regional analysis of the central Great Plains. *Bioscience,* **4,** 685-692.

Burtsev, I.I., 1974: Hail suppression. *WMO Scientific Conf. Weather Modification,* Boulder, Colorado, 2nd, 1976, 217-222.

Butzer, K.W., 1980: Adaptation to global environmental change. *Professional Geographer,* **32,** 269-278.

Byers, H.R., 1974: History of weather modification. In *Weather and Climate Modification*, W.N. Hess, Ed., John Wiley and Sons, Inc., New York, 3-44.

Callendar, G.S., 1938: The artificial production of carbon dioxide and its influence on temperature. *Quart. J. Roy. Meteor. Soc.,* **64,** 223-240.

Carlson, T.N., and S.G. Benjamin, 1980: Radiative heating rates for Saharan dust. *J. Atmos. Sci.,* **37,** 193-213.

Cess, R.D., G.L. Potter, S.J. Ghan, and W.L. Gates, 1985: The climatic effects of large injections of atmospheric smoke and dust: A study of climate feedback mechanisms with one- and three-dimensional climate models. *J. Geophys. Res.,* **90,** 12,937-12,950.

Cess, R.D., G.L. Potter, J.P. Blanchet, G.J. Boer, A.D. Del Genio, M. Déqué, V. Dymnikov, V. Galin, W.L. Gates, S.J. Ghan, J.T. Kiehl, A.A. Lacis, H. Le Treut, Z.-X. Li, X.-Z. Liang, B.J. McAvaney, V.P. Meleshko, J.F.B. Mitchell, J.-J. Morcrette, D.A. Randall, L. Rikus, E. Roeckner, J.F. Royer, U. Schlese, D.A. Sheinin, A. Slingo, A.P. Sokolov, K.E. Taylor, W.M. Washington, R.T. Wetherald, I. Yagai, and M.-H. Zhang, 1990: Intercomparison and interpretation of climate feedback processes in 19 atmospheric general circulation models. *J. Geophys. Res.*, **95**, 16,601-16,615.

Cess, R.D., G.L. Potter, M.-H. Zhang, J.-P. Blanchet, S. Chalita, R. Colman, D.A. Dazlich, A.D. Del Genio, V. Dymnikov, V. Galin, D. Jerrett, E. Keup, A.A. Lacis, H. Le Treut, Z.-Z. Liang, J.-F. Mahfouf, B.J. McAvaney, V.P. Meleshko, J.F.B. Mitchell, J.-J. Morcrette, P.M. Norris, D.A. Randall, L. Rikus, E. Roeckner, J.-F. Royer, U. Schlese, D.A. Sheinin, J.M. Slingo, A.P. Sokolov, K.E. Taylor, W.M. Washington, R.T. Wetherald, and I. Yagai, 1991: Interpretation of snow-climate feedback as produced by 17 general circulation models. *Science,* **253,** 888-892.

Chang, J., and P.J. Wetzel, 1991: Effects of spatial variations of soil moisture and vegetation on the evolution of a prestorm environment: A numerical case study. *Mon. Wea. Rev.*, **119**, 1368-1390.

Changnon, S.A., Jr., 1968: The La Porte anomaly—fact or fiction? *Bull. Amer. Meteor. Soc.,* **49,** 4-11.

Changnon, S.A., Jr., 1981a: Midwestern cloud, sunshine and temperature trends since 1901: Possible evidence of jet contrail effects. *J. Appl. Meteor.,* **20,** 496-508.

Changnon, S.A., Jr., 1981b: METROMEX: A review and summary. Stanley A. Changnon, Jr., Ed., American Meteorological Society, Meteorological Monographs, **18,** No. 40, 181 pp.

Changnon, S.A., Jr., and F.A. Huff, 1977: La Porte again: A new anomaly. *Bull. Amer. Meteor. Soc.,* **58,** 1069-1072.

Changnon. S.A., Jr., and F.A. Huff, 1986: The urban-related nocturnal rainfall anomaly at St. Louis. *J. Climate Appl. Meteor.,* **25,** 1985-1995.

Changnon, S.A., Jr., and J.L. Ivens, 1981: History repeated: The forgotten hail cannons of Europe. *Bull. Amer. Meteor. Soc.,* **62,** 368-375.

Changnon, S.A., Jr., and W.H. Lambright, 1987: The rise and fall of federal weather modification policy. *J. Wea. Mod.,* **19,** 1-12. October 1981.

Charlock, T.P., 1981: Cloud optics as a possible stabilizing factor in climate change. *J. Atmos. Sci.,* **38,** 661-663.

Charlock, T.P., 1982: Cloud optical feedback and climate stability in a radiative-convective equilibrium. *Tellus,* **34,** 245-254.

Charlson, R.J., J.E. Lovelock, M.O. Andreae, and S.G. Warren, 1987: Oceanic phytoplankton, atmospheric sulfur, cloud albedo and climate. *Nature,* **326,** 655-661.

Charlson, R.J., S.E. Schwartz, J.M. Hales, R.D. Cess, J.A. Coakley, Jr., J.E. Hansen, D.J. Hofmann, 1992: Climate forcing by anthropogenic aerosols. *Science,* **255,** 423-430.

Charney, J.G., 1975: Dynamics of desert and drought in the Sahel. *Quart. J. Roy. Meteor. Soc.,* **101,** 193-202.

Charney, J.G., W.J. Quirk, S.-H. Chow, and J. Kornfield, 1977: A comparative study of the effects of albedo change on drought in semiarid regions. *J. Atmos. Sci.*, **34**, 1366-1385.

Chemical Manufacturers Association, 1983: 1982 world production and sales of fluorocarbons FC-11 and FC-12. Washington, D.C., 29 pp.

Chen, C.H., and H.D. Orville, 1980: Effects of mesoscale on cloud convection. *J. Appl. Meteor.*, **19**, 256-274.

Chisnell, R.F., and J. Latham, 1976a: Ice multiplication in cumulus clouds. *Quart. J. Roy. Meteor. Soc.*, **102**, 133-156.

Chisnell, R.F., and J. Latham, 1976b: Comments on the paper by Mason, Production of ice crystals by riming in slightly supercooled cumulus. *Quart. J. Roy. Meteor. Soc.*, **102**, 713-715.

Coakley, J.A., and R.D. Cess, 1985: Response of the NCAR community climate model to the radiative forcing by the naturally occurring tropospheric aerosol. *J. Atmos. Sci.,* **42,** 1677-1692.

Coakley, J.A., R.L. Bernstein, and P.A. Durkee, 1987: Effect of shipstack effluents on cloud reflectivity. *Science,* **237,** 1020-1022.

Cobb, W.E., and H.J. Wells, 1970: The electrical conductivity of oceanic air and its correlation to global atmospheric pollution. *J. Atmos. Sci.,* **27,** 814-819.

Cohen, S., 1991: Pork in the sky. *The Washington Post Magazine*, **November 10**, 15-17, 36-39.

Committee on Scholarly Communication with the People's Republic of China (CSCPRC), 1992: Grasslands and grassland sciences in northern China. National Academy Press, Washington, DC, 214 pp.

Cooper, C., 1988: Controls on global biomass. Presented at the Chapman Conference on Gaia Hypothesis, March 7-11, 1988.

Cooper, W.A., and R.P. Lawson, 1984: Physical interpretation of results from the HIPLEX-1 experiment. *J. Climate Appl. Meteor.,* **23,** 523-540.

Cotton, W.R., 1972a: Numerical simulation of precipitation development in supercooled cumuli, Part I. *Mon. Wea. Rev.,* **100,** 757-763.

Cotton, W.R., 1972b: Numerical simulation of precipitation development in supercooled cumuli, Part II. *Mon. Wea. Rev.,* **100,** 764-784.

Cotton, W.R., 1985: Atmospheric convection and nuclear winter. *American Scientist,* **73,** 275-280.

Cotton, W.R., 1990: *Storms.* Geophysical Science Series, Vol. 1. ASTeR Press, Fort Collins, CO, 158 pp.

Cotton, W.R., and R.A. Anthes, 1989: *Storm and Cloud Dynamics.* Academic Press, Inc., San Diego. International Geophysics Series, Vol. 44., 883 pp.

Cotton, W. R., R.L. Walko, and G.L. Stephens, 1992: Cirrus cloud generation by smoke injections. Submitted to *Theor. Appl. Climatol.*

Covey, C., 1985: Climatic effect of nuclear war. *BioScience,* **35,** 563.

Covey, C., 1987: Protracted climatic effects of massive smoke injection into the atmosphere. *Nature,* **325,** 701-703.

Covey, C., S.H. Schneider, and S.L. Thompson, 1984: Global atmospheric effects of massive smoke injections from a nuclear war: Results from general circulation model simulations. *Nature*, **308,** 21-25.

Covey, C., S.L. Thompson, and S.H. Schneider, 1985: Nuclear winter: Diagnosis of atmospheric general circulation model simulations. *J. Geophys. Res.,* **90,** 5615-5628.

Craig, H., and C.C. Chou, 1982: Methane: The record in polar ice cores. *Geophys. Res. Lett.,* **9,** 1221-1224.

Crutzen, P.J., and J.W. Birks, 1982: The atmosphere after a nuclear war: Twilight at noon. *Ambio,* **11,** 114-125.

Crutzen, P.J., I.E. Galbally,and C. Brühl, 1984: Atmospheric effects from post-nuclear fires. *Climatic Change,* **6,** 323-364.

Cullis, C.F., and M.M. Hirschler, 1980: Atmospheric sulphur: Natural and man-made sources. *Atmos. Environ.,* **14,** 1263-1278.

Dalu, G.A., R.A. Pielke, R. Avissar, G.Kallos, and A. Guerrini, 1991: Linear impact of subgrid-scale thermal inhomogeneities on mesoscale atmospheric flow with zero synoptic wind. *Ann. Geophys.*, **9**, 641-647.

DAR Bulletin, 1990. Division of Atmospheric Research, CSIRO, AUSTRALIA, Issue #11, 16 pp.

Davis, F.W., D.S. Schimel, M.A. Friedl, J.C. Michaelsen, T.G.F. Kittel, R. Dubayah, and J. Dozier, 1992: Covariance of biophysical data with digital topographical and landuse maps over the FIFE site. *J. Geophys. Res.*, (in press).

Delmas, R., and C. Boutron, 1980: Are the past variations of the stratospheric sulfate burden recorded in Central Antarctic snow and ice layers? *J. Geophys. Res.,* **85,** 5645-5649.

Dennis, A.S., and A. Koscielski, 1972: Height and temperature of first echoes in unseeded and seeded convective clouds in South Dakota. *J. Appl. Meteor.,* **11,** 994-1000.

Dennis, A.S., and D.J. Musil, 1973: Calculations of hailstone growth and trajectories in a simple cloud model. *J. Atmos. Sci.,* **30,** 278-288.

Dennis, A.S, A. Koscielski, D.E. Cain, J.H. Hirsch, and P.L. Smith, Jr., 1975: Analysis of radar observations of a randomized cloud seeding experiment. *J. Appl. Meteor.,* **14,** 897-908.

Dickinson, R.E., Editor, 1987: *The Geophysiology of Amazonia*, Wiley for the United Nations University, New York, 526 pp.

Dickinson, R.E., 1991: Global change and terrestrial hydrology – A review. *Tellus*, **43AB**, 176-181.

Dickinson, R.E., and A. Henderson-Sellers, 1988: Modelling tropical deforestation: A study of GCM land-surface parameterizations. *Quart. J. Roy. Meteor. Soc.*, **114**, 439-462.

Dickinson, R.E., R.M. Errico, F. Giorgi, and G.T. Bates, 1989: A regional climate model for the western United States. *Climatic Change,* **15,** 383-422.

Doran, J.C., F.J. Parnes, R.L. Coulter, and T.L. Crawford, 1992a: A field study of the effects of inhomogeneities of surface sensible and latent heat fluxes. *Proc. Third Symposium of Global Change Studies*, Atlanta, Georgia. American Meteorological Society, Boston, January 5-10, 1992.

Doran, J.C., J.M. Hubbe, W.J. Shaw, D.D. Baldocchi, T.L. Crawford, R.J. Dobosy, and T.J. Meyers, 1992b: Comparisons of sensible and latent heat fluxes using surface and aircraft data over adjacent wet and dry surfaces. *Proc. Third Symposium of Global Change Studies*, Atlanta, Georgia. American Meteorological Society, Boston, January 5-10, 1992.

Drummond, A.J., and J.R. Hickey, 1971: Large-scale reflection and absorption of solar radiation by clouds as influencing earth radiation budget: New aircraft measurements. *Preprints, Int. Conf. on Weather Modification*, Canberra, Australian Acadamy of Science and American Meteorological Society, 267-276.

Dudal, R., 1987: Land resources for plant production. In *Resources and World Development*, D.J. McLaren and B.J. Skinner, Eds., John Wiley and Sons, Limited, New York, 659-670.

Dyson, F.J., 1988: *Infinite in all directions.* Gillford lectures given at Aberdeen, Scotland, April-November 1985, Harper and Row, New York.

Ellsaesser, H.W., M.C. MacCracken, J.J. Walton, and S.L. Grotch, 1986: Global climatic trends as revealed by the recorded data. *Rev. Geophys.,* **24,** 745-792.

Emanuel, K.A., 1987: The dependence of hurricane intensity on climate. *Nature,* **326,** 483-485.

Emanuel, K.A., 1988: Toward a general theory of hurricanes. *Amer. Sci.,* **76,** 371-379.

Enger, L., and M. Tjernström, 1991: Estimating the effects on the regional precipitation climate in a semiarid region caused by an artificial lake using a mesoscale model. *J. Appl. Meteor.*, **30**, 227-250.

English, M., 1973: Alberta hailstorms. Part II: Growth of large hail in the storm. *Meteor. Monogr.,* **14,** 37-98.

Farmer, G., T.M.L. Wigley, P.D. Jones, M. Salmon, 1989: Documenting and explaining recent global-mean temperature changes. Climatic Research Unit, Norwich, Final Report to NERC, UK, Contract GR3/6565.

Federer, B., A. Waldvogel, W. Schmid, H.H. Schiesser, F. Hampel, M. Schweingruber, W. Stahel, J. Bader, J.F. Mezeix, N. Doras, G. d'Aubigny, G. DerMegreditchian, and D. Vento, 1986: Main results of Grossversuch IV. *J. Climate Appl. Meteor.*, **25**, 917-957.

Folland, C.K., T.R. Karl, K.Ya. Vinnikov, 1990: Observed climate variations and change. In *Climate Change. The IPCC Scientific Assessment.* J.T. Houghton, G.J. Jenkins, and J.J. Ephraums, Eds., Cambridge University Press, Cambridge, MA., 193-238.

Fourier, J.B., 1827: Remarques générales sur les températures du globe terrestre et des espaces planétaries. *Paris, Mém. Acad. Sci.*, **VII,** 570-604.

Fournier d'Albe, E.M., and P. Mosino Aleman, 1976: A large-scale cloud seeding experiment in the Rio Nazas Catchment Area, Mexico. *Proc., Second WMO Scientific Conference on Weather Modification*, Boulder, CO, 2-6 August 1976, 143-149.

Fraser-Smith, A.C., 1989: Forum: Space science strategy. *Eos*, **December 26**, 1569.

Friis-Christensen, E., and K. Lassen, 1991: Length of the solar cycle: An indicator of solar activity closely associated with climate. *Science*, **254**, 698-700.

Fritz, S., 1954: Scattering of solar energy by clouds of "large drops." *J. Meteor.*, **11**, 291-300.

Fritz, S., 1958: Absorption and scattering of solar energy in clouds of "large water drops"- II. *J. Meteor.*, **15**, 51-58.

Fung, I. and Michael Prather, 1990: Greenhouse Gas Trends. In *Policy Options for Stabilizing Global Climate.* D.A. Lashof and D.A. Tirpak, Eds., U.S. Environmental Protection Agency, Office of Policy, Planning and Evaluation. Hemisphere Publishing Corp. New York.

Gagin, A., and J. Neumann, 1981: The second Israeli randomized cloud seeding experiment: Evaluation of the results. *J. Appl. Meteor.*, **20,** 1301-1311.

Gagin, A., D. Rosenfeld, and R.E. Lopez, 1985: A relationship between height and precipitation characteristics of summertime convective cells in South Florida. *J. Atmos. Sci.,* **42,** 84-94.

Gaian Science, 1991: James Lovelock on 20+ years of Gaian ideas. **May-July 1991**, 1-4.

Garratt, J.R., 1992: On the sensitivity of climate simulations to land-surface and atmospheric boundary layer treatments in GCMs – A review. *J. Climate*, (in press).

Garstang, M. B.E. Kelbe, G.D. Emmitt, and W.B. London, 1987: Generation of convective storms over the escarpment of northeastern South Africa. *Mon. Wea. Rev.*, **115**, 429-443.

Garstang, M., S. Ulanski, S. Greco, J. Scala, R. Swap, D. Fitzjarrald, D. Martin, E. Browell, M. Shipman, V. Connors, R. Harriss, and R. Talbot, 1990: The Amazon Boundary-Layer Experiment (ABLE 2B): A meteorological perspective. *Bull. Amer. Meteor. Soc.*, **71**, 19-32.

Gast, P.R., A.S. Jursa, J. Castelli, S. Basu, and J. Aarons, 1965: Solar electromagnetic radiation. In *Handbook of Geophysics and Space Environments*, S.L. Valley, Ed., McGraw-Hill, New York, 16-1–16-38.

Gentry, R.C., 1974: Hurricane modification. In *Climate and Weather Modification*, W.N. Hess, Ed., John Wiley and Sons, Inc. New York, 497-521.

Gerard, J-C., D. Delcourt, and L. M. Francois, 1990: The maximum entropy production principle in climate models: Application to the fain young sun paradox. *Quart. J. Roy. Meteor. Soc.,* **116,** 1123-1132.

Ghan, S.J., M.C. MacCracken, and J.J. Walton, 1988: Climatic response to large atmospheric smoke injections: Sensitivity studies with a tropospheric general circulation model. *J. Geophys. Res.,* **93,** 8315-8337.

Ghuman, B.S., and R. Lal, 1986: Effects of deforestation on soil properties and microclimate of a high rain forest in southern Nigeria. In *The Geophysiology of Amazonia*, R.E. Dickinson, Ed., John Wiley and Sons, New York, 225-244.

Giorgi, F., 1990: Simulation of regional climate using a limited area model nested in a general circulation model. *J. Climate*, **3**, 941-963.

Giorgi, F., and L.O. Mearns, 1991: Approaches to the simulation of regional climate change: A review. *Rev. Geophys.*, **29**, 191-216.

Giorgi, F., and G. Visconti, 1989: Two-dimensional simulations of possible mesoscale effects of nuclear war fires. 2. Model results. *J. Geophys. Res.,* **94,** 1145-1163.

Glatzmaier, G.A., and R.C. Malone, 1988: Global climate simulations of the ENUWAR case studies. Presented at the SCOPE-ENUWAR Moscow Workshop, 21-25 March 1988, Moscow, USSR.

Glenn, E., C. Hodges, H. Lieth, R.A. Pielke, and L. Pitelka, 1992: Halophytes to remove carbon from the atmosphere. *Environment,* **34,** 40-43.

Global Change, 1990: Terrestrial biosphere exchange with global atmospheric chemistry, terrestrial biosphere perspective of the IGAC Report. Companion to the Dookie Report. IGBP Report No. 13., Stockholm, 103 pp.

Golding, B.W., P. Goldsmith, N.A. Machin, and A. Slingo, 1986: Importance of local mesoscale factors in any assessment of nuclear winter. *Nature,* **319,** 301-303.

Graedel, T.E., and P.J. Crutzen, 1989: The changing atmosphere. *Sci. Amer.*, **September**, 58-68.

Graetz, R.D., 1991: The nature and significance of the feedback of changes in terrestrial vegetation on global atmospheric and climate changes. *Climate Change*, **18**, 147-173.

Grant, L.O., 1986: Hypotheses for the Climax Wintertime Orographic Cloud Seeding Experiments. In *Precipitation Enhancement–A Scientific Challenge*, R. Braham, Ed. *Meteor. Monogr.,* **21,** 105-108.

Grant, L.O., and R.E. Elliot, 1974: The cloud seeding temperature window. *J. Appl. Meteor.,* **13,** 355-363.

Gray, W.M., 1990: Strong association between West African rainfall and U.S. landfall of intense hurricanes. *Science,* **249**, 1251-1256.

Gray, W.M., 1991: Florida's hurricane vulnerability – Hostage to African rainfall. Presented at the 5th Governor's Hurricane Conference, Tampa, Florida, June 7, 1991, 24 pp.

Griffiths, J.F., and D.M. Driscoll, 1982: *Survey of climatology.* Charles E. Merrill Publishing Co., Columbus, Ohio.

Grotch, S.L., 1988: Regional intercomparisons of general circulation model predictions and historical climate data. Contract Report Contract No. DOE/NBB-0084, 291 pp.

Haberle, R.M., T.P. Ackerman, and O.B. Toon, 1985: Global transport of atmospheric smoke following a major nuclear exchange. *Geophys. Res. Lett.,* **12,** 405-408.

Hallett, J., 1981: Ice crystal evolution in Florida summer cumuli following AgI seeding. *Preprints, Eighth Conf. on Inadvertent and Planned Weather Modification*, Reno, American Meteorological Society, 114-115.

Hallett, J., and S. C. Mossop, 1974: Production of secondary ice particles during the riming process. *Nature,* **249,** 26-28.

Hänel, R.A., B.J. Conrath, V.G. Kunde, C. Prabhakara, I. Revah, V.V. Salomonson, and G. Wolford, 1972: The Nimbus 4 infrared spectroscopy experiment, 1. Calibrated thermal emission spectra. *J. Geophys. Res.,* **11,** 2629-2641.

Hansen, J.E., W. Wong, and A.A. Lacis, 1978: Mount Agung eruption provides test of a global climatic perturbation. *Science,* **199,** 1065-1068.

Hansen, J.E., A.A. Lacis, P. Lee, and W-C. Wang, 1980: Climatic effects of atmospheric aerosols. *Ann. N.Y. Acad. Sci.,* **338,** 575-587.

Hansen, J., D. Johnson, A. Lacis, S. Lebedeff, P. Lee, D. Rind, and G. Russell, 1981: Climate impact of increasing atmospheric carbon dioxide. *Science,* **213,** 957-966.

Hansen, J., A. Lacis, D. Rind, G. Russell, P. Stone, I. Fung, R. Ruedy, and J. Lerner, 1984: Climate sensitivity: Analysis of feedback mechanisms. In *Climate Processes and Climate Sensitivity, Geophys. Monogr. Ser.,* Vol. 29, J.E. Hansen and T. Takahashi, Eds., American Geophysical Union, Washington, D.C., 130-163.

Hansen, J., A. Lacis, D. Rind, G. Russell, I. Fung, P. Ashcraft, S. Lebedeff, R. Ruedy, and P. Stone, 1986: The greenhouse effect: Projections of global climate change. In *Effects of Changes in Stratospheric Ozone and Global Climate,* J.G. Titus, Ed., U.S. Environmental Protection Agency, Washington, DC, 379 pp.

Hansen, J., I. Fung, A. Lacis, D. Rind, S. Lebedeff, R. Reudy, G. Russell, and P. Stone, 1988: Global climate changes as forecast by Goddard Institute for Space Studies three-dimensional model. *J. Geophys. Res.,* **93,** 9341-9364.

Hartmann, D.L., 1984: On the role of global-scale waves in ice-albedo and vegetation-albedo feedback. In *Climate Processes and Climate Sensitivity,* J.E. Hansen and T. Takahashi, Eds., Geophysical Monogr., **5,** No. 29, 18-28, American Geophysical Union.

Harvey, L.D.D., and S.H. Schneider, 1985: Transient climate response to external forcing on $10^0 - 10^4$ year time scales. *J. Geophys. Res.,* **90,** 2191-2222.

Harwell, M.A., and T.C. Hutchinson, 1985: *Environmental Consequences of Nuclear War, Volume II: Ecological and Agricultural Effects.* SCOPE 28, John Wiley and Sons, Chichester, 374 pp.

Havens, B. S., J. E. Jiusto, and B. Vonnegut, 1978: Early history of cloud seeding. Project Cirrus Fund, SUNY-ES/328, 1400 Washington Ave., Albany, NY 12203, 75 pp.

Hayden, B.P., 1992: Desert ecosystems, greenhouse gases and climate control. *Science,* (submitted).

Henderson-Sellers, A., 1990: Getting GARP to grow grass: Land surfaces in global climate models. *Aust. Meteor. Mag.,* **38,** 245-254.

Henderson-Sellers, A., and K. McGuffie, 1989: Sulphate aerosols and climate. *Nature,* **340,** 436-437.

Herman, G.F., 1977: Solar radiation in summertime Arctic stratus clouds. *J. Atmos. Sci.,* **34,** 1423-1432.

Herron, M.M, 1982: Impurity sources of F^-, $C1^-$, NO_3 and SO_4^{2-} in Greenland and Antarctic precipitation. *J. Geophys. Res.,* **87,** 3052-3060.

Hindman, E.E., II, P.V. Hobbs, and L.F. Radke, 1977a: Cloud condensation nuclei from a paper mill. Part I: Measured effects on clouds. *J. Appl. Meteor.,* **16,** 745-752.

Hindman, E.E., II, P.M. Tag, B.A. Silverman, and P.V. Hobbs, 1977b: Cloud condensation nuclei from a paper mill. Part II: Calculated effects on rainfall. *J. Appl. Meteor.,* **16,** 753-755.

Hitchcock, D.R., and J.E. Lovelock, 1967: Life detection by atmospheric analysis. *Icarus,* **7,** 149-159.

Hjelmfelt, M.R., 1980: Numerical simulation of the effects of St. Louis on boundary layer airflow and convection. Ph.D. Dissertation, University of Chicago, Chicago, IL, 185 pp.

Hobbs, P.V., and A.L. Ragno, 1985: Ice particle concentrations in clouds. *J. Atmos. Sci.,* **42,** 2523-2549.

Hobbs, P.V., L.F. Radke, and S.E. Shumway, 1970: Cloud condensation nuclei from industrial sources and their apparent influence on precipitation in Washington State. *J. Atmos. Sci.,* **27,** 81-89.

Hobbs, P.V., T.J. Matejka, P.H. Herzegh, J.D. Locatelli, and R.A. Houze, 1980: The mesoscale and microscale structure and organization of clouds and precipitation in midlatitude cyclones. I. A case study of a cold front. *J. Atmos. Sci.,* **37,** 568-596.

Hoffert, M.I., A.J. Callegari, and C.-T. Hsieh, 1980: The role of deep sea heat storage in the secular response to climatic forcing. *J. Geophys. Res.,* **85,** (C11), 6667-6679.

Houghton, J.T., G.J. Jenkins, and J.J. Ephraums, 1990: *Climate Change. The IPCC Scientific Assessment.* Cambridge University Press, Cambridge, MA, 365 pp.

Houghton, R., 1987: Biotic changes consistent with the increased seasonal amplitude of atmospheric CO^2 concentrations. *J. Geophys. Res.,* **92,** 4223-4230.

Houghton, R., 1988: Reply to S. Idso, Comment on 'Biotic changes consistent with the increased seasonal amplitude of atmospheric CO^2 concentrations.' *J. Geophys. Res.,* **93,** 1747-1748.

Hulstrom, R.L., and T.L. Stoffel, 1990: Some effects of the Yellowstone fire smoke cloud on incident solar irradiance. *J. Climate,* **3,** 1485-1490.

Hummel, J.R., 1982: Surface temperature sensitivities in a multiple cloud radiative-convective model with a constant and pressure-dependent lapse rate. *Tellus,* **34,** 203-208.

Hummel, J.R., and W.R. Kuhn, 1981a: Comparison of radiative-convective models with constant and pressure-dependent lapse rates. *Tellus,* **33,** 254-261.

Hummel, J.R., and W.R. Kuhn, 1981b: An atmospheric radiative-convective model with interactive water vapor transport and cloud development. *Tellus,* **33,** 372-381.

Hummel, J.R., and R.A. Reck, 1981: Carbon dioxide and climate: The effects of water transport in radiative-convective models. *J. Geophys. Res.,* **86,** 12,035-12,038.

Hunt, B.G., 1981: An examination of some feedback mechanisms in the carbon dioxide climate problem. *Tellus,* **33,** 78-88.

Hunt, B.G., and N.C. Wells, 1979: An assessment of the possible future climate impact of carbon dioxide increases based on a coupled one-dimensional atmospheric-oceanic model. *J. Geophys. Res.,* **84,** 787-791.

Idso, S.B, 1988: Me and the modelers: Perhaps not so different after all. *Climatic Change,* **12,** 93.

Imbrie, J., and K.P. Imbrie, 1979: Ice Ages. Solving the mystery. Enslow Publishers, Short Hills, NJ, 224 pp.

IPCC, 1991: Climate Change, The IPCC Scientific Assessment Intergovernmental Panel on Climate Change. WMO/UNEP, xi-xxxiii.

Iribarne, J.V., and R.G. de Pena, 1962: The influence of particle concentration on the evolution of hailstones. *Nubila,* **5,** 7-30.

Karl, T.R., and P.D. Jones, 1989: Urban bias in area-averaged surface air temperature trends. *Bull. Amer. Meteor. Soc.,* **70,** 265-270.

Karl, T.R., J.D. Tarpley, R.G. Quayle, H.F. Diaz, D.A. Robinson, and R.S. Bradley, 1989: The recent climate record: What it can and cannot tell us. *Rev. Geophys.*, **27**, 405-430.

Kaufman, Y.J., R.S. Fraser, and R.L. Mahoney, 1991: Fossil fuel and biomass burning effect on climate. *J. Climate*, **4**, 578-588.

Keeling, C.D., R.B. Bacastow, and T.P. Whort, 1982: Measurements of the concentration of carbon dioxide at Mauna Loa observatory, Hawaii, in *Carbon Dioxide Review: 1982.* W.C. Clark, Ed., Oxford University Press, New York, 377-385.

Keller, V.W., and R.I. Sax, 1981: Microphysical development of a pulsating cumulus tower: A case study. *Quart. J. Roy. Meteor. Soc.,* **107,** 679-697.

Kellogg, W.W., 1977: Effects of human activities on global climate. WMO Tech. Note No. 156, World Meteorological Organization, Geneva, Switzerland.

Kellogg, W.W., 1982: Precipitation trends on a warmer earth. In *Interpretation of Climate and Photochemical Models, Ozone, and Temperature Measurements*, R.A. Reck and J.R. Hummel, Eds., Institute of Physics, New York, 35-46.

Kellogg, W.W., 1983: Feedback mechanisms in the climate system affecting future levels of carbon dioxide. *J. Geophys. Res.,* **88,** 1263-1269.

Kellogg, W.W., 1991: Response to skeptics of global warming. *Bull. Amer. Meteor. Soc.*, **72**, 499-511.

Kerr, R.A., 1982: Cloud seeding: One success in 35 years. *Science,* **217,** 519-521.

Kerr, R.A., 1991: Could the sun be warming the climate. *Science*, **254**, 652-653.

King, D.A., G.E. Bingham, and J.R. Kercher, 1985: Estimating the direct effect of CO_2 on soybean yield. *J. Environ. Mgt.,* **20,** 51-62.

Knight, C.A., and P. Squires, 1982: *Hailstorms of The Central High Plains, Parts I and II,* C.A. Knight and P. Squires, Eds., Colorado Associated University Press, Boulder, CO.

Koenig, L.R., 1977: The rime-splintering hypothesis of cumulus glaciation examined using a field-of-flow cloud model. *Quart. J. Roy. Meteor. Soc.,* **103,** 585-606.

Koenig, L.R., and F.W. Murray, 1976: Ice-bearing cumulus cloud evolution: Numerical simulations and general comparison against observations. *J. Appl. Meteor.*, **15**, 747-762.

Kondrat'yev, K. Ya., V.I. Binenko, and O.P. Petrenchuck, 1981: Radiative properties of clouds influenced by a city. *Izv., Atmos. Oceanic Phys.,* **17,** 122-127.

Kramer, M.L., D.E. Seymour, M.E. Smith, R.W. Reeves, and T.T. Frankenberg, 1976: Snowfall observations from natural-draft cooling tower plumes. *Science,* **193,** 1239-1241.

Kratzer, A., 1956: *The Climate of Cities.* Translation, American Meteorological Society, 221 pp.

Kuhn, P.M., 1970: Airborne observations of contrail effects on the thermal radiation budget. *J. Atmos. Sci.,* **27,** 937-942.

Kump, L.R., 1988: Terrestrial feedback in atmospheric oxygen regulation by fire and phosphorus. *Nature*, **335**, 152-154.

La Brecque, M., 1989a: Detecting climate change, I: Taking the world's shifting temperature. *MOSAIC*, **20**, No. 4, 2-9.

La Brecque, M., 1989b: Detecting climate change, II: The impact of the water budget. *MOSAIC*, **20**, 10-16.

Labitzke, K., 1987: Sunspots, the QBO, and the stratosphere temperature in the North Polar region. *Geophys. Res. Lett.,* **14,** 535-537.

Lal, M., and V. Ramanathan, 1984: The effects of moist convective and water vapor radiative processes on climate sensitivity. *J. Atmos. Sci.,* **41,** 2238-2249.

Lamb, D., J. Hallet, and R.I. Sax, 1981: Mechanistic limitations to the release of latent heat during the natural and artificial glaciation of deep convective clouds. *Quart. J. Roy. Meteor. Soc.*, **107**, 935-954.

Landsberg, H.E., 1956: The climate of towns. *Man's Role in Changing the Face of the Earth,* University of Chicago Press, 584-603.

Landsberg, H.E., 1970: Man-made climatic changes. *Science,* **170,** 1265-1274.

Langmuir, I., 1948: The production of rain by chain reaction in cumulus clouds at temperatures above freezing. *J. Meteor.* **5,** 175.

Langmuir, I., 1953: Final Report, Project Cirrus, Contract W-36-039-SC-32427, RL-785. General Electric Company, Research Laboratory, Schnectady, NY.

Lanicci, J.M., T.N. Carlson, and T.T. Warner, 1987: Sensitivity of the Great Plains severe-storm environment to soil-moisture distribution. *Mon. Wea. Rev.*, **115**, 2660-2673.

Lashof, D.A., 1989: The dynamic greenhouse: Feedback processes that may influence future concentrations of atmospheric trace gases and climatic change. *Climatic Change,* **14,** 213-242.

Lashof, D.A., and D.A. Tirpak, Eds., 1989: *Policy Options for Stabilizing Global Climate*, U.S. Environmental Protection Agency, Office of Policy, Planning and Evaluation.

Lean, J., and D.A. Warrilow, 1989: Simulation of the regional climatic impact of Amazon deforestation. *Nature*, **342**, 411-413.

Ledley, T.S., and S.L. Thompson, 1986: Potential effect of nuclear war smokefall on sea ice. *Climatic Change,* **8,** 155-171.

Lee, R., 1978: *Forest Micrometeorology.* Columbia University Press, 276 pp.

Lee, T.J., 1992: Formulation of the Land Ecosystem-Atmosphere Feedback Model. Ph.D. Dissertation, Colorado State University, Department of Atmospheric Science, In preparation.

Lee, T.J., R.A. Pielke, T.G.F. Kittel, and J.F. Weaver, 1991: Atmospheric modeling and its spatial representation of land surface characteristics. *Proc., First International Conference/Workshop on Integrating Geographic Information Systems and Environmental Modeling*, Invited contribution, September, 1991, Boulder, Colorado.

Leith, H., 1963: The role of vegetation in the carbon dioxide content of the atmosphere. *J. Geophys. Res.*, **68**, 3887-3898.

Levy, G., and W.R. Cotton, 1984: A numerical investigation of mechanisms linking glaciation of the ice-phase to the boundary layer. *J. Climate Appl. Meteor.*, **23**, 1505-1519.

Lewis, W., 1951: On a seven day periodicity. *Bull. Amer. Meteor. Soc.*, **32**, 192.

Lian, M.S., and R.D. Cess, 1977: Energy balance climate models: A reappraisal of ice-albedo feedback. *J. Atmos. Sci.*, **34**, 1058-1062.

Lindzen, R.S., 1990: Some coolness concerning global warming. *Bull. Amer. Meteor. Soc.*, **71**, 288-299.

Lindzen, R.S., A.Y. Hou, and B.F. Farrell, 1982: The role of convective model choice in calculating the climate impact of doubling CO_2. *J. Atmos. Sci.*, **39**, 1189-1205.

Loose, T., and R.D. Bornstein, 1977: Observations of mesoscale effects on frontal movement through an urban area. *Mon. Wea. Rev.*, **105**, 562-571.

Lovelock, J.E., 1972: Gaia as seen through the atmosphere. *Atmos. Environ.*, **6**, 579-580.

Lovelock, J.E., 1988: *The Ages of Gaia.* W.W. Norton, New York, 252 pp.

Lovelock, J.E., 1990: Hands up for the Gaia hypothesis. *Nature*, **344**, 100-102.

Lovelock, J.E., and M. Whitfield, 1982: Life span of the biosphere. *Nature*, **296**, 561-563.

Lovelock, J.E., R.J. Maggs, and R.A. Rasmussen, 1972: Atmospheric dimethyl sulphide and the natural sulphur cycle. *Nature*, **237**, 452-453.

Luoma, J.R., 1991: Gazing into our greenhouse future. *Audubon*, **March**, 52-59, 124-125.

MacCracken, M.C., 1983: Nuclear war: Preliminary estimates of the climatic effects of a nuclear exchange. Lawrence Livermore National Laboratory Report UCRL-89770, Livermore, CA.

MacCracken, M.C., 1985: Carbon dioxide and climate change: Background and overview. In *The Potential Climatic Effects of Increasing Carbon Dioxide*, M.C. MacCracken, and F.M. Luther, Eds., U.S. Department of Energy, Washington, DC (DOE/ER-0237), 381 pp.

MacCracken, M.C., and J.S. Chang, 1975: Preliminary study of the potential chemical and climatic effects of atmospheric nuclear explosions. Lawrence Livermore Lab. Rep. UCRL-51653.

MacCracken, M.C., and J.J. Walton, 1984: The effects of interactive transport and scavenging of smoke on the calculated temperature change resulting from large amounts of smoke. Lawrence Livermore National Laboratory Report UCRL-91446, Livermore, CA.

Mahrer, Y., and R.A. Pielke, 1976: The numerical simulation of the air flow over Barbados. *Mon. Wea. Rev.*, **104**, 1392-1402.

Mahrer, Y., and R.A. Pielke, 1978: The meteorological effect of the changes in surface albedo and moisture. *Israel Meteorological Soc. (IMS)*, 55-70.

Mahfouf, J.F., E. Richard, and P. Mascart, 1987: The influence of soil and vegetation on the development of mesoscale circulation. *J. Climate Appl. Meteor.*, **26**, 1483-1495.

Malm, W.C., 1989: Atmospheric haze: Its sources and effects on visibility in rural areas of the continental United States. *Environ. Monitoring Assessment,* **12,** 203-225.

Malone, R.C., L.H. Auer, G.A. Glatzmaier, M.C. Wood, and O.B. Toon, 1985: Influence of solar heating and precipitation scavenging on the simulated lifetime of post-nuclear war smoke. *Science,* **230,** 317.

Malone, R.C., L.H. Auer, G.A. Glatzmaier, M.C. Wood, and O.B. Toon, 1986: Three-dimensional simulations including interactive transport, scavenging, and solar heating of smoke. *J. Geophys. Res.,* **91,** 1039-1053.

Manabe, S., 1971: Estimate of future changes in climate due to increase of carbon dioxide concentration in the air. In *Man's Impact on the Climate,* W.H. Mathews, W.W. Kellogg, and G.D. Robinson, Eds., MIT Press, Cambridge, MA, 249-264.

Manabe, S., and R. Stouffer, 1979: Sensitivity of a global climate model to an increase of CO_2 concentration in the atmosphere. *J. Geophys. Res.,* **85,** 491-493.

Manabe, S., and R. Stouffer, 1980: Sensitivity of a global climate model to an increase of CO_2 concentration in the atmosphere. *J. Geophys. Res.,* **85,** 5529-5554.

Manabe, S., and R.T. Wetherald, 1967: Thermal equilibrium of the atmosphere with a given distribution of relative humidity. *J. Atmos. Sci.,* **24,** 241-259.

Manabe, S., and R.T. Wetherald, 1975: The effects of doubling the CO_2 concentration on climate of a general circulation model. *J. Atmos. Sci.,* **32,** 3-15.

Manabe, S., and R.T. Wetherald, 1980: On the distribution of climate change resulting from an increase in CO_2-content of the atmosphere. *J. Atmos. Sci.,* **37,** 99-118.

Mann, C., 1991: Lynn Margulis: Science's unruly earth mother. *Nature,* **252**, 378-381.

Marland, G., 1991. CO_2 emissions. In Boden, T.A., P. Kanciruk, and M.P. Farrell, *Trends '90: A Compendium of Data on Global Change,* Carbon Dioxide Information Analysis Center, Oak Ridge National Laboratory, Oak Ridge, Tennessee, 92-133.

Marwitz, J.D., 1972: The structure and motion of severe hailstorms. Part I: Supercell storms. *J. Appl. Meteor.,* **11,** 166-179.

Mascart, P., O. Taconet, J.-P. Pinty, and M.B. Mehrez, 1991: Canopy resistance formulation and its effect in mesoscale models: A HAPEX perspective. *Agric. Forest Meteor.*, **54**, 319-351.

Mason, B.J., 1971: *The Physics of Clouds,* 2nd Ed., Clarendon Press, Oxford, 671 pp.

Mather, G.K., 1991: Coalescence enhancement in large multicell storms caused by the emissions from a Kraft paper mill. *J. Appl. Meteor.*, **30,** 1134-1146.

Mayewski, P.A., W.B. Lyons, M.J. Spencer, M. Twickler, W. Dansgaard, B. Koci, C.I. Davidson, and R.E. Honrath, 1986: Sulfate and nitrate concentrations from a South Greenland ice core. *Science,* **232,** 975-977.

Mayewski, P.A., W.B. Lyons, M.J. Spencer, M.S. Twickler, C.F. Buck, and S. Whitlow, 1990: An ice-core record of atmopsheric response to anthropogenic sulphate and nitrate. *Nature*, **346**, 554-556.

McNider, R.T., M.D. Moran, and R.A. Pielke, 1988: Influence of diurnal and inertial boundary layer oscillations on long-range dispersion. *Atmos. Environ.*, **22**, 2445-2462.

McPherson, E.G., and G.C. Woodard, 1989: The case for urban reLeaf: Tree planting pays. *Arizona's Economy*, December, 1989.

McPherson, E.G., and G.C. Woodard, 1990: Cooling the urban heat island with water-and-energy efficient landscapes. *Arizona Review*, Karl Eller Graduate School of Management, College of Business and Public Administration, The University of Arizona, Tucson, Arizona, 1-8.

Meehl, G.A., 1990: Development of global coupled ocean-atmosphere general circulation models. *Climate Dyn.,* **5,** 19-33

Meitín, J.G., W.L. Woodley, and J.A. Flueck, 1984: Exploration of extended-area treatment effects in FACE-2 using satellite imagery. *J. Climate Appl. Meteor.,* **23,** 63-83.

Merrill, R.T., 1988: Environmental influences on hurricane intensification. *J. Atmos. Sci.,* **45,** 1678-1687.

Mesinger, F., and N. Mesinger, 1992: Has hail suppression in Eastern Yugoslavia led to a reduction in the frequency of hail? *J. Appl. Met.,* **31,** 104-111.

Meszaros, E., 1988: On the possible role of the biosphere in the control of atmospheric clouds and precipitation. *Atmos. Environ.,* **22,** 423-424.

Michaels, P.J., D.E. Sappington, D.E. Stooksbury, and B.P. Hayden, 1990: Regional 500 mb heights and U.S. 1000-500 mb thickness prior to the radiosonde era. *Theor. Appl. Climatol.*, **42**, 149-154.

Mielke, P.W., Jr., G.W. Brier, L.O. Grant, G.J. Mulvey, and P.N. Rosensweig, 1981: A statistical re-analysis of the replicated Climax I and II wintertime orographic cloud seeding experiments. *J. Appl. Meteor.,* **20,** 643-660.

Miller, W.F., R.A. Pielke, M. Garstang, and S. Greco, 1988: Simulations of the mesoscale circulation in the Able II region. *Proc., Fifth Brazilian Congress of Meteorology.* November 5-11, 1988, Rio de Janeiro, Brazil.

Mitchell, J.F.B., 1989: The "greenhouse" effect and climate change. *Rev. Geophys.,* **27,** 115-139.

Mitchell, J.F.B., C.A. Senior, and W.J. Ingram, 1989: CO_2 and climate: A missing feedback? *Nature,* **341,** 132-134.

Mitchell, J.F.B., and A. Slingo, 1988: Climatic effects of nuclear war: The role of atmospheric stability and ground heat fluxes. *J. Geophys. Res.,* **93,** 7037-7045.

Möller, F., 1963: On the influence of changes in the CO_2 concentration in air on the radiation balance of the earth's surface and on the climate. *J. Geophys. Res.,* **68,** 3877-3886.

Mossop, S.C., and J. Hallett, 1974: Ice crystal concentration in cumulus clouds: Influence of the drop spectrum. *Science,* **186,** 632-634.

Muhs, D.R., 1985: Age and paleoclimatic significant of Holocene sand dunes in northeastern Colorado. *Ann. Assoc. Amer. Geogr.*, **75**, 566-582.

Murcray, W.B., 1970: On the possibility of weather modification by aircraft contrails. *Mon. Wea. Rev.*, **98**, 745-748.

Murray, F.W., and L.R. Koenig, 1979: Simulation of convective cloudiness, rainfall, and associated phenomena caused by industrial heat released directly to the atmosphere. Report, Rand Corporation, R-2456-DOE, 93 pp.

Murty, Bh. V.R., and K.R. Biswas, 1968: Weather modification in India. *Proc., First National Conference on Weather Modification,* Albany, N.Y., 28 April – 1 May, 71-80.

Namais, J., 1980: Some concomitant regional anomalies associated with hemispherically averaged temperature variations. *J. Geophys. Res.*, **85**, 1585-1590.

National Academy of Sciences, 1975: *Long-Term Worldwide Effects of Multiple Nuclear-Weapon Detonations.* Washington, D.C.

National Academy of Sciences, 1980: *The Atmospheric Sciences: National Objectives for the 1980's.* Washington, D.C., 130 pp.

National Academy of Sciences, 1991: *Policy Implications of Greenhouse Warming – Synthesis Panel Report.* National Academy Press, Washington, D.C., 127 pp.

Neftel, A., J. Beer, H. Oeschger, F. Zürcher, and R.C. Finkel, 1985a: Sulfate and nitrate concentrations in snow from South Greenland. *Nature,* **314**, 611-613.

Neftel, A., E. Moor, H. Oeschger, and B. Stauffer, 1985b: Evidence from polar ice cores for the increase in atmospheric CO_2 in the past two centuries. *Nature,* **315**, 45-47.

Nemani, R.R., and S.W. Running, 1989: Estimation of regional surface resistance to evapotranspiration from NDVI and Thermal-IR AVHRR Data. *J. Appl. Meteor.*, **28**, 276-284.

Neumann, J., and S. Parpola, 1987: Climatic change and the eleventh-tenth-century eclipse of Assyria and Babylonia. *JNES*, **46**, 161-182.

Neumann, J., and R.M. Sigrist, 1978: Harvest dates in ancient Mesopotamia as possible indicators of climatic variations. *Climatic Change*, **1**, 239-252.

Nickerson, E.C., E. Richard, R. Rosset, and D.R. Smith, 1986: The numerical simulation of clouds, rain, and airflow over the Vosges and Black Forest mountains: A meso-β model with parameterized microphysics. *Mon. Wea. Rev.*, **114**, 398-414.

NOAA Environmental Digest, 1990: Selected environmental indicators of the United States and the Global Environment. September 1990, Office of the Chief Scientist, 66 pp. plus appendices.

Nobre, C.A., P.J. Sellers, and J. Shukla, 1991: Amazonian deforestation and regional climate change. *J. Climate*, **4**, 957-988.

Notes on the State of Virginia, 1976. Thomas Jefferson. Originally published in 1861. Introduction to the Torchbook edition by Thomas Perkins Abernethy, Gloucester, Massachusetts, 228 pp.

O'Leary, J.W., 1984: The role of halophytes in irrigated agriculture. In *Plants: Strategies for Crop Improvement*, R.C. Staples, Ed., John Wiley and Sons, Inc., New York.

Odingo, R.S., 1990: The definition of desertification: Its programmatic consequences for UNEP and the international community. Desertification Control Bulletin, United Nations Environment Programme, November 18, 1990.

Ohtake, T., and P.J. Huffman, 1969: Visual range of ice fog. *J. Appl. Meteor.*, **8**, 499-501.

Ookouchi, Y., M. Segal, R.C. Kessler, and R.A. Pielke, 1984: Evaluation of soil moisture effects on the generation and modification of mesoscale circulations. *Mon. Wea. Rev.*, **112**, 2281-2292.

Orville, H.D., and J-M. Chen, 1982: Effects of cloud seeding, latent heat of fusion, and condensate loading on cloud dynamics and precipitation evolution: A numerical study. *J. Atmos. Sci.*, **39**, 2807-2827.

Orville, H.D., P.A. Eckhoff, J.E. Peak, J.H. Hirsch, and F.J. Kopp, 1981: Numerical simulation of the effects of cooling tower complexes on clouds and severe storms. *Atmos. Environ.,* **15**, 823-836.

Orville, H.D., J.H. Hirsch, and L.E. May, 1980: Application of a cloud model to cooling tower plumes and clouds. *J. Appl. Meteor.,* **19,** 1260-1272.

Otterman, J., 1974: Baring high-albedo soils by overgrazing. A hypothesized desertification mechanism. *Science*, **86**, 531-533.

Otterman, J., 1977: Anthropogenic impact on the albedo of the earth. *Climatic Change*, **1**, 137-155.

Otterman, J., and C.J. Tucker, 1985: Satellite measurements of surface albedo and temperatures in semi-desert. *J. Climate Appl. Meteor.*, **24**, 228-235.

Otterman, J., A. Manes, S. Rubin, P. Alpert, and D.O'C. Starr, 1990: An increase of early rains in southern Israel following land-use change? *Bound.-Layer Meteor.*, **53**, 333-351.

Paltridge, G.W., 1980: Cloud-radiation feedback to climate. *Quart. J. Roy. Meteor. Soc.,* **106,** 895-899.

Paltridge, G.W., and C.M.R. Platt, 1976: *Radiative processes in meteorology and climatology.* Developments in Atmospheric Science, 5. Elsevier Science Publishers, New York.

Pecker, J.-C., and S.K. Runcorn, Eds., 1990: *The Earth's Climate and Variability of the Sun over Recent Millennia.* The Royal Society, London, 287 pp.

Penner, J.E., 1986: Uncertainties in the smoke source term for 'nuclear winter' studies. *Nature,* **324,** 222-226.

Peterson, T.C., 1991: The relationships between SST anomalies and clouds, water vapor, and their radiative effects. Ph.D. dissertation, Department of Atmospheric Science, Colorado State University, Fort Collins, CO 80523, 259 pp.

Petukhov, V.K., Ye. M. Feygelson, and N.I. Manuylova, 1975: The regulating role of clouds in the heat effects of anthropogenic aerosols and carbon dioxide. *Izv. Atmos. Oceanic Phys.,* **11,** 802-808.

Philander, S.G., 1990: *El Niño, La Niña & the Southern Oscillation.* Academic Press, London.

Phulpin, T., J.P. Jullien, and D. Lasselin, 1989: AVHRR data processing to study the surface canopies in temperate regions. First results of HAPEX. *Int. J. Remote Sens.*, **10**, 869-884.

Pielke, R.A., 1974a: A three-dimensional numerical model of the sea breezes over South Florida. *Mon. Wea. Rev.*, **102**, 115-134.

Pielke, R.A., 1974b: A comparison of three-dimensional and two-dimensional numerical predictions of sea breezes. *J. Atmos. Sci.*, **31**, 1577-1585.

Pielke, R.A., 1984: *Mesoscale meteorological modeling.* Academic Press, New York, N.Y., 612 pp. (Translated into Chinese with corrections by the Chinese Meteorological Press, 1990.)

Pielke, R.A., 1990: *The Hurricane*, Routledge Press, London, England, 228 pp.

Pielke, R.A., 1991: Overlooked scientific issues in assessing hypothesized greenhouse gas warming. *Environ. Software*, **6**, 100-107.

Pielke, R.A., and R. Avissar, 1990: Influence of landscape structure on local and regional climate. *Landscape Ecology*, **4**, 133-155.

Pielke, R.A., G. Dalu, J.R. Garratt, T.G.F. Kittel, R.A. Stocker, T.J. Lee, and J.S. Snook, 1990d: Influence of mesoscale landuse on weather and climate and its representation for use in large-scale models. *Proc., Indo-U.S. Seminar on "The Parameterization of Subgrid-Scale Processes in Dynamical Models of Medium-Range Prediction and Global Climate"*, Pune, India. August 6-10, 1990.

Pielke, R.A., G. Dalu, M.D. Moran, M. Uliasz, T.J. Lee, and R.A. Stocker, 1991a: Impacts of land surface characteristics on atmospheric dispersion. *Preprints, Seventh Joint Conference on Applications of Air Pollution Meteorology with AWMA*, New Orleans, Louisiana, AMS, January 14-18, 1991, 302-307.

Pielke, R.A., G. Dalu, J.S. Snook, T.J. Lee, and T.G.F. Kittel, 1991b: Nonlinear influence of mesoscale landuse on weather and climate. *J. Climate*, **4**, 1053-1069.

Pielke, R.A., G. Dalu, J. Weaver, J. Lee, and J. Purdom, 1990c: Influence of landuse on mesoscale atmospheric circulation. *Extended Abstracts Volume, Fourth Conference on Mesoscale Processes*, Boulder, Colorado, June 25-29, 1990, AMS, Boston, 226-227.

Pielke, R.A., and E. Kennedy, 1980: Mesoscale terrain features, January 1980. Report # UVA-ENV SCI-MESO-1980-1, 29 pp.

Pielke, R.A., T.J. Lee, J. Weaver, and T.G.F. Kittel, 1990b: Influence of vegetation on the water and heat distribution over mesoscale sized areas. *Preprints, 8th Conference on Hydrometeorology*, Kananaskis Provincial Park, Alberta, Canada, October 22-26, 1990, 46-49.

Pielke, R.A. and M. Uliasz, 1992: Influence of landscape variability on atmospheric dispersion. *J. Air Waste Mgt.*, (in press).

Pielke, R.A., J. Weaver, T. Kittel, and J. Lee, 1990a: Use of NDVI for mesoscale modeling. *Proc. Workshop on the "Use of Satellite-Derived Vegetation Indices in Weather and Climate Prediction Models"*, Camp Springs, Maryland. February 26-27, 1990, 83-85.

Pielke, R.A., and X. Zeng, 1989: Influence on severe storm development of irrigated land. *Natl. Wea. Dig.*, **14**, 16-17.

Pinty, J.-P., P. Mascart, E. Richard, and R. Rosset, 1989: An investigation of mesoscale flows induced by vegetation inhomogeneities using an evapotranspiration model calibrated against HAPEX-MOBILHY data. *J. Appl. Meteor.*, **28**, 976-992.

Pittock, A.B., T.P. Ackerman, P.J. Crutzen, M.C. MacCracken, C.S. Shapiro, and R.P. Turco, 1986: *Environmental Consequences of Nuclear War: Vol. I., Physical and Atmospheric Effects*, SCOPE 28, John Wiley and Sons, Chichester, 374 pp.

Policy Statement of the American Meteorological Society on Global Climate Change. *Bull. Amer. Meteor. Soc.*, **72**, 57-59.

Porch, W.M., C.-Y. J. Kao, and R.G. Kelley, Jr., 1990: Ship trails and ship induced cloud dynamics. *Atmos. Environ.,* **24A,** 1051-1059.

Post, W.M., W.R. Emanuel, P.J. Zinke, and A.G. Stangenberger, 1982: Soil carbon pools and world life zones. *Nature*, **298**, 156-159.

Prahm, L.P., U. Torp, and R.M. Stern 1976: Deposition and transformation rates of sulfur oxides during atmospheric transport over the Atlantic. *Tellus,* **28,** 355-372.

Pruppacher, H.R., and J.D. Klett, 1978: *Microphysics of clouds and precipitation.* Reidel, Dordrecht.

Rabin, R.M., S. Stadler, P.J. Wetzel, D.J. Stensrud, and M. Gregory, 1990: Observed effects of landscape variability of convective clouds. *Bull. Amer. Meteor. Soc.*, **71**, 272-280.

Radke, L.F., J.A. Coakely, Jr., and M.D. King, 1989: Direct and remote sensing observations of the effects of ships on clouds. *Science,* **246,** 1146-1148.

Ramanathan, V., 1981: The role of ocean-atmosphere interactions in the CO_2 climate problem. *J. Atmos. Sci.,* **38,** 918-930.

Ramanathan, V., and J.A. Coakley, 1978: Climate modeling through radiative-convective models. *Rev. Geophys. Space Phys.,* **16,** 465-489.

Ramanathan, V., and W. Collins, 1991: Thermodynamic regulation of ocean warming by cirrus clouds deduced from observations of the 1987 El Niño. *Nature,* **351,** 27-32.

Ramanathan, V., L. Callis, R. Cess, J. Hansen, I. Isaksen, W. Kuhn, A. Lacis, F. Luther, J. Mahlman, R. Reck, and M. Schlesinger, 1987: Climate chemical interactions and effects of changing atmospheric trace gases. *Rev. Geophys.,* **25,** 1441-1482.

Ramanathan, V., B.R. Barkstrom, and E.F. Harrison, 1989a: Climate and the earth's radiation budget. *Phys. Today,* **42,** 22-32.

Ramanathan, V., R.D. Cess, E.F. Harrison, P. Minnis, B.R. Barkstrom, E. Ahmad, and D. Hartmann, 1989b: Cloud-radiative forcing and climate: Results from the earth radiation budget experiment. *Science,* **243,** 1-140.

Ramaswamy, V., 1988: Aerosol radiative forcing and model responses. In *Aerosols and Climate*, P.V. Hobbs and M.P. McCormick, Eds., A. Deepak Publishing, 349-372.

Ramaswamy, V., and J.T. Kiehl, 1985: Sensitivities of the radiative forcing due to large loadings of smoke and dust aerosols. *J. Geophys. Res.,* **90,** 5597-5613.

Randall, D.A., 1992: Global climate models: What and how. In *Global Warming: Physics and Facts,* G.B. Levi, D. Hafemeister, and R. Scribner, Eds., American Physical Society, pp. 24-35.

Randall, D.A. and S. Tjemkes, 1991: Clouds, the earth's radiation budget, and the hydrologic cycle. *Paleogeography, Paleoclimatology, Paleoecology (Global and Planetary Change Section),* **90**, 3-9.

Randall, D., T. Carlson, and Y. Mintz, 1984: The sensitivity of a general circulation model to Saharan dust heating. In *Aerosols and Their Climatic Effects,* H.E. Gerber and A. Deepak, Eds., A. Deepak Publishing, 123-132.

Randall, D.A., Harshvardhan, D.A. Dazlich, and T.G. Corsetti, 1989: Interactions among radiation, convection, and large-scale dynamics in a general circulation model. *J. Atmos. Sci.,* **46,** 1943-1970.

Randall, D.A., R.D. Cess, J.P. Blanchet, G.J. Boer, D.A. Dazlich, A.D. Del Genio, M. Deque, V. Dymnikov, V. Galin, S.J. Ghan, A.A. Lacis, H. Le Treut, Z.-X. Li, X.-Z. Liang, B.J. McAvaney, V.P. Meleshko, J.F.B. Mitchell, J.-J. Morcrette, G.L. Potter, L. Rikus, E. Roeckner, J.F. Royer, U. Schlese, D.A. Sheinin, J. Slingo, A.P. Sokolov, K.E. Taylor, W.M. Washington, R.T. Wetherald, I. Yagai, and M.-H. Zhang, 1992: Intercomparison and interpretation of surface energy fluxes in atmospheric general circulation models. *J. Geophys. Res.,* **97,** 3711-3724.

Ratner, B., 1962: Climatological effect of changeover to hygrothermometer. *Mon. Wea. Rev.,* **90,** 89-96.

Raynaud, D., and J.M. Barnola, 1985: An Antarctic ice core reveals atmospheric CO_2 variations over the past few centuries. *Nature,* **315,** 309-311.

Reynolds, D.W., 1988: A report on winter snowpack-augmentation. *Bull. Amer. Meteor. Soc.,* **69,** 1290-1300.

Reynolds, D.W., and A.S. Dennis, 1986: A review of the Sierra Cooperative Pilot Project. *Bull. Amer. Meteor. Soc.,* **67,** 513-523.

Reynolds, D.W., T.H. VonderHaar, and S.K. Cox, 1975: The effect of solar radiation absorption in the tropical troposphere. *J. Appl. Meteor.,* **14,** 433-443.

Richards, J.F., 1986: Chapter 2: World environmental history and economic development. In *Sustainable Development of the Biosphere,* W.C. Clark and R.E. Munn, Eds., International Institute for Applied Systems Analysis, Laxenburg, Austria, Cambridge University Press, 53-74.

Rind, D., 1984: The influence of vegetation on the hydrologic cycle in a global climate model. In *Climate Processes and Climate Sensitivity.*, Geophysical Monograph No. 29, J. Hanson and C. Takahashi, Eds., American Geophysical Union, Washington, D.C., 73-91.

Rind, D., E.W. Chiou, W. Chu, J. Larsen, S. Oltmans, J. Lerner, M.P. McCormick, and L.M. McMaster, 1991: Positive water vapour feedback in climate models confirmed by satellite data. *Nature,* **349,** 500-503.

Robinson, G.D., 1958: Some observations from aircraft of surface albedo and the albedo and absorption of cloud. *Arch. Meteor. Geophys. Bioklim.,* **B9,** 28-41.

Robuck, A., 1978: Internally and externally caused climate change. *J. Atmos. Sci.,* **35,** 1111-1122.

Robuck, A., 1979: The "Little Ice Age": Northern Hemispheric average observations and model calculations. *Science,* **206,** 1402-1404.

Robuck, A., 1981: A latitudinally dependent volcanic dust veil index, and its effect on climatic simulations. *J. Volcan. Geotherm. Res.,* **11,** 67-80.

Robuck, A., 1984a: Climate model simulations of the effects of the El Chichon eruption. *Geofis. Int.,* **23,** 403-414.

Robock, A., 1984b: Snow and ice feedbacks prolong effects of nuclear winter. *Nature,* **310,** 667-670.

Rodriguez, H., 1992: Influence of mesoscale variations in land surface albedo on large-scale averaged heat fluxes. M.S. Thesis, Department of Atmospheric Science, Colorado State University.

Roeker, E., U. Schlese, J. Biercamp, and P. Loewe, 1987: Cloud optical depth feedbacks and climate modelling. *Nature,* **329,** 138-140.

Rowntree, P.R., and J. Walker, 1978: The effects of doubling the CO_2 concentration on radiative-convective equilibrium. In *Carbon Dioxide, Climate and Society,* J. Williams, Ed., Pergamon, New York, 181-191.

Roy, A.K., Bh. V. Ramana Murty, R.C. Srivastava, and L.T. Khemani, 1961: Cloud seeding trails at Delhi during monsoon months, July to Sept. (1957-1959), *Indian J. Meteor. Geophys.,* **12,** 401-412.

Rozenberg, G.V., M.S. Malkevich, V.S. Malkova, and V.L. Syachinov, 1974: Determination of the optical characteristics of clouds from measurements of reflected solar radiation by KOSMOS 320. *Izv. Acad. Sci USSR, Atmos. Oceanic Phys.,* **10,** 14-24.

Running, S.W., 1990: Estimating terrestrial primary productivity by combining remote sensing and ecosystem simulation. In *Remote Sensing of Biosphere Functioning*, R. Hobbs and H. Mooney, Eds., Springer-Verlag, New York, 65-86.

Running, S.W., and J.C. Coughlan, 1988: A general model of forest ecosystem processes for regional applications. I. Hydrologic balance, canopy gas exchange and primary production processes. *Ecol. Modelling*, **42**, 125-154.

Sagan, C., 1983: Nuclear war and climatic catastrophe: Some policy implications. *Foreign Affairs,* **62,** (Winter 1983/84), 257-292.

Saito, T., 1981: The relationship between the increase rate of downward long-wave radiation by atmospheric pollution and the visibility. *J. Meteor. Soc.,* **59,** 254-261.

Salati, E., J. Marques, and L.C.B. Molion, 1978: Origem e distribuicão das chuvas na Amazônia. *Interciencia*, **3**, 200-205.

Sax, R.I., 1976: Microphysical response of Florida cumuli to AgI seeding. *Second WMO Scientific Conf. on Weather Modification*, Boulder, WMO, 109-116.

Sax, R.I., and V.W. Keller, 1980: Water-ice and water-updraft relationships near -10° C within populations of Florida cumuli. *J. Appl. Meteor.,* **19,** 505-514.

Sax, R.I., J. Thomas, and M. Bonebrake, 1979: Ice evolution within seeded and nonseeded Florida cumuli. *J. Appl. Meteor.,* **18,** 203-214.

Schaefer, V.J., 1948a: The production of clouds containing supercooled water droplets or ice crystals under laboratory conditions. *Bull. Amer. Meteor. Soc.,* **29**, 175.

Schaefer, V.J., 1948b: The natural and artificial formation of snow in the atmosphere. *Trans. Amer. Geophys. Union,* **29,** 492.

Schaefer, V.J., 1966: Condensed water in the free atmosphere in air colder than –40° C. *J. Appl. Meteor.,* **1,** 481-488.

Schaefer, V.J., 1969: The inadvertent modification of the atmosphere by air pollution. *Bull. Amer. Meteor. Soc.,* **50,** 199-206.

Scheraga, Joel and Irving Mintzer, 1990: Introduction. In *Policy Options for Stabilizing Global Climate,* D.A. Lashof and D.A. Tirpak, Eds., U.S. Environmental Protection Agency, Office of Policy, Planning and Evaluation. Hemisphere Publishing Corp. New York.

Schickendanz, P.T., 1976: Effects of irrigation on precipitation in the Great Plains. Final report to NSF, RANN, Illinois State Water Survey, University of Illinois, Urbana, 105 pp.

Schlesinger, M.E., 1983: A review of climate models and their simulation of CO_2-induced warming. *Int. J. Environ. Studies,* **20,** 103-114.

Schlesinger, M.E., and J.F.B. Mitchell, 1987: Climate model simulations of the equilibrium climatic response to increased carbon dioxide. *Rev. Geophys.,* **25,** 760-798.

Schneider, S.H., 1989: The changing climate. *Scientific American,* **September**, 70-79.

Schneider, S.H., 1990: The global warming debate heats up: An analysis and perspective. *Bull. Amer. Meteor. Soc.,* **71**, 1291-1304.

Schneider, S.H., and S.L. Thompson, 1988: Simulating the climatic effects of nuclear war. *Nature,* **333,** 221-227.

Schwartz, S.E., 1988: Are global cloud albedo and climate controlled by marine phytoplankton? *Nature,* **336,** 441-445.

Schwartz, S.E., 1989: Sulphate aerosols and climate. *Nature,* **340,** 515-516.

Schwartz, M.D. and T.R. Karl, 1990: Spring phenology: Nature's experiment to detect the effect of "green-up" on surface maximum temperatures. *Mon. Wea. Rev.*, **118**, 883-890.

Schwartzman, D.W., and T. Volk, 1989: Biotic enhancement of weathering and the habitability of earth. *Nature*, **340**, 457-459.

Scorer, R.S., 1987: Ship trails. *Atmos. Environ.,* **21,** 1417-1425.

Scott, B.C., and P.V. Hobbs, 1977: A theoretical study of the evolution of mixed-phase cumulus clouds. *J. Atmos. Sci.*, **34**, 812-826.

Seaver, W.L., and J.E. Lee, 1987: A statistical examination of sky cover changes in the contiguous United States. *J. Climate Appl. Meteor.,* **26,** 88-95.

Segal, M., R.A. Pielke, and Y. Mahrer, 1983: On climatic changes due to a deliberate flooding of the Qattara depression (Egypt). *Climatic Change*, **5**, 73-83.

Segal, M., R. Avissar, M.C. McCumber, and R.A. Pielke, 1988: Evaluation of vegetation effects on the generation and modification of mesoscale circulations. *J. Atmos. Sci.*, **45**, 2268-2292.

Segal, M., W. Schreiber, G. Kallos, R.A. Pielke, J.R. Garratt, J. Weaver, A. Rodi, and J. Wilson, 1989: The impact of crop areas in northeast Colorado on midsummer mesoscale thermal circulations. *Mon. Wea. Rev.*, **117**, 809-825.

Sellers, P.J., 1991: Modeling and observing land-surface-atmosphere interactions on large scales. Chapter 4 in *Surveys in Geophysics*, **12**, Kluwer Academic Publishers, 85-114.

Sellers, P.J. and F.G. Hall, Editors, 1992: FIFE JGR Special Issue. (In preparation).

Sheets, R.C., 1981: Tropical cyclone modification: The Project STORMFURY Hypothesis. Miami, Florida, NOAA Technical Report ERL 414-AOML30, 52 pp.

Shine, K.P., 1991: On the cause of the relative greenhouse strength of gases such as the halocarbons. *J. Atmos. Sci.*, **48**, 1513-1518.

Shuttleworth, W.J., 1988: Macrohydrology – The new challenge for process hydrology. *J. Hydrol.*, **100**, 31-56.

Simon, C., 1981: Clues in the clay. *Sci. News*, Nov. 14, 1981, 314-315.

Simpson, J., 1980: Downdrafts as linkages in dynamic cumulus seeding effects. Notes. *J. Appl. Meteor.,* **19,** 477-487.

Simpson, J., and A.S. Dennis, 1972: Cumulus clouds and their modification. NOAA Tech. Memo., ERLOD-14, Washington, D.C., 148 pp.

Simpson, R.H., and J.S. Malkus, 1964: Experiments in hurricane modification. *Sci. Amer.,* **211,** 27-37.

Simpson, R.H., M.R. Ahrens, and R.D. Decker, 1963: A cloud seeding experiment in Hurricane Esther, 1961. *National Hurricane Research Project Report No. 60,* U.S. Department of Commerce, Weather Bureau, Washington, D.C., 30 pp.

Simpson, R.H., et al., 1978: TYMOD: Typhoon Modification Final Report. Prepared for the Government of the Philippines. Virginia Technology, Arlington, Virginia.

Simpson, J., N.E. Westcott, R.J. Clerman, and R.A. Pielke, 1980: On cumulus mergers. *Arch. Meteor. Geophys. Bioklim., Series A*, **29**, 1-40.

Simpson, J., N.E. Westcott, R.J. Clerman, and R.A. Pielke, 1980: On cumulus mergers. *Arch. Meteor. Geophys. Bioklim.*, Series A, **29**, 1-40.

Singer, S. Fred, Ed., 1992: *Global Climate Change: Human and Natural Influences*, Chapter 19, Paragon House Publishers, New York, New York, 424 pp.

Slingo, T., 1988: Can plankton control climate? *Nature,* **336,** 421.

Small, R.D., B.W. Bush, and M.A. Dore, 1989: Initial smoke distribution for nuclear winter calculations. *Aerosol Sci. Tech.,* **10,** 37.

Smith, E.A., A.Y. Hsu, W.L. Crosson, R.T. Field, L.J. Fritschen, R.J. Gurney, E.T. Kanemasu, W. Kustas, D. Nie, W.J. Shuttleworth, J.B. Stewart, S.B. Verma, H. Weaver, and M. Wesley, 1992: Area averaged surface fluxes and their time-space varaiblity over the FIFE experimental domain. *J. Geophys. Res.*, (in press).

Somerville, R.C.J., and L.A. Remer, 1984: Cloud optical thickness feedbacks in the CO_2 climate problem. *J. Geophys. Res.,* **89,** 9668-9672.

Spanner, M.A., L.L. Pierce, S.W. Running, and D.L. Peterson, 1990: The seasonality of AVHRR data of temperate coniferous forests: Relationship with leaf area index. *Remote Sens. Environ.*, **33**, 97-112.

Squires, P., 1966: An estimate of the anthropogenic production of cloud nuclei. *J. Rech. Atmos.,* **2/3,** 297-308.

Stenchikov, G.L., and P. Carl, 1985: Climatic consequences of nuclear war: Sensitivity against large-scale inhomogeneities in the initial atmospheric pollutions. Central Inst. Electron Physics, Berlin, GDR.

Stephens, G.L., and S-C. Tsay, 1990: On the cloud absorption anomaly. *Quart. J. Roy. Meteor. Soc.,* **116,** 671-704.

Stephens, G.L., G.W. Paltridge, and C.M.R. Platt, 1978: Radiation profiles in extended water clouds, III. Observations. *J. Atmos. Sci.,* **35,** 1837-1848.

Stockston, C.W., 1990: Climatic variability on the scale of decades to centuries. *Climatic Change*, **16**, 173-183.

Stouffer, R.J., S. Manabe, and K. Bryan, 1989: Interhemispheric asymmetry in climate response to a gradual increase of atmospheric CO_2. *Nature,* **342,** 660-662.

Sulakvelidze, G.K., B.I. Kiziriya, and V.V. Tsykunov, 1974: Progress of hail suppression work in the U.S.S.R. In *Climate and Weather Modification*, W.N. Hess, Ed., Wiley, New York, 410-431.

Sundqvist, H., 1978: A parameterization scheme for nonconvective condensation including prediction of cloud water content. *Quart. J. Roy. Meteor. Soc.*, **104**, 667-690.

Sundqvist, H., 1981: Prediction of stratiform clouds: Results from a 5-day forecast with a global model. *Tellus*, **33**, 242-253.

Super, A.B., and B.A. Boe, 1988: Microphysical effects of wintertime cloud seeding with silver iodide over the Rocky Mountains. Part III: Observations over the Grand Mesa, Colorado. *J. Appl. Meteor.,* **27,** 1166-1182.

Super, A.B., and J.A. Heimbach, 1988: Microphysical effects of wintertime cloud seeding with silver iodide over the Rocky Mountains. Part II: Observations over the Bridger Range, Montana. *J. Appl. Meteor.,* **27,** 1152-1165.

Super, A.B., B.A. Boe, and E.W. Holroyd, III, 1988: Microphysical effects of wintertime cloud seeding with silver iodide over the Rocky Mountains. Part I: Experimental design and instrumentation. *J. Appl. Meteor.,* **27,** 1145-1151.

Tanre, D., J.F. Geleyn, and J. Slingo, 1984: First results of the introduction of an advanced aerosol-radiation interaction in the ECMWF low resolution global model. In *Aerosols and Their Climatic Effects*, H.E. Gerber and A. Deepak, Eds., A. Deepak Publishing, 133-177.

Thatcher, M., 1990: On long term climate prediction. *J. Air Waste Mgt.*, **40**, 1086-1087.

Thompson, S.L., 1985: Global interactive transport simulations of nuclear war smoke. *Nature,* **317**, 35-39.

Thompson, S.L., V. Ramaswamy, and C. Covey, 1987: Atmospheric effects of nuclear war aerosols in general circulation model simulations: Influences of smoke optical properties. *J. Geophys. Res.,* **92,** 10942-10960.

Thompson, S.L., and S.H. Schneider, 1986: Nuclear winter reappraised. *Foreign Affairs,* **65,** 981.

Titus, J.G., 1990a: Strategies for adapting to the greenhouse effect. *APA Journal*, 331-323.

Titus, J.G., 1990b: Greenhouse effect, sea level rise and barrier islands: Case study of Long Beach Island New Jersey. *Coastal Management*, **18**, 65-90.

Tjernström, M., 1987a: A study of flow over complex terrain using a three-dimensional model – A preliminary model evaluation focusing on fog and stratus. *Ann. Geophys.*, **5**, 469-486.

Tjernström, M., 1987b: A three-dimensional meso-γ-scale model for studies of stratiform boundary layer clouds. A model description. Department of Meteorology, Uppsala University, Sweden, Report No. 85.

Trabalka, J.R., Ed., 1985: *Atmospheric Carbon Dioxide and the Global Carbon Cycle*, U.S. Department of Energy, Washington, DC. (Available as NTIS DOE/ER-0239 from Natl. Tech. Inf. Serv., Springfield, VA.)

Tripoli, G.J., 1986: Nucleation scavenging in smoke plumes induced by large urban fires: Some preliminary results. Presentation to Global Effects Technical Meeting, NASA Ames Research Center, Moffett Field, California, 25-27 February 1986.

Tripoli, G.J., and W.R. Cotton, 1980: A numerical investigation of several factors contributing to the observed variable intensity of deep convection over South Florida. *J. Appl. Meteor.*, **19**, 1037-1063.

Tukey, J.W., D.R. Brillinger, and L.V. Jones, 1978: The role of statistics in weather resources management. The Management of Weather Resources, Vol. II. *Final Report of Weather Modification Advisory Board,* Department of Commerce, Washington, D.C. [U.S. Government Printing Office No. 003-0180-0091-1.]

Turco, R.P., O.B. Toon, T.P. Ackerman, J.B. Pollack, and C. Sagan, 1983: Nuclear winter: Global consequences of multiple nuclear explosions. *Science,* **222,** 1283-1292.

Turco, R.P., O.B. Toon, T.P. Ackerman, J.B. Pollack, and C. Sagan, 1990: Climate and smoke: An appraisal of nuclear winter. *Science,* **247,** 166-176.

Turner, C.L., T.R. Seastedt, M.I. Dyer, T.G.F. Kittel, and D.S. Schimel, 1992: Effect of management and topography on the radiometric response of tallgrass prairie. *J. Geophys. Res.*, (in press).

Twomey, W., 1974: Pollution and the planetary albedo. *Atmos. Environ.,* **8,** 1251-1256.

Twomey, S., 1977: The influence of pollution on the shortwave albedo of clouds. *J. Atmos. Sci.,* **34,** 1149-1152.

Twomey, S., and T.C. Cocks, 1982: Spectral reflectance of clouds in the near-infrared: comparison of measurements and calculations. *J. Meteor. Soc. Japan,* **60,** 583-592.

Twomey, S., M. Piepgrass, and T.L. Wolfe, 1984: An assessment of the impact of pollution on global albedo. *Tellus,* **36B,** 356-366.

UCAR/NOAA, 1991: Reports to the Nation. Winter 1991. Published by the UCAR Office for Interdisciplinary Earth Studies and the NOAA Office of Global Programs, 21 pp.

Ulanski, S., and M. Garstang, 1978a: The role of surface divergence and vorticity in the lifecycle of convective rainfall, Part I: Observation and analysis. *J. Atmos. Sci.*, **35**, 1047-1062.

Ulanski, S., and M. Garstang, 1978b: The role of surface divergence and vorticity in the life cycle of convective rainfall. Part II: Descriptive model. *J. Atmos. Sci.*, **35**, 1063-1069.

U.S. National Research Council, 1985: *The Effects on the Atmosphere of a Major Nuclear Exchange.* National Academy Press, Washington.

Vogelmann, A.M., A. Robock, and R.G. Ellingson, 1988: Effects of dirty snow in nuclear winter simulations. *J. Geophys Res.,* **93,** 5319-5332.

von Foerster, H., P.M. Mora, L.W. Amiot, 1960: Doomsday: Friday, 13 November, A.D. 2026. *Science,* **132,** 1291.

Vonnegut, B., 1947: The nucleation of ice formation by silver iodide. *J. Appl. Phys.,* **18,** 593.

Vukovich, F.M., J.W. Dunn, III, and B.W. Crissman, 1976: A theoretical study of the St. Louis heat island: the wind and temperature distribution. *J. Appl. Meteor.,* **15,** 417-440.

Vupputuri, R.K.R., 1986: Effect of ozone photochemistry on atmospheric and surface temperature changes due to large atmospheric injections of smoke and NO_x by a large-scale nuclear war. *Atmos. Environ.,* **20,** 665-680.

Wahl, E., 1951: On a seven-day periodicity in weather in the United States during April, 1950. *Bull. Amer. Meteor. Soc.,* **32,** 193.

Wang, W.-C., and P.H. Stone, 1980: Effect of ice-albedo feedback on global sensitivity in a one-dimensional radiative-convective climate model. *J. Atmos. Sci.,* **37,** 545-552.

Wang, W.-C., W.B. Rossow, J.S. Yao, and M. Wolfson, 1981: Climate sensitivity of a one-dimensional radiative-convective model with cloud feedback. *J. Atmos. Sci.,* **38,** 1167-1178.

Warner, J., 1968: A reduction in rainfall associated with smoke from sugar-cane fires—An inadvertent weather modification? *J. Appl. Meteor.,* **7,** 247-251.

Warner, J., 1971: Smoke from sugar-cane fires and rainfall. *Proc., International Conference on Weather Modification*, Canberra, Australia, Sept. 6-11, 1971.

Warner, J., 1973: The microstructure of cumulus clouds. Part V. Changes in droplet size distribution with cloud age. *J. Atmos. Sci.,* **30,** 1724-1726.

Warner, J., and S. Twomey, 1967: The production of cloud nuclei by cane fires and the effects on cloud droplet concentration. *J. Atmos. Sci.,* **24,** 704-706.

Warren, S.G., and W.J. Wiscombe, 1985: Dirty snow after nuclear war. *Nature,* **313,** 467.

Warren, S.G., C.J. Hahn, J. London, R.M. Chervin, and R.L. Jenne, 1988: Global distribution of total cloud cover and cloud type amounts over the ocean. Prepared for the United States Department of Energy, Washington, DC and the National Center for Atmospheric Research, Boulder, CO.

Washington, W.M., and G.A. Meehl, 1984: Seasonal cycle experiment on the climate sensitivity due to a doubling of CO_2 with an atmospheric general circulation model coupled to 'a single mixed-layer ocean model. *J. Geophys. Res.,* **89,** 9475-9503.

Washington, W.M., and G.A. Meehl, 1989: Climate sensitivity due to increased CO_2: Experiments with a coupled atmosphere and ocean general circulation model. *Climate Dynamics,* **4,** 1-38.

Watson, A.J., and J.E. Lovelock, 1983: Biological homeostasis of the global environment: The parable of Daisyworld. *Tellus*, **35B**, 284-289.

Weickmann, H.K., 1964: The language of hailstorms and hailstones. *Nubila,* **6,** 7-51.

Welch, R.M., and W.G. Zdunkowski, 1976: A radiation model of the polluted atmospheric boundary layer. *J. Atmos. Sci.*, **33**, 2170-2184.

Welch, R.M., J. Paegle, and W.G. Zdunkowski, 1978: Two-dimensional numerical simulation of the effects of air pollution upon the urban-rural complex. *Tellus*, **30**, 136-150.

Wetherald, R.T., and S. Manabe, 1986: An investigation of cloud cover change in response to thermal forcing. *Climatic Change,* **8,** 5-23.

Wetherald, R.T., and S. Manabe, 1988: Cloud feedback processes in a general circulation model. *J. Atmos. Sci.,* **45,** 1397-1415.

Wexler, H., 1951: Periodicity of weather. *Chem. Eng. News,* **29,** 3933.

Whitten, R.C., W.J. Borucki, and R.P. Turco, 1975: Possible ozone depletions following nuclear explosions. *Nature,* **257,** 38-39.

Wigley, T.M.L., 1988: The climate of the past 10,000 years and the role of the Sun. In *Secular solar and geomagnetic variations in the last 10,000 years*, F.R. Stephenson and A.W. Wolfendale, Eds., Kluwer Academic Publishers, Hingham, MA, 209-244.

Wigley, T.M.L., 1989: Possible climate change due to SO_2-derived cloud condensation nuclei. *Nature,* **339,** 365-367.

Wigley, T.M.L., and P.M. Kelly, 1990: Holocene climatic change, ^{14}C wiggles and variations in solar irradiance. *Phil. Trans. Roy. Soc. London*, **330**, 547-559.

Wigley, T.M.L., and S.C.B. Raper, 1987: Thermal expansion of sea water associated with global warming. *Nature,* **330,** 127-131.

Wigley, T.M.L., P.D. Jones, and P.M. Kelly, 1979: Scenario for a warm, high-CO_2 world. *Nature,* **283,** 17-21.

Wigley, T.M.L., P.D. Jones, and P.M. Kelly, 1986: Empirical climate studies. In *The greenhouse effect, climate change and ecosystems*, B. Bolin, B.R. Döös, J. Jäger and R.A. Warrick, Eds., John Wiley and Sons, Chichester, U.K.

Williams, J., 1979: Anomalies in temperature and rainfall during warm Arctic season as a guide the the formulation of climate scenarios. *Climatic Change,* **2**, 249.

Willson, R.C., and H.S. Hudson, 1988: Solar luminosity variations in solar cycle 21. *Nature,* **332**, 810-812.

Willson, R.C., H.D. Hudson, C. Frohlich, and R.W. Brusa, 1986: Long-term downward trend in total solar irradiance. *Science,* **234**, 1114-1117.

Wilson, C.A., and J.F.B. Mitchell, 1987: A doubled CO_2 climate sensitivity experiment with a global climate model including a simple ocean. *J. Geophys. Res.,* **92**, 13,315-13,343.

Wiscombe, W.J., R.M. Welch, and W.D. Hall, 1984: The effects of very large drops on cloud absorption, Part I: parcel models. *J. Atmos. Sci.,* **41**, 1336-1355.

Wolff, G.T., 1986: Measurements of SO_x, NO_x and aerosol species off Bermuda. *Atmos. Environ.,* **20**, 1229-1239.

Woodcock, A.H., and R.H. Jones, 1970: Rainfall trends in Hawaii. *J. Appl. Meteor.,* **9**, 690-696.

Woodley, W.L., 1970: Precipitation results from a pyrotechnic cumulus seeding experiment. *J. Appl. Meteor.,* **9**, 242-257.

Woodley, W., J. Flueck, R. Biondini, R.I. Sax, J. Simpson, and A. Gagin, 1982a: Clarification of confirmation in the FACE-2 experiment. *Bull. Amer. Meteor. Soc.,* **63**, 273-276.

Woodley, W., J. Jordan, J. Simpson, R. Biondini, J. Flueck, and A. Barnston, 1982b: Rainfall results of the Florida Area Cumulus Experiment, 1970-1976. *J. Appl. Meteor.,* **21**, 139-164.

Woodley, W., A. Barnston, J.A. Flueck, and R. Biondini, 1983: The Florida Area Cumulus Experiment's Second Phase (FACE-2). Part II: Replicated and confirmatory analyses. *J. Climate Appl. Meteor.,* **22,** 1529-1540.

Xian, Z., and R.A. Pielke, 1991: The effects of width of land masses on the development of sea breezes. *J. Appl. Meteor.*, **30**, 1280-1304.

Yan, H., and R.A. Anthes, 1987: The effect of latitude on the sea breeze. *Mon. Wea. Rev.*, **115**, 939-956.

Yoshikado, H., 1981: Statistical analyses of the sea breeze pattern in relation to general weather patterns. *J. Meteor. Soc. Japan*, **59**, 98-107.

Young, K.C., 1977: A numerical examination of some hail suppression concepts. In *Hail: A review of hail science and hail suppression*, G.B. Foote and C.A. Knight, Eds., *Meteorol. Monogr.*, **16,** No. 38, 195-214.

Young, G.S., and R.A. Pielke, 1983: Application of terrain height variance spectra to mesoscale modeling. *J. Atmos. Sci.*, **40**, 2555-2560.

Young, G.S., R.A. Pielke, and R.C. Kessler, 1984: A comparison of the terrain height variance spectra of the front range with that of a hypothetical mountain. *J. Atmos. Sci.*, **41**, 1249-1250.

Zeng, X., R.A. Pielke, and R. Eykholt, 1990: Chaos in Daisyworld. *Tellus*, **42B**, 309-318.

Index